Children's Games in Street and Playground

CHASING · CATCHING · SEEKING · HUNTING
RACING · DUELLING · EXERTING
DARING · GUESSING · ACTING · PRETENDING

BY

IONA AND PETER OPIE

Oxford   New York
OXFORD UNIVERSITY PRESS
1984

Oxford University Press, Walton Street, Oxford OX2 6DP
London Glasgow New York Toronto
Delhi Bombay Calcutta Madras Karachi
Kuala Lumpur Singapore Hong Kong Tokyo
Nairobi Dar es Salaam Cape Town
Melbourne Auckland
and associated companies in
Beirut Berlin Ibadan Mexico City Nicosia

Oxford is a trade mark of Oxford University Press

© Iona and Peter Opie 1969

First published 1969
First issued as an Oxford University Press paperback 1984

British Library Cataloguing in Publication Data
Opie, Iona
Children's games in street and playground.—
(Oxford paperbacks)
1. Games—Great Britain
2. Outdoor recreation—Great Britain
I. Title II. Opie, Peter
796'.01'922 GV1204.43
ISBN 0-19-281489-3

Library of Congress Cataloging in Publication Data
Opie, Iona Archibald.
Children's games in street and playground.
(Oxford paperbacks)
Includes index.
1. Games—Great Britain. 2. Games—Great Britain—
History. 3. Games—History. I. Opie, Peter. II. Title.
GV1204.43.O65 1984 796'.0941 84-3919
ISBN 0-19-281489-3 (pbk.)

Printed in Great Britain by
Richard Clay (The Chaucer Press) Ltd
Bungay, Suffolk

3,94,3
Office use.

Preface

DURING the past fifty years shelf-loads of books have been written instructing children in the games they ought to play, and some even instructing adults on how to instruct children in the games they ought to play, but few attempts have been made to record the games children in fact play. It seems to be presumed that children today (unlike those in the past) have few diversions of their own, that they are incapable of self-organization, have become addicted to spectator amusements, and will languish if left to rely on their own resources. It is felt that the enlightened adult is one who thinks up ideas for them, provides them with 'play materials', and devotes time to playing with them. Certainly our attitude to the young has changed since the nineteenth century, when Herbert Spencer could complain that men of education felt the rearing of fine animals to be a fitter subject for attention than the bringing up of fine human beings. Yet our vision of childhood continues to be based on the adult–child relationship. Possibly because it is more difficult to find out about, let alone understand, we largely ignore the child-to-child complex, scarcely realizing that however much children may need looking after they are also people going about their own business within their own society, and are fully capable of occupying themselves under the jurisdiction of their own code.

In the present study we are concerned solely with the games that children, aged about 6–12, play of their own accord when out of doors, and usually out of sight. We do not include, except incidentally, party games, scout games, team games, or any sport that requires supervision; and we concentrate for the most part on the rough-and-tumble games which, though they may require energy and sometimes fortitude, do not need even the elementary equipment of bat and ball. We are interested in the simple games for which, as one child put it, 'nothing is needed but the players themselves'. It is intended that the intricacies of ball-bouncing and other ball games, of skipping, marbles, fivestones, hopscotch, tipcat, and gambling, shall be described in a further volume, where they can be treated more fully, together with the singing games.

Our accounts of these games are based on information from more than 10,000 children attending local-authority schools in England, Scotland, and the eastern part of Wales; and this survey in the 1960s has been supplemented by the similar investigation we undertook in the 1950s when preparing *The Lore and Language of Schoolchildren* (1959). It should be remarked, however, that although our survey has extended from the Channel Islands to the Shetland Islands, and takes notice of a number of remote areas, the vast majority of children who have contributed have spent their lives in great cities such as London, Liverpool, Bristol, and Glasgow, and are the children who are ordinarily supposed to be the least satisfied with simple pleasures.

The fact is that there is no town or city known to us where street games do not flourish. Indeed, during twenty years' inquiry, we have not met a child who was unable to tell us something interesting, and who did not unwittingly increase the size of our files. However, the research-worker who is blessed with an unending flow of information can be in as embarrassing a position as he whose sources are limited, when he comes to making his report. Our problem editorially has been to present our findings in a form in which the facts are readily available, yet not so prosaically that the spirit of street play, its zest, variety, contradictions, and disorderliness, is entirely lost. Street play is not a social activity that lends itself to neat analysis and tabulation. The robustness of juvenile society allows of no more than partial agreement on what each game is called, let alone on how it is played. Yet long acquaintance with the games has sometimes shown that the differences between one place and another are more apparent than real. Having the benefit of several hundred descriptions of the more popular games, from perhaps fifty to sixty places, we have, we think, usually been able to understand the basic principle of a game; and have thereafter been able to identify the more significant regional variations and idiosyncrasies. We have then, where necessary, appended a list of its colloquial and dialect names. Thus no matter under what name a game is known to the reader, its whereabouts in the volume should be readily located by reference to the index, which includes every one of the 2,500 names of games appearing in the text.

We have also given thought to the order in which the games appear, arrangement being by the basic motif of the game; and it is hoped that with the aid of the pages of Contents and Analysis (pp. xvii–xxvi), it will be possible to find a game even when its name is not known, as when, for instance, the British equivalent is sought to a game played elsewhere in the world. For this reason we have not felt it necessary to record every name of every parallel game played in other countries. Reference is made to a game's counterparts in other languages only when the comparison is significant; or when, as happens not infrequently, the sport was described or depicted at an earlier date on the Continent than in Britain.

It was at first thought that it would be possible to give the histories of only a minority of present-day street games; and that these few, being well-known antiques (e.g. 'Leapfrog', 'Blind Man's Buff', and 'Piggy-back Fighting'), would be so well documented already that an appropriate reference to where the history could be found would suffice. Unhappily the more we examined these histories and checked the original sources, the more unsatisfactory were previous works on children's games found to be. Further, it became increasingly evident that the way a game was played in the past is sometimes highly relevant when an explanation is required of a peculiarity in the way it is played today. The majority of games are therefore accompanied by a historical note, following the sign *⁎*, and we hope that the reader who is interested only in the activities of the contemporary schoolchild will not be too dismayed at finding his amusements compared with those in medieval France or ancient Greece.

In saying that we find the previous histories of children's games unsatisfactory it should not be thought that we are ungrateful for them. The pioneer work in Britain, Strutt's *Glig-Gamena Angel Deod, or, The Sports and Pastimes of the People of England*, 1801, is a remarkable undertaking that embodies considerable research, and can scarcely be faulted in what it says, only in what it does not say, for it allows small space to the 'sports of children'. For the past seventy years the standard work has been Alice Bertha Gomme's *Traditional Games of England, Scotland, and Ireland*, 2 vols., 1894–8. Lady Gomme, wife of George Laurence Gomme, a pioneer folklorist and first secretary of the Folk-Lore Society (founded 1878), was attracted to the subject of children's play by the singing games, and these remained her abiding interest.[1] In consequence her work is a vivid reflection of the exciting years towards the end of the Victorian era when English folksong was being discovered, and children, too, were found to be filling the air with unwritten poetry. However, these wonders on the lips of the people, whether milkmaids or mudlarks, were looked upon almost uniformly as relics of antiquity. They were examined as if they were archaeological remains, rather than living organisms which are constantly evolving, adapting to new situations, and renewing themselves or being replaced. Alice Gomme, under the influence of her husband (author of *Folk-lore Relics of Early Village Life*), was fascinated by the possibility that the

---

[1] In her early appeals for assistance in collecting she asked only for singing games; and years later, when she contributed the article on 'Children's Games' to the *Encyclopaedia Britannica*, all games other than singing and dramatic games were disposed of in three lines.

games she was collecting were the survivals of primitive custom and belief; and this somewhat romantic attitude, as it now seems, materially affected the quality of her research. Lovers of English folksong will ever be indebted to her for stimulating the collection of children's songs in the nineteenth century, and for inducing several excellent folklorists (notably Addy and Gregor) to contribute their collectings to her pages. But since she thought of change chiefly in terms of decay, she had little interest in working upon historical principles; and she had, apparently, no curiosity to study early or foreign literature. Her literary references are almost without exception taken at second hand, are not always correctly transcribed or understood, and her notes, other than those for the singing games, are seldom the labour of love they appear to be.

A more scholarly and perceptive work to our mind, and attractively written as well, is W. W. Newell's *Games and Songs of American Children*, 1883, augmented 1903. Newell, like everyone else in his time, thought the games were fast disappearing; but he had a sense of history, a good knowledge of languages, and an understanding of the operation of oral tradition. He realized that any game or rite, wherever collected, was likely to be only a local version of a general stock, and with modestly presented references he drew attention over and over again to the European versions of games he was discovering on his side of the Atlantic. In this he has been well supported in recent times by his countryman Paul G. Brewster in *American Nonsinging Games*, 1953; and it has been a pleasure to refer on occasion to both these works as places where additional analogues are to be found.

Many folklorists, however, seem to be magically insulated from literature not strictly within their discipline; and neither Newell, one of the founders of the American Folklore Society, nor Lady Gomme, seems to have been aware of the juvenile books on children's games which had long existed, even in their day, both in Britain and in the United States. This extensive literature has been examined systematically for the present work, apparently for the first time. In addition full use has been made of the signposts, clearly marked for those willing to look for them, that are provided by *The Oxford English Dictionary* and *The English Dialect Dictionary*. And the work has also benefited from a happy twenty years of general reading. Yet our greatest asset as historians has been fortuitous. It has been the knowledge we already happened to possess about the ways in which the games are played today. It is remarkable how much guesswork has been expended on classical, medieval, and Tudor pastimes, simply because the learned commentators in the eighteenth and nineteenth centuries, closeted in their studies, lacked knowledge of the games that their own children were playing in the sunshine outside their windows. And this in itself is an illustration of the gap there always is between the generations. For although some games have disappeared, and many, naturally, have altered over the years, it can still be said that the way to understand the 'wanton sports' of Elizabethan days, and the horseplay of even earlier times, is to watch the contemporary child engrossed in his traditional pursuits on the metalled floor of a twentieth-century city.

West Liss in Hampshire   I.O.& P.O.
1959-1969

IT is fifteen years since *Children's Games in Street and Playground* was first published, but I believe the picture it gives of children's unofficial games is substantially the same today as it was in 1969. Certainly my weekly visits to Liss playground confirm that the old favourites are still being played—'Om Pom', 'British Bulldog', 'Egg Bacon and Chips', 'Mr Wolf', 'Queenie, Queenie, Who's Got the Ball?', 'Stick in the Mud', and 'Farmer, Farmer, May We Cross Your Golden River?'; and when a seven-year-old excitedly fetches me to 'Come and see our new game' I am reminded that the games are new to each generation of children, and have no need to be replaced. Playgrounds in other parts of the country, where I have in recent years been recording singing games for a companion volume, *The Singing Game,* have provided further evidence that the competitive games are being played much as before, and even the repertoire of dipping rhymes seems scarcely to have altered. Correspondents have reported variations and elaborations, and occasionally a new local name; but such real innovations as have occurred have been in the daring and pretending games, which are most affected by modern technology, television, and current events. As well as continuing to play 'Last Across' in front of trains and across busy streets, children now ride up and down on top of lifts in tower blocks, and hurl rolls of metal foil at high-voltage electricity cables, holding the end of the foil as long as they dare before the two make contact in a lethal flurry of sparks. Pretending games, though mostly based on family life and wars between unspecified opponents, now often enact episodes from galactic sagas such as 'Star Wars'; and world news, as seen on television, continues to inspire its quota of games—'Hostages' was in vogue during the siege of the Iranian Embassy in 1980, and 'Falklands War' throughout the summer of 1982.

West Liss, 1984   I.O.

Acknowledgements

A WORK of this kind cannot be accomplished by two people on their own; and it has again been our good fortune that many perceptive and energetic people, strategically placed in different parts of Britain, and indeed in different parts of the world, have been willing to help us. Our hope is that they have never felt we were taking their assistance for granted, any more than have they, we hope, taken for granted our ability to do justice to the wealth of material they have made available to us.

The following are the chief schools, together with the teachers responsible at the time, who, in addition to those listed in *The Lore and Language of School-children*, pp. xi–xiv, have contributed to this work or who have made further contributions since then.

ABERDEEN	Powis Secondary School (Mr. H. W. Valentine, Head Teacher; Miss Netta Y. Dick, Woman Adviser)
ACCRINGTON	Peel Park County Junior School (Mr. John Heaton)
BACUP	St. Saviour's County Primary School (Mrs. Margaret Dearden)
BANBURY	Easington School for Girls (Miss M. H. Dawson, Head Mistress)
BARROW-IN-FURNESS	St. Aloysius Roman Catholic Secondary School (Miss M. Morris)
BERRY HILL	Berry Hill Secondary School (Mr. K. R. E. Farmer, Head Master)
BISHOP AUCKLAND	Cockton Hill Junior School (Mr. Harold Guthrie, Head Master)
BRIGHTLINGSEA	St. James' Junior School (Mr. W. Wilcox, Head Master)
BRISTOL	Embleton Junior School (Mr. W. H. James, Head Master) Portway Secondary Girls' School (Miss B. O. Draper, Head Mistress) Westbury Park Junior Mixed and Infants' School (Mr. W. J. Lander, Head Master)
BUTE	Kingarth School (Miss C. M. McFarlane)
CRICKHOWELL	Crickhowell Junior School (Miss Joyce Short)
CUMNOCK	Cumnock Academy (Mr. A. L. Taylor; Dr. John Strawhorn; Mr. Charles McLeod)

CUMNOCK, NEW	Bank Junior Secondary School (The English Staff)
DUNOON	Dunoon Grammar School (Mr. Alasdair J. S. Sinclair)
EDINBURGH	James Gillespie's Boys' School (Miss Ella Henderson)
	John Watson School (Miss A. M. Ireland)
ENFIELD	Grange Park Primary School (Mrs. Olive Sellick)
	Ponders End Boys' School (Mr. A. H. Sellick, Head Master)
	Ponders End Secondary School (Mr. A. H. Sellick, Head Master)
FRODSHAM	Frodsham Church of England Junior School (through Mrs. L. R. Stanton)
GLASGOW	The Grammar School, Uddingston (Mr. Ian C. MacLeod)
GLASTONBURY	St. John's Boys' School (Mr. G. C. White)
GRIMSBY	Harold Secondary Modern Boys' School (Mr. D. H. Potts, Head Master)
GUERNSEY	Castel Primary School (Mr. Laurence Adkins, Head Master; Mr. R. J. Gill)
	The Grammar School for Boys, St. Peter Port (Mr. Laurence Adkins)
	Vale Junior School (Mr. Laurence Adkins)
HELENSBURGH	Hermitage Senior Secondary School (Dr. Ian J. Simpson)
HELSTON	Helston County Primary School (Miss J. Martin)
INVERNESS	Dalneigh Primary School (Miss Katherine Mackintosh)
IPSWICH	Priory Heath Junior School (Mr. H. A. James, Head Master)
	Rushmere St. Andrew County Primary School (Miss Helen Southgate, Head Teacher)
	Springfield Junior School (Mr. Leslie Stow, Head Master)
JERSEY	The Convent F.C.J., St. Helier (Miss Lynda Hodkinson)
KINLOCHLEVEN	Kinlochleven Junior Secondary School

KNIGHTON   Knighton County Secondary School (Mr. Frank Noble; Mr. and Mrs. H. J. Williams)

LANGHOLM   Langholm Academy (Miss Eva Smart)

LEWIS   Nicholson Primary School, Stornoway (Mr. John Angus Maciver)

LISS   Liss County Junior School (Mr. D. M. Dolman, Head Master, and successively Mr. I. W. Cutler, Head Master)

LIVERPOOL   Waterloo County Secondary School (Mr. G. H. Roberts)

LONDON   Beaufort House Junior Mixed School, Fulham (Mr. J. R. Bevan, Head Master)

Godwin County Primary School, West Ham (Mr. R. S. Richards, Head Master)

Munster Junior Mixed School, Fulham (Head Master, Mr. I. H. W. Haines)

Netley Junior Mixed School, Camden Town (Mr. D. D. Mackay, Head Master)

St. John's Parochial School, Kilburn (Mr. A. Kinsman, Head Master)

Walworth Secondary School (Miss Valerie J. Avery)

LYDEARD ST. LAWRENCE   Lydeard St. Lawrence County Primary School (Miss Mona Penny, Head Mistress)

MANCHESTER (SALE)   Sale Grammar School (Dr. G. G. Urwin)

MARKET RASEN   The Modern School (Mr. S. B. Vickers, Head Master)

MIDDLETON CHENEY   Middleton Cheney County Primary School (Mr. George Stevens, Head Master)

NORWICH   Colman Junior School (Miss Antonia Steedman)

OFFHAM   Offham County Primary School (Mrs. E. M. Melville, Head Mistress)

ORKNEY   Kirkwall Grammar School (Mr. H. G. MacKerron, Rector)

The Academy, Stromness (Mr. W. Groundwater, Rector)

OXFORD   Headington Secondary Modern School (Miss Margaret Hornsey)

PERTH   Kinnoull School (Mr. J. S. Soane, Head Master)

PLYMOUTH	Pennycross Primary School (Mr. F. Uglow, Head Master)
ST. ANDREWS	Madras College (Dr. J. Thompson, Rector)
ST. LEONARDS-ON-SEA	St. Mary Star of the Sea School (through Mrs. L. R. Stanton)
SCARBOROUGH	Gladstone Road Junior School (Miss R. Horsman; Mr. H. Weale)
SHETLAND	Livister School, Whalsay (Mrs. J. Grant Wood, Head Teacher)
	Scalloway Junior Secondary School (Mr. Robert N. Hutchison, Head Master, and successively Mr. John Gray, Head Master)
SPENNYMOOR	Spennymoor Grammar Technical School (Mr. D. Cockburn, Head Master; Mr. Matthew Walton; Mr. E. W. Ashton)
SWANSEA	Brynmill Junior Mixed School (Mr. Reginald Gammon)
	Glanmor Secondary School for Girls (Miss Joyce A. Terrett)
	Waunarlwydd Infants' School (Miss Dorothy Lloyd, Head Mistress)
TWICKENHAM	Twickenham Technical School (Mr. T. P. Stanton)
WELSHPOOL	Ardwyn Nursery and Infant School (Miss L. N. Griffiths, Head Mistress)
	The High School (Miss F. H. Rosser)
	Welshpool Secondary Modern School (Mr. J. Spergeon)
WIDECOMBE-IN-THE-MOOR	Widecombe-in-the-Moor County Primary School (Mr. F. A. Baldock, Head Master)
WIGAN	St. Cuthbert's Junior School (Mr. B. J. Murphy, Head Master)
WILMSLOW	Chancel Lane Church of England Controlled School (Mr. Roy H. Couchman, Head Master)
WINDERMERE	Windermere Endowed School for Girls (Miss Brakewell, Head Mistress)
WOLSTANTON	Ellison Street County Primary School (through Miss Fay Prendergast)
YORK	Tang Hall Primary School (Miss Theodora Ross)

Moreover, Mrs. L. R. Stanton, Head of the Education Department, Maria Assumpta College of Education, encouraged her students to acquire information for us in more than forty schools, chiefly but not exclusively in the Greater London area. Miss A. E. Osmond and students of Bedford College of Education again went to much trouble on our behalf tackling specific problems; and Mr. P. C. Sheppard of Brentwood School, Miss E. M. Wright, Head Mistress, Huyton College Preparatory School, and Mr. Alex Helm of Congleton followed up particular lines of inquiry.

A number of teachers other than those named above have either allowed us to visit them and their pupils, or have supplied us with tape recordings or other information, amongst them: Mr. Harvey Macpherson (to whom we are also grateful for a copy of Dean's *Alphabet of Sports*) when at St. Leonard's Primary School, Banbury; Mr. K. Bumstead, Bramford Primary School; Mr. Bryn Jones, Fairfield Grammar School, Bristol; Mr. K. M. Allan, Head Master, Ellesmere County Primary School; Mrs. Margaret Hope Luff, Milton House School, Edinburgh; Miss G. R. Brooker, Head Mistress, Greatham County Primary School; Mr. Herbert Brelsford, Head Master, St. Martin's School, Guernsey; Mr. F. H. Le Poidevin, Head Master, Amherst Junior Mixed School, Guernsey; Mr. J. I. H. Fleming, successively Head Teacher of Flotta Public School, and Head Master of Stenness Public School, in Orkney; Miss Anita Colquhoun, Sand School, Garderhouse; Mrs. Inga I. A. Thomson, Skeld Public School, Shetland; Miss G. L. Landon, Sneyd Green Junior School; and Mr. and Mrs. Frank L. Pinfold, Swine Church of England School. No tally has been kept of the number of children spoken to in the street when we have been travelling around, but we would like to remark on the courtesy children have always shown us when we approached them, and on the way they have never seemed surprised by our questions.

Individual adults who have answered queries, or who have provided us with specialist, background, or local information, have included: Mr. F. O'Brien Adams (recollections of Bembridge, Isle of Wight); the late Mrs. Barbara Aitken (who also collected for Lady Gomme); Mrs. Cornelia Baké (assistance with Dutch material); Mrs. Barbara Bagshaw (Bristol); Miss Gertrude M. Black (Stornoway); Miss Theo Brown, Folklore Recorder, The Devonshire Association; Mrs. Aldyth M. Cadoux, who collected for us in Moscow; Miss Kathleen Crawshaw, who sent Gillington's *Hampshire Singing Games*; Miss K. H. Crofts (Kennington); Mr. Gilbert H. Fabes (recollections of Vauxhall Bridge Road); Mr. Alec Forster (Anglesey); Mrs. Gande (Luccombe); Mr. and Mrs. Fritz Gasch; Mr. H. W. Harwood of Halifax, whose passing was a sad loss to us; Mr. G. E. Hawkins (recollections of Hull); Miss M. Hewson (Louth); Mr. G. H. Hobbs (Petersfield); Mr. Philip Howell (Birmingham and Cheltenham); Mrs. Kathleen Hunt (Bicester and Newcastle upon Tyne); Mrs. C. C. Hurst (recollections of Earl Shilton, Leicestershire); Mr. P. G. Inwood (Wrecclesham in the nineteenth century); Mr. Chris Kilkenny (Newcastle upon Tyne); Mr. John MacInnes (translations of children's papers written in Gaelic); Miss Maureen MacMillan (Ballachulish); Miss H. Milne; Mr. Pat Page (Dundee); Mr. Julian Pilling (Nelson); Mrs. Vivien Pope; Mrs. Lilian M. C. Randall (references to games in medieval manuscripts); Dr. and Mrs. Rossell Hope Robbins; Mr. F. C.

Rowlands (Eccleshill); Mr. Stewart Sanderson; Mr. Frank Shaw (Liverpool); Mrs. Margaret Smallwood (recollections of Salisbury); Mr. Alexander L. Stark (Stirling); Mr. George Stratton (Birkenhead); Miss Wendy Twine (Liss); Mr. Andrew Weir of *The Yorkshire Post*; Miss Dora M. Wild (Douglas, Isle of Man); and The Revd. J. Stafford Wright.

We are also indebted to well-wishers in different places who have often been to much trouble arranging that the survey should have a good coverage, amongst them: Miss R. Beresford; Mrs. Kate Bone; Miss J. B. T. Christie; Mr. and Mrs. L. Gilbert; Dr. Hugh Marwick; Mr. Roger Mayne; Mr. Dennis Potter; Mrs. Bruce Proudfoot; Mr. R. U. Sayce; Mr. J. H. Spence, Director of Education, Zetland; Miss K. I. Stevenson; Miss Ruth L. Tongue; Mr. R. A. Wake; Mr. D. Woodward; and Mrs. Michael Young (Sasha Moorsom). In addition, we would like to put on record that it is no coincidence that references are numerous in this volume to the localities where Mr. Laurence Adkins, Mr. Roy H. Couchman, Mr. Robert S. Richards, Miss Yvonne K. Rodwell, Mr. A. H. Sellick, Mr. Alasdair J. S. Sinclair, Mr. A. L. Taylor, Mr. S. B. Vickers, and Mr. and Mrs. Matthew Walton have been active; while many as are the references to Swansea we can be certain there would have been yet more had it not been for the untimely death in 1965 of Miss Joyce A. Terrett who had been collecting for us continuously for fourteen years.

Our information has also been supplemented, most happily, by some extensive collections of games made independently of our survey. We are indebted to Mrs. Violet Brewser for presenting us with the important collection of children's manuscripts and related material on games formed in Somerset in the early twenties by her father, the late A. S. Macmillan; to Dr. Francis Celoria for the folklore record sheets concerning games assembled under his auspices in the Department of Extramural Studies, Keele University; to Mr. Cuthbert Graham and *The Press and Journal*, Aberdeen, for the large collection of games and bairns' rhymes contributed by readers in 1959; to Mr. David Holbrook for kindly giving us the safe-keeping of Lady Gomme's manuscripts subsequent to the publication of her book; and to Miss G. Johnson, Chief Librarian and Curator, Camberwell Public Libraries, for a further 489 children's essays, this time on 'Games I play with my friends', entered for the annual essay competition sponsored by the Libraries Committee.

A feature of this work which has given us particular pleasure has been the amount of contemporary material we have had for comparative purposes through the co-operation of collectors and correspondents overseas. We are grateful to Larer Helge Børseth for her book *Min mann Mass*, and for other Norwegian material; Dr. Paul G. Brewster for his invaluable *American Nonsinging Games*; Mrs. Kenneth Carpenter (Patricia Evans) of Reno for her book *Rimbles* and other help; Miss Nina Demurova for a report from Moscow; Professore and Signora V. Gargiulo for assistance in and around Capri; Miss Joan Hunt of Edmonton, Alberta, and through her Mr. F. W. Wootton, The Principal, Hazeldean Elementary School, Edmonton, for the contributions of 130 pupils; Signora Matizia Maroni Lumbroso, La Presidente, Fondazione Ernesta Besso, Rome, for her invaluable books *Giochi descritti e illustrati dai bambini* and *Conte, cantilene e filastrocche*, and other kindnesses; Mrs. Philip Martin and Miss Barbara

Martin for information and demonstrations of games played in Montreal; Miss Elisabeth Nielsen for Danish games; Professeur Roger Pinon for the further parts as they appeared of his great work *La Nouvelle Lyre Malmédienne*; Professor Brian Sutton-Smith of the Teachers College, Columbia University, for his volume *The Games of New Zealand Children*, and for numerous off-prints; Mme M.-M. Rabecq-Maillard, Conservateur du Musée d'Histoire de l'Éducation, Paris, for following up inquiries; Mme H. A. Rasheed for descriptions of games played in Egypt; Dr. Ian Turner of Monash University, Victoria, for Australian material collected in preparation for his book *Cinderella Dressed in Yella*; Mr. Carl Withers, editor of the Dover edition of Newell's *Games and Songs of American Children*, for many books and kindnesses; Frl. Barbara Zeller for information from Zurich; and Dr. Dorothy Howard, editor of the Dover edition of Lady Gomme's *Traditional Games*, whose manuscript collections of games in Australia and the United States we have again consulted. Further, Mrs. Cecily Hancock (Storrs, Connecticut), Miss Myra Iwagami (Chicago), Mr. and Mrs. Nelles (Ottawa), and Professor Morris Silverman (New York), have kept continually alert to our interests sending books, press cuttings, and other materials; while Professor Archer Taylor of Berkeley, who remains at the centre of the folklore scene, has not only kept us informed of new European publications, but has presented us with a number of valuable works including Otto Kampmüller's *Oberösterreichische Kinderspiele*.

Likewise friends in Britain such as Lady Archibald, Miss Joan Ford, Miss Joan Hassall, Miss Carrol Jenkins, Mr. Roland Knaster, Miss B. A. Kneller, Mrs. Margaret Opie, Miss Jean C. Rodger, and Mr. and Mrs. Tom Todd, have continually kept their eyes open for us, sending press cuttings, children's books, and playthings; and just because Mr. James Opie, Mr. Robert Opie, and Miss Letitia Opie are our children, we should not, and do not, take for granted their continued collaboration now they are away from home.

We have also to thank Father Damian Webb O.S.B., the vividness of whose photographs is the result not just of technical skill, but of a deep understanding of children and their games; Mrs. Robert Berk (when Miss Jenny King) whose strenuous year on the survey was as good for our morale as it certainly was for our files; Mrs. Gillian Davies who, during her free time, was also industrious at our third desk; and Miss Gwen Twine who could not be with us a day let alone eight years without being a friend to this work.

Finally we are, as ever, deeply indebted to Miss F. Doreen Gullen of Scarborough who as our mentor examined our pages as they were written, and who as our friend somehow made it so much easier that they should be written.

Contents and Analysis

3. CATCHING GAMES

Games in which a player attempts to intercept other players
who are obliged to move from one designated place to another
(often from one side of a road to another), and who if caught
either take the catcher's place or, more often, assist him

4. SEEKING GAMES

Games in which a player tries to find others, who obtain safety by remaining out of sight or by getting back to the starting place

5. HUNTING GAMES

Games in which there are no boundaries, in which both pursuers and pursued generally operate in teams, and in which the pursued generally have to give some assistance to their pursuers

6. RACING GAMES

Races, and chases over set courses, in which fleetness of foot is not necessarily the decisive factor

§ *Races in which the Progress of Those Taking Part is Dependent on Their Fulfilling a Condition or Possessing a Particular Qualification*

7. DUELLING GAMES

Games in which two players place themselves in direct conflict with each other

§ *Contests Mainly Requiring Strength*

§ Contests Requiring Nerve and Skill

§ Contests Requiring Fortitude

§ Duels by Proxy

8. EXERTING GAMES

Games in which the qualities of most account are physical strength and stamina

9. DARING GAMES

Games in which players incite each other to show their mettle

10. GUESSING GAMES

Games in which guessing is a necessary prelude or climax to physical action

11. ACTING GAMES

Games in which particular stories are enacted with set dialogue

12. PRETENDING GAMES

Children make-believe they are other people, or in other situations, and extemporize accordingly

Distribution Maps

Introduction

'And the streets of the city shall be full of boys and girls
playing in the streets thereof.'

Zechariah, viii. 5

WHEN children play in the street they not only avail themselves of one of
the oldest play-places in the world, they engage in some of the oldest and
most interesting of games, for they are games tested and confirmed by
centuries of children, who have played them and passed them on, as
children continue to do, without reference to print, parliament, or adult
propriety. Indeed these street games are probably the most played, least
recorded, most natural games that there are. Certainly they are the most
spontaneous, for the little group of boys under the lamp-post may not
know, until a minute before, whether they are to play 'Bish Bash' or
'Poison' or 'Cockarusha', or even know that they are going to play.

A true game is one that frees the spirit. It allows of no cares but those
fictitious ones engendered by the game itself. When the players commit
themselves to the rhythm and incident of 'Underground Tig' or 'Witches
in the Gluepots' they opt out of the ordinary world, the boundary of their
existence becomes the two pavements this side of a pillar-box, their only
reality the excitement of avoiding the chaser's touch. Yet it is not only the
nature of the game that frees the spirit, it is the circumstances in which it is
played. The true game, as Locke recognized years ago, is the one that
arises from the players themselves.

'Because there can be no *Recreation* without Delight, which depends not
alway on Reason, but oftener on Fancy, it must be permitted Children not only
to divert themselves, but to do it after their own Fashion; provided it be
innocently, and without Prejudice to their Health.'

It may even be argued that the value of a game as recreation depends on
its inconsequence to daily life. In the games which adults organize for
children, or even merely oversee in a playground, the outside world is ever
present. Individual performances tend to become a matter for congratula-
tion or shame; and in a team game, paradoxically, individual responsibility presses hardest. The player who 'lets down his side' can cheer
himself only with the sad reflection that those who speak loudest about the
virtues of organized sport are the people who excel in it themselves, never

the duffers. He is not likely to have been told that such a man as Robert Louis Stevenson felt that cricket and football were colourless pastimes, scarcely play at all, compared with the romance of Hide and Seek.

THE APPEAL OF THE GAMES

Play is unrestricted, games have rules. Play may merely be the enactment of a dream, but in each game there is a contest. Yet it will be noticed that when children play a game in the street they are often extraordinarily naïve or, according to viewpoint, highly civilized. They seldom need an umpire, they rarely trouble to keep scores, little significance is attached to who wins or loses, they do not require the stimulus of prizes, it does not seem to worry them if a game is not finished. Indeed children like games in which there is a sizeable element of luck, so that individual abilities cannot be directly compared. They like games which restart almost automatically, so that everybody is given a new chance. They like games which move in stages, in which each stage, the choosing of leaders, the picking-up of sides, the determining of which side shall start, is almost a game in itself. In fact children's games often seem laborious to adults who, if invited to join in, may find themselves becoming impatient, and wanting to speed them up. Adults do not always see, when subjected to lengthy preliminaries, that many of the games, particularly those of young children, are more akin to ceremonies than competitions. In these games children gain the reassurance that comes with repetition, and the feeling of fellowship that comes from doing the same as everyone else. Children may choose a particular game merely because of some petty dialogue which precedes the chase:

> 'Sheep, sheep, come home.'
> 'We're afraid.'
> 'What of?'
> 'The wolf.'
> 'Wolf has gone to Devonshire
> Won't be here for seven years,
> Sheep, sheep, come home.'

As Spencer remarked, it is not only the amount of physical exercise that is taken that gives recreation, 'an agreeable mental excitement has a highly invigorating influence'. Thus children's 'interest' in a game may well be the incident in it that least appeals to the adult: the opportunity it affords to thump a player on the back, as in 'Strokey Back', to behave stupidly and be applauded for the stupidity, as in 'Johnny Green', to say aloud the colour of someone's panties, as in 'Farmer, Farmer, may we cross your Golden River?' And in a number of games, for instance 'Chinese Puzzle', there may be little purpose other than the ridiculousness of the experiment itself:

'Someone as to be on. The one who is on as to turn round while the others hold hands and make a round circul. Then you get in a muddle, one persun could clime over your arms or under your legs or anything ales that you could make a muddle, then when they have finished they say "Chinese Puzzle we are all in a muddle", then the persun turns round and goes up to them and gets them out of the muddle without breaking their hands, and then the persun who was on choose someone ales, and then it goes on like that. It is fun to play Chinese Puzzle.'

Here, indeed, is a British game where little attempt is made to establish the superiority of one player over another. In fact the function of such a game is largely social. Just as the shy man reveals himself by his formalities, so does the child disclose his unsureness of his place in the world by welcoming games with set procedures, in which his relationships with his fellows are clearly established. In games a child can exert himself without having to explain himself, he can be a good player without having to think whether he is a popular person, he can find himself being a useful partner to someone of whom he is ordinarily afraid. He can be confident, too, in particular games, that it is his place to issue commands, to inflict pain, to steal people's possessions, to pretend to be dead, to hurl a ball actually at someone, to pounce on someone, or to kiss someone he has caught. In ordinary life either he never knows these experiences or, by attempting them, makes himself an outcast.

It appears to us that when a child plays a game he creates a situation which is under his control, and yet it is one of which he does not know the outcome. In the confines of a game there can be all the excitement and uncertainty of an adventure, yet the young player can comprehend the whole, can recognize his place in the scheme, and, in contrast to the confusion of real life, can tell what is right action. He can, too, extend his environment, or feel that he is doing so, and gain knowledge of sensations beyond ordinary experience. When children are small, writes Bertrand Russell, 'it is biologically natural that they should, in imagination, live through the life of remote savage ancestors'. As long as the action of the game is of a child's own making he is ready, even anxious, to sample the perils of which this world has such plentiful supply. In the security of a game he makes acquaintance with insecurity, he is able to rationalize absurdities, reconcile himself to not getting his own way, 'assimilate reality' (Piaget), act heroically without being in danger. The thrill of a chase is accentuated by viewing the chaser not as a boy in short trousers, but as a bull. It is not a classmate's back he rides upon but a knight's fine charger. It is not a party of other boys his side skirmishes with but Indians, Robbers, 'Men from Mars'. And, always provided that the environment is of his own choosing, he—or she—is even prepared to meet the 'things that happen in the dark', playing games that would seem strange amusement if it was thought they were being taken literally: 'Murder in the Dark', 'Ghosties in the Garret', 'Moonlight, Starlight, Bogey won't come out Tonight'. And yet, within the context of the game, these alarms *are* taken literally.

THE AGE OF THE PLAYERS

When generalizing about children's play it is easy to forget that each child's attitude to each game, and his way of playing it, is constantly changing as he himself matures; his preferences moving from the fanciful to the ritualistic, from the ritualistic to the romantic (i.e. the free-ranging games, 'Hide and Seek', 'Cowboys and Indians'), and from the romantic to the severely competitive. The infants, 5–7 years old, may play some of the same games in their playground that the juniors do across the way, but in a more personal, less formalized style. Their chasing game, in which they clutch the railings to be safe, is called, perhaps, 'Naughty Boys'. 'We're playing naughty boys, we've run away from home.' ('Touch Iron' or 'Touch Green' is only emerging from make-believe.) The boys who are moving on hands and feet, stomachs upward, in another part of the playground, say they are being Creatures, 'horrible creatures in the woods'. The juniors, in the next playground, would not play like this, not move about publicly on hands and feet, unless, that is, it was part of a 'proper' game, one in which they were *chasing* each other.

Today, in an increasingly integrated society, children become self-conscious about the games they play on their own more quickly than they used to do. They discard them two, and even three years earlier than they did in the days before the introduction of organized sport. When Lord John Russell, aged 11, started his school-days at Westminster in 1803, he recorded in his diary that 'the boys play at hoops, peg-tops, and pea-shooters'. At Eton in 1766 the sports in vogue included Hopscotch, Headimy, Peg in the Ring, Conquering Lobs (marbles), Trap-ball, Chuck, Steal Baggage, and Puss in the Corner. At Sedgley Park School in Stafford-shire, about 1805, the boys were content with Kites, Marloes (marbles),

Peg-tops, Hoops, Backs (leap-frog), Beds (hopscotch), Cat, Rounders, Skipping, and even with 'playing horses'. Today few boys at grammar school would contemplate such sports. The experts are aged eight, nine, and ten. Even girls, who it might be thought would hold out against being organized, now look down upon informal games once they have a chance of taking part in the recognized sports:

'When I was five I played at Beddies. At seven I learned to play E.I.O. At nine I played "Alla Baba who's got the ball-a". Now I am fourteen I play tennis and netball, not the games I used to play when I was smaller.'

Indeed, a game which at one time was the breath of life to a child a short while later may be cast aside and become an embarrassment to remember. Thus a 10-year-old's description of 'Queeny' starts as follows:

'My favourite game is a game that lots and lots of children can play at once. It is a lovely game, children of two years old can play it right up to the age of twelve. It is very enjoyable. It has also a rhyme to it. It goes like this:

> Queeny, Queeny, who's got the ball,
> Is she fat or is she small,
> Is she big or is she thin,
> Does she play the violin,
> Yes or no?'

But a 14-year-old girl says of the game:

'When I was smaller I used to play with my friends games which seem very silly and babyish to me now . . . for example Queeny Ball. Now that was babyish! Today I would never dream of standing out in the street chanting "Queeny ball, ball, ball" but then I simply wallowed in such fun as I called it.'

Thus one child's attitude to a game may vary from another's to a greater extent than does either of them from that of the adult spectator. In fact when a child enters his teens (earlier if in the 'A' stream) a curious but genuine disability may overtake him. He may, as part of the process of growing up, actually lose his recollection of the sports that used to mean so much to him. As a 13-year-old wrote when describing 'Aunts and Uncles', 'King Ball', 'Kick the Can', 'Hide and Seek', and 'I Draw a Snake':

'All these games are quite common round Glenzier, that is how I can remember them. There are other games played once in a while but these I cannot remember. The games I can remember are a little too young for me to play, although I play King Ball sometimes.'

Older children can thus be remarkably poor informants about the games. Twelve-year-olds may be heard talking like old men and women ('Our street used to be very lively but children don't play the games like we used to do'). A 15-year-old in Liverpool, where 'Queeny' is popular, was certain the game was no longer known. Fourteen-year-olds, re-met in the street, from whom we wanted further information about a game they had showed us proudly a year before, have listened to our queries with blank incomprehension. Paradoxical as it may appear, a 5-year-old in his first term at school may well be aware of more self-organized games than a 15-year-old about to leave school.

THE AGE OF THE GAMES

These games that 'children find out for themselves when they meet', as Plato put it, seem to them as new and surprising when they learn them as the jokes in this week's comic. Parties of schoolchildren, at the entrance to the British Museum, secretly playing 'Fivestones' behind one of the columns as they wait to go in, little think that their pursuits may be as great antiquities as the exhibits they have been brought to see. Yet, in their everyday games, when they draw straws to see who shall take a disliked role, they show how Chaucer's Canterbury pilgrims determined which of them should tell the first tale. When they strike at each other's plantains, trying to decapitate them, they play the game a medieval chronicler says King Stephen played with his boy-prisoner William Marshal to humour him. When they jump on a player's back, and make him guess which finger they hold up ('Husky-bum, Finger or Thumb?') they perpetuate an amusement of ancient Rome. When they hit a player from behind, in the game 'Stroke the Baby', and challenge him to name who did it, they unwittingly illuminate a passage in the life of Our Lord. And when they enter the British Museum they can see Eros, clearly depicted on a vase of 400 B.C., playing the game they have just been told to abandon.

There was no need for Plato to urge that boys be forbidden to make alterations in their games, lest they be led to disobey the laws of the State in later life. Boys are such sticklers for tradition that after 2,000 years they have not yet given up Epostrakismos ('Ducks and Drakes'), Schoenophilinda ('Whackem'), Apodidraskinda (running-home 'Hide and Seek'), Ostrakinda (a form of 'Crusts and Crumbs'), Chytrinda ('Frog in the Middle'), Strombos ('Whipping top'), Dielkustinda ('Tug of War'), and at least two forms of Muinda ('Blind Man's Buff' and 'How Far to London'). Even the limping witch Empusa seems to have survived to this day in the guise of Limpety Lil.

If a present-day schoolchild was wafted back to any previous century he would probably find himself more at home with the games being played than with any other social custom. If he met his counterparts in the Middle Ages he might enjoy games of Prisoners' Base, Twos and Threes, street-football, Fox and Chickens, Hunt the Hare, Pitch and Toss, and marbles, as well as any of the games from classical times; and judging by the illuminations in the margins of manuscripts, he would be a prince among his fellows if he was good at piggyback fighting.

The Elizabethans played Bowls (one of their most common games), Barley-break, Stoolball, 'King by your Leaue' (a form of running-home Hide and Seek), 'Sunne and Moone' (Tug of War), and 'Crosse and Pile' (Heads and Tails). Shakespeare himself mentions 'All hid, all hid', 'Cherrie-pit', 'Fast and loose', 'Handy-dandy' (see *Oxford Dictionary of Nursery Rhymes*, pp. 197–8), 'Hide Fox and all after', 'Hoodman-blinde', 'Leape-frogge', 'Push-pin' (a game now played with pen-nibs), and 'Span-counter'.

Amongst games 'used by our countrey Boys and Girls', named by the Cheshire antiquary Randle Holme in 1688, were 'Battle-dore or Shuttle cock', 'Bob Apple' (see *Lore and Language of Schoolchildren*, pp. 272–3), 'Chase Fire', 'Drop Glove', 'Hare and Hound', 'Hide and seech', 'Hop skotches', 'Hornes Hornes', 'Fives', 'Jack stones', 'King I am', 'Long Larrance', 'Pi[t]ch and Hussle', and 'Puss in the corner'. And the 'inno-cent Games that Good Boys and Girls' diverted themselves with in *A Little Pretty Pocket-Book*, 1744, one of the first books to be published for juvenile amusement, included: Cricket, Base-Ball, 'Chuck-Farthing', 'Peg-Farthing', 'Taw' (marbles), 'Knock out and Span', 'Hop-Hat', 'Thread the Needle', 'All the Birds in the Air', and 'I sent a Letter to my Love'.

Even more revealing, perhaps, than the age of the games, is the persis-tence of certain practices during the games. The custom of turning round a blindfold player *three times* before allowing him to begin chasing seems already to have been standard practice in the seventeenth century. The quaint notion that a player becomes 'warm' when nearing the object he is seeking was doubtless old when Silas Wegg adopted it (*Our Mutual Friend*, III. vi). The stratagem of making players choose one of two objects, such as an 'orange' or a 'lemon', to decide which side they shall take in a pulling match, was almost certainly employed by the Elizabethans. The rule that a special word and finger-sign shall give a player respite in a game appears to be a legacy of the age of chivalry. The convention that the player who does worst in a game shall be punished, rather than that he who does best shall be rewarded, has an almost continuous history stretching from classical antiquity. And the ritual confirmation that a player has been caught, by crowning him or by tapping him three times, prevalent today even in such sophisticated places as Ilford and Enfield, was mentioned by Cromek in his *Remains of Nithsdale and Galloway Song* in 1810 ('If the intruder be caught on the hostile ground he is *taend*, that is, clapped three times on the head, which makes him a prisoner'), and is also the rule—as are other of these conventions—amongst children in France, Germany, Austria, Italy, and the United States.

VARIATION IN THE GAMES

If children played their games invariably in the way the previous generation played them, the study of youthful recreation could be a matter merely of antiquarian scholarship. But they do not. Despite the motherly influence of tradition, of which we have seen examples, children's play is like every other social activity, it is subject to continual change. The fact that the games are played slightly differently in different places, and may even vary in name, is itself evidence that mutation takes place. ('Chinese Puzzle', for instance, is also known as 'Chinese Muddle', 'Chinese Puddle', 'Jigsaw Puzzle', 'Chinese Knots', 'French Knots', 'Chain Man', 'Tangle Man', 'Policeman', and 'Cups and Saucers'.) In addition, as is well known, new sports emerge that may or may not in the course of time become traditional. (During the past decade there has been the 'Hula Hoop', 'Scoobeedoo', 'Split the Kipper', 'American Skipping', and 'Ippyop' or 'Belgian Skipping'.) And for reasons that are usually social or environmental, some games become impracticable (e.g. games played with caps are fast disappearing), while others are overlaid or replaced by new versions that are found to be more satisfactory. ('Conkers', played with horse chestnuts, which became possible with the introduction of the horse-chestnut tree, *Aesculus hippocastanum*, has now displaced the centuries-old contest with cobnuts.)

Yet the most fundamental kind of change that takes place is less obvious, although continual. This is the variation that occurs over the years in the relative popularity of individual games. At any one time some games are gaining in popularity; some, presumably, are at their peak; and others are in marked decline; and this variation affects not only the frequency with which each game is played but its actual composition. Thus games that are approaching their peak of popularity are easily recognizable, just as are customs and institutions that are nearing their zenith and about to decay.

A game enjoying absolute favour fatally attracts additional rules and formalities; the sport becomes progressively more elaborate, the playing of it demands further finesse, and the length of time required for its completion markedly increases. (In our day 'Statues' was a simple amusement of seeing who, after a sharp pull, could be the best statue; today it is a procedure-ridden pastime incorporating at least four additional operations.) Indeed, as a game grows in popularity its very name may grow. (Thirty years ago in Liss the old game known as 'Stoop' was already being called 'Three Stoops and Run for Your Life'; today, still more popular, it is 'Three Stoops, Three Pokers, and Run for Your Life'.) On the other hand, games which are in a decline lose their trimmings; the players become disdainful of all but the actual contest; the time-taking preliminaries and poetic formulas which gave the game its quality are discarded; and fragments of the game may even be taken over by another game that is on the up-grade (part of the introductory formula of 'Hickety, Bickety', for instance, is now repeated in the seeking game, 'North, South, East, West').

The identification and listing of games that have been declining in popularity over the past fifty years, and those that have been most noticeably gaining in popularity, may help to show the factors that currently affect children's choice of games.

Games diminishing in popularity	*Games growing in popularity*
Anything under the Sun	Bad Eggs (a ball game)
Baste the Bear (virtually obsolete)	Bar the Door
Blind Man's Buff	Block
Bull in the Ring	British Bulldog
Cat and Mouse	Budge He
Crusts and Crumbs	Donkey (a ball-bouncing game)
Duckstone (an aiming game)	Fairies and Witches
Finger or Thumb?	Farmer, Farmer, may we cross your
Fool, Fool, Come to School	Golden River?
French and English	Film Stars
Hide and Seek (the simple form)	Hi Jimmy Knacker
Honey Pots	I Draw a Snake upon your Back
I Sent My Son John	Jack, Jack, Shine a Light
King of the Castle	Kerb or Wall
Kiss in the Ring	Kingy
Knifie (Mumbletypeg)	Kiss Chase
Leapfrog	May I?
Odd and Even (a gambling game)	Off-Ground He
Old Man in the Well	Peep Behind the Curtain

Prisoners' Base
Sardines
Stag
Stroke the Baby (Hot Cockles)
Territories
Tipcat
Tom Tiddler's Ground
Touch Iron and Touchwood
Tug of War
Twos and Threes
Warning

Poison
Queenie (in its new form)
Relievo
Split the Kipper
Statues
Stuck in the Mud
Three Stoops and Run For Ever
Tin Can Tommy
Touch Colour
Truth, Dare, Promise, or Opinion
What's the time, Mr. Wolf?

There has been a marked decrease in the playing of games in which one player is repeatedly buffeted by the rest. (It is apparently not now felt as amusing as it used to be that one player should remain at a disadvantage indefinitely.) There has been an increase (possibly a corresponding increase) in the playing of games in which children fight each other on roughly equal terms. Above all, we feel it is no coincidence that the games whose decline is most pronounced are those which are best known to adults, and therefore the most often promoted by them; while the games and amusements that flourish are those that adults find most difficulty in encouraging (e.g. knife-throwing games and chases in the dark), or are those sports, such as ball-bouncing and long-rope skipping, in which adults are ordinarily least able to show proficiency.

WHERE CHILDREN PLAY: PLAYING IN THE STREET

Where children are is where they play. They are impatient to be started, the street is no further than their front door, and they are within call when tea is ready. Indeed the street in front of their home is seemingly theirs, more theirs sometimes than the family living-room; and of more significance to them, very often, than any amenity provided by the local council. When a young coloured boy from Notting Hill was being given a week's holiday in a Wiltshire village, and was asked how he liked the country, he promptly replied, 'I like it—but you can't play in the road as you can in London.'

Yet, as we know, Zechariah's vision of the new Jerusalem is not as splendid in practice as he makes it appear. Windows are liable to be broken, caretakers appear from blocks of flats telling the children to keep off the grass, obstinate car-drivers insist on making their way down the street, and, nightly, little dramas are enacted between 10-year-olds and the tendentious:

'We were having a lovely game of Relievo when a man across the road came out and moved us. My friend Ann said, "Oh shut up you're always moaning". Then the man said, "I will see your father about this it is going too far. Someone is trying to have a sleep." Then my other friend said "So is my dad". Then the man shouted for his dog Flash and sent him after us.'

What is curious about these embroilments is that children always do seem to have been in trouble about the places where they played. In the nineteenth century there were repeated complaints that the pavements of London were made impassable by children's shuttlecock and tipcat.[1] In Stuart times, Richard Steele reported, the vicinity of the Royal Exchange was infested with uninvited sportsmen, and a beadle was employed to whip away the 'unlucky Boys with Toys and Balls'. Even in the Middle Ages, when it might be supposed a meadow was within reach of every Jack and Jill in Britain, the young had a way of gravitating to unsuitable places. In 1332 it was found necessary to prohibit boys and others from playing in the precincts of the Palace at Westminster while Parliament was sitting. In 1385 the Bishop of London was forced to declaim against the ball-play about St. Paul's; and in 1447, away in Devonshire, the Bishop of Exeter was complaining of 'yong peple' playing in the cloister, even during divine service, such games as 'the toppe, queke, penny prykke, and most atte tenys, by the which the walles of the saide Cloistre have be defowled and the glas wyndowes all to brost'.

Should such persistent choice of busy and provocative play-places alert us that all is not as appears in the ghettos of childhood? Children's deepest pleasure, as we shall see, is to be away in the wastelands, yet they do not care to separate themselves altogether from the adult world. In some forms of their play (or in certain moods), they seem deliberately to attract attention to themselves, screaming, scribbling on the pavements, smashing milk bottles, banging on doors, and getting in people's way. A single group of children were able to name twenty games they played which involved running across the road. Are children, in some of their games, expressing something more than high spirits, something of which not even they, perhaps, are aware? No section of the community is more rooted to where it lives than the young. When children engage in 'Last Across' in front of a car is it just devilment that prompts the sport, or may it be some impulse of protest in the tribe? Perhaps those people will appreciate this question most who have asked themselves whether the convenience of motorists thrusting through a town or village is really as important as the well-being

[1] 'This mania for playing at cat', commented *Punch*, 23 April 1853, 'is no less absurd than dangerous, for it is a game at which nobody seems to win, and which, apparently, has no other aim than the windows of the houses, and the heads of the passengers.'

of the people whose settlement it is, and who are attempting to live their
lives in it.

PLAY IN RESTRICTED ENVIRONMENT

'It is a pleasant sight to see the young play with those of their own age
at tick, puss in the corner, hop-scotch, ring-taw, and hot beans ready
buttered: and in these boyish amusements much self-denial and good
nature may be practised. This, however, is not always the case . . .'

The Boy's Week-Day Book, 1834

The places specially made for children's play are also the places where
children can most easily be watched playing: the asphalt expanses of
school playgrounds, the cage-like enclosures filled with junk by a local
authority, the corners of recreation grounds stocked with swings and
slides. In a playground children are, or are not, allowed to make chalk
diagrams on the ground for hopscotch, to bounce balls against a wall, to
bring marbles or skipping ropes, to play 'Conkers', 'Split the Kipper',
'Hi Jimmy Knacker'. Children of different ages may or may not be kept
apart; boys may or may not be separated from girls. And according to the
closeness of the supervision they organize gangs, carry out vendettas, place
people in Coventry, gamble, bribe, blackmail, squabble, bully, and fight.
The real nature of young boys has long been apparent to us, or so it has
seemed. We have only to travel in a crowded school bus to be conscious of
their belligerency, the extraordinary way they have of assailing each other,
verbally and physically, each child feeling—perhaps with reason—that it is
necessary to keep his end up against the rest. We know from accounts of
previous generations with what good reason the great boarding schools,
and other schools following, limited boys' free time, and made supervised
games a compulsory part of the curriculum. As Sydney Smith wrote in
1810, it had become an 'immemorial custom' in the public schools that
every boy should be alternately tyrant and slave. The tyranny of the
monitors at Christ's Hospital, wrote Lamb, was 'heart-sickening to call to
recollection'. Southey's friend who went to Charterhouse was nearly killed
by the cruelty of the other boys who 'used to lay him before the fire till he
was scorched, and shut him in a trunk with sawdust till he had nearly
expired with suffocation'. Even at Marlborough, not founded until 1843,
a new boy might be branded with an anchor by means of a red-hot poker.
And at so tranquil-seeming an establishment as Harnish Rectory, run by
the Revd. Robert Kilvert, some of the boys were 'a set of little monsters' in
their depravity. 'The first evening I was there,' recalled Augustus Hare, 'at
nine years old, I was compelled to eat Eve's apple quite up—indeed, the

Tree of Knowledge of Good and Evil was stripped absolutely bare: there was no fruit left to gather.'

Such accounts, which can usually be reinforced by personal experience of school life, have increasingly influenced educational practice over the past hundred years, leading us to believe that a *Lord of the Flies* mentality is inherent in the young. Yet there is no certainty that this judgement is well founded. In one respect we remain as perverse as we were in Spencer's day, devoting more time to observing the ways of animals than of our own young. Thus recent extensive studies of apes and monkeys have shown, perhaps not unexpectedly, that animal behaviour in captivity is not the same as in the wild. In the natural habitat the welfare of the troop as a whole is paramount, the authority of the experienced animal is accepted, the idiosyncrasies of members of the troop are respected. But when the same species is confined and overcrowded the toughest and least-sensitive animal comes to the top, a pecking order develops, bullying and debauchery become common, and each creature when abused takes his revenge on the creature next weakest to himself. In brief, it appears that when lower primates are in the wild, and fending for themselves, their behaviour is 'civilized', certainly in comparison with their behaviour when they are confined and cared for, which is when they most behave 'like animals'.

Our observations of children lead us to believe that much the same is true of our own species. We have noticed that when children are herded together in the playground, which is where the educationalists and the psychologists and the social scientists gather to observe them, their play is markedly more aggressive than when they are in the street or in the wild places. At school they play 'Ball He', 'Dodge Ball', 'Chain Swing', and 'Bull in the Ring'. They indulge in duels such as 'Slappies', 'Knuckles', and 'Stinging', in which the pleasure, if not the purpose, of the game is to dominate another player and inflict pain. In a playground it is impracticable to play the free-ranging games like 'Hide and Seek' and 'Relievo' and 'Kick the Can', that are, as Stevenson said, the 'well-spring of romance', and are natural to children in the wastelands.

Often, when we have asked children what games they played in the playground we have been told 'We just go round aggravating people.' Nine-year-old boys make-believe they are Black Riders and in a mob charge on the girls. They play 'Coshes' with knotted handkerchiefs, they snatch the girls' ties or hair ribbons and call it 'Strip Tease', they join hands in a line and rush round the playground shouting 'Anyone who gets in our way will get knocked over', they play 'Tweaking', running behind a person and tweaking the lobe of his ear as they run off. One teacher, who asked her own 6-year-old what game he really enjoyed at school, was surprised to find it was 'getting gangs on to people'. He said, 'We get in a line and slap our sides as we run, and push down or bump a child.'

Such behaviour would not be tolerated amongst the players in the street or the wasteland; and for a long time we had difficulty reconciling these accounts with the thoughtfulness and respect for the juvenile code that we had noticed in the quiet places. Then we recollected how, in our own day, children who had seemed unpleasant at school (whose term-time behaviour at boarding school had indeed been barbarous), turned out to be surprisingly civilized when we met them in the holidays. We remembered hearing how certain inmates of institutions, and even people in concentration camps during the war, far from having a feeling of camaraderie, were liable to seek their pleasure in making life still more intolerable for those who were confined with them (see, for instance, Pierre d'Harcourt, *The Real Enemy*). It seems to us that something is lacking in our understanding of the child community, that we have forgotten Cowper's dictum that 'Great schools suit best the sturdy and the rough', and that in our continual search for efficient units of educational administration we have overlooked that the most precious gift we can give the young is social space: the necessary space—or privacy—in which to become human beings.

THE WASTELANDS

> 'There is no doubt that the first world war and the coming of the motor car killed, I suppose for ever, the playing of street games in this country.'
> H. E. Bates

Children are all about us, living in our own homes, eating at our tables, and it might be wondered how we ever supposed (along with H. E. Bates, J. B. Priestley, Richard Church, Howard Spring, and other professional observers of the social scene) that they had stopped playing in the way we ourselves used to play. Yet the belief that traditional games are dying out is itself traditional; it was received opinion even when those who now regret the passing of the games were themselves vigorously playing them. We overlook the fact that as we have grown older our interests have changed, we have given up haunting the places where children play, we no longer have eyes for the games, and not noticing them suppose them to have vanished. We forget that children's amusements are not always ones that attract attention. They are not prearranged rituals for which the players wear distinctive uniforms, freshly laundered. Unlike the obtrusive sports of grown men, for which ground has to be permanently set aside and perpetually tended, children's games are ones which the players adapt to their surroundings and the time available. In fact most street games are as happily played in the dark as in the light. To a child 'sport is sweetest when there be no spectators'. The places they like best for play are the secret places 'where no one else goes'.

The literature of childhood abounds with evidence that the peaks of a child's experience are not visits to a cinema, or even family outings to the sea, but occasions when he escapes into places that are disused and over-grown and silent. To a child there is more joy in a rubbish tip than a flowering rockery, in a fallen tree than a piece of statuary, in a muddy track than a gravel path. Like Stanley Spencer he may 'see more in a dustbin in his village than in a cathedral abroad'. Yet the cult amongst his elders is to trim, to pave, to smooth out, to clean-up, to prettify, to convert to economic advantage—as if 'the maximum utilisation of surrounding amenities' had become a line of poetry.

Ironically the bombing of London was a blessing to the youthful generations that followed. 'We live facing a bombsite where boys throw stones, light fires, make camps and roast potatoes', writes an 11-year-old. 'In my neighbourhood,' wrote a Peckham child in 1955, 'the sites of Hitler's bombs are many, and the bigger sites with a certain amount of rubble provide very good grounds for Hide and Seek and Tin Can Tommy.' To a child the best parts of a park are the parts that are the least maintained. It is his nature to be attracted to the slopes, the bushes, the long grass, the waterside. 'Ours is a good park there are still places in it that are wild', observed a 10-year-old. But what do the authorities do? They exploit our wealth to make improvements for the worse. They invade the parks, erecting kiosks and tea gardens, and side-shows for those who require their entertainments ready-made. It is not only Battersea Park (the enchanted garden of our childhood) that has been turned into a honky-tonk. The trend is universal. In one small town we know there are some municipal gardens, the only place where children can play, and on the largest lawn they have laid-out and fenced off an immense bowling green for the summer pleasure of the middle-aged. The centre of our own home town possessed, miraculously, until two years ago, a small dark wood adjoining a car park. If an adult entered its shade he might imagine he was alone, unless he became aware that the trees above his head were a play-ground for Lilliputians. Now the trees have been cut down, the ground levelled, a stream canalized, and the area flooded with asphalt to make an extension to the car park. Should we be surprised if children play around the cars, if cars get damaged, if sometimes boys are tempted to more serious offences? Having cleared away the places that are naturally wild it is becoming the fashion to set aside other places, deposit junk in them, and create 'Adventure Playgrounds', so called, the equivalent of creating Whipsnades for wild life instead of erecting actual cages. The next need is to advertise in *The Times Educational Supplement* (for example 8 February 1963) for Play Leaders at 32s. for 2½ hours: apply Chief Officer (A/B/197/2) L.C.C. Parks Department, County Hall, S.E. 1. (WAT 5000 Ext. 7621) P.K. A2. Or, more recently (24 January 1969) for 'Senior Play Leader' at 40s. 6d. for 2½ hours: apply Parks Department, Cavell House, 2a Charing Cross Rd., W.C. 2, 836 5464, Ext. 144. The provision of playmates for the young has become an item of public expenditure.

In the past, traditional games were thought to be dying out, few people cared, and the games continued to flourish. In the present day we assume children to have lost the ability to entertain themselves, we become concerned, and are liable, by our concern, to make what is not true a reality. In the long run, nothing extinguishes self-organized play more effectively than does action to promote it. It is not only natural but beneficial that there should be a gulf between the generations in their choice of recreation. Those people are happiest who can most rely on their own resources; and it is to be wondered whether middle-class children in the United States will ever reach maturity 'whose playtime has become almost as completely organized and supervised as their study' (Carl Withers). If children's games are tamed and made part of school curricula, if wastelands are turned into playing-fields for the benefit of those who conform and ape their elders, if children are given the idea that they cannot enjoy themselves without being provided with the 'proper' equipment, we need blame only ourselves when we produce a generation who have lost their dignity, who are ever dissatisfied, and who descend for their sport to the easy excitement of rioting, or pilfering, or vandalism. But to say that children should be allowed this last freedom, to play their own games in their own way, is scarcely to say more than John Locke said almost three centuries ago:

'Children have as much a Mind to shew that they are free, that their own good Actions come from themselves, that they are absolute and independent, as any of the proudest of your grown Men.'

And speaking of their recreation he observed how it was freedom 'they extreamly affect'; it was 'that Liberty alone which gives the true Relish and Delight to their ordinary Play Games'.

I

Starting a Game

'Zig zag zooligar
Zim zam bum.'
A Manchester 'dip'

SUCH is the capacity of the young for turning whatever they do into
a sport, that collecting players for a game can be a game in itself. Two or
three children, arms round each other's shoulders, reel across the playground chanting in the way they have heard other children chant before
them: 'Who wants a game of *Sticky Toffee*? Who wants a game of *Sticky
Toffee*?' Those who want to play attach themselves to the line, the line
becomes unwieldy, the chant becomes a roar, there are more than enough
players for the game, yet nobody now seems in the least inclined to break
the line and begin playing. 'Sticky Toffee', or whatever the game proposed, has been forgotten in the very success of their summons to play it,
a summons which varies according to local prescription, for instance: 'All
join-ee join-ee up', or 'Arly-arly-in, who's a playing?', or 'All in, all in—
hands in the dip'. In Pontypool:

> All in, all in, a bottle of gin,
> All out, all out, a bottle of stout.

In Brighouse:

> All here who's laiking [playing],
> Mary Ann's baking.

And in Whalsay, Shetland, when a game of 'Aggie Waggie' is proposed:

> 'Whā's cŏmĭn ĭ wīr fŭn āggĭe wāggĭe?'

('They seem able to fit any name into the rhythm', commented their
teacher.)

Sometimes, however, playground recruitment is less voluntary. Two
or three youngsters form a ring round a solitary child and threaten: 'Have
a smack or join in the ring', or 'Pinch, punch, or join the bunch', or

> Pinch, punch, join in the ring,
> Pinch, punch, no girls in.

('Please, miss,' said a little girl to the teacher on playground duty, 'that boy keeps hitting me and says it's a game.')

In some places the press-gang think they will be successful if they demand, 'Join the ring or tell us your sweetheart's name.' In Chingford:

> Pinch or punch or join in the ring,
> If you don't you'll have to sing
> Or tell us your true love's name.

And in East Dulwich:

> Pinch or punch or join in the ring,
> Or tell us the name of your sweetheart,
> Or do you believe in Santa Claus,
> Or have you a house of your own?

While in Hinckley, Leicestershire, they ask—with what counts in the playground as craft—'Eggs or Bacon or join the ring?' If the child chooses either eggs or bacon he is told they are 'Not done yet', and he is, they feel, left with no option but to join the ring.

AVOIDANCE OF DISLIKED ROLE

On occasion a game starts in a flash, the players themselves hardly knowing how it began. More often children feel a game is not a proper one if such matters have not first been settled as who is to be allowed to play, what the boundaries are to be, and whether dropping-out is to be permitted. Yet the chief impediment to a swift start is the fact that in most games one player has to take a part that is different from the rest; and all children have, or affect to have, an insurmountable objection to being the first one to take this part. Tradition, if not inclination, demands that they do whatever they can to avoid being the chaser, or the seeker, or the one who, as they express it, is 'on', 'on it', 'he', or 'it'. Thus it is recognized throughout Britain (by everyone except the slowcoach) that the last person to exempt himself shall be the first to be 'on'; and this rule is so embedded in children's minds that their immediate response to the proposal of a game is to cry out 'Bags no on', 'I bags not on it', 'Me fains first', 'Foggy not on', or whatever is the locally accepted term of exemption. (In Banbury it is 'Baggy laggy', in Bishop Auckland 'Nanny on', in Forfar 'Chap no out', in Wigan 'Brit'.) On occasion, even the person suggesting the game may feel he must safeguard himself by saying in one gulp, 'Let's-play-Tig-fains-I-be-on-it'.

Yet, in some places, words are not enough to give exemption. In Norfolk the person who becomes the chaser is the one who last bobs down to the ground saying 'Vains'. In Putney it is he or she who last touches the ground, turns around, and says 'Bags I'm not on it'. In Swansea it is the child who last says and acts upon the words 'Not on this tippits', touching his shoe-cap; or 'Tippits, touch the ground, turn around, no back answers, one, two, three'. In Ruthin one child holds out his arms in a circle in front of him, shouting 'Hands in the bucket', and the last person to put his hands in the circle is 'out'. In many places one child will cry 'Last off ground is it' or 'Last on high is it'. Alternatively they have a race to a lamp-post or to the end of the playground and the last to arrive is made the chaser; for it is apparently felt not unsatisfactory that the slowest runner should be the one who has to pursue them in the game. And sometimes, according to time-honoured precedent, the last person to arrive for the game finds himself welcomed with genuine warmth, for it is he who has to take the unwanted role.[1]

Such methods of settling who shall be 'on' appear eminently fair to the alert; and they will try any device to make someone other than themselves be he. They say, 'A, B, C, D, F, G, H. What have I missed out?' and when some half-awake replies 'E', he is told 'That's right, you are!' They say 'Whose shoes are the shiniest?' and when a player claims the distinction he is promptly given the important role. They shout 'Cannon' or 'Quick Fire', and the player who is dolt enough to inquire what this means will have talked himself into a job. In parts of Lincolnshire (e.g. Cleethorpes and Market Rasen) the children who are to play a game silently form a circle, giving each other instructions by signs, and even pushing each other into place without saying anything, for 'the first person to speak is *it*'. In some places, indeed, a player may require fortitude if he is to avoid the obnoxious role. In Sale, after the boys have formed a circle, one of them walks round pretending to kick each lad on the shins, and 'the one who flinches is on'. At Chapeltown, near Sheffield, the boys make a circle, putting one foot forward into the ring, and a player takes round

---

[1] That this was also the custom in the seventeenth century seems evident from John Suckling's allegorical description of 'Barley-break'. In this game there were three pairs of players, of whom one pair were the catchers and had to stand in mid-field, designated 'Hell' (see hereafter p. 129). Suckling's account, printed in *Fragmenta Aurea*, 1646, p. 24, begins:

> Love, Reason, Hate, did once bespeak
> Three mates to play at barley-break;
> Love, Folly took; and Reason, Fancy;
> And Hate consorts with Pride; so dance they:
> Love coupled last, and so it fell
> That Love and Folly were in hell.

a brick which he pretends to drop on their foot to make one of them move. And at Lydney, says a 13-year-old boy, there is no pretending about it:

'When you want to start a game you pick a stone and throw it at the people's feet. The first to jump back or say anything is on it.'

HE-NAMES

Children's dislike of being the player with a certain power in a game, of being the one, that is, from whom the others flee or hide, has led some folklorists to suppose that this player originally represented a being who was evil or supernatural. They observe that in France and Germany such a player is sometimes termed a wolf, that in Spain he is 'El Dimoni', and in Japan, likewise, 'Oni'. In some games in Britain, too, the chaser is 'Old Mr. Wolf', 'Old Mother Witch', or 'the Devil'; yet in the majority of instances there seems little reason to think these roles have, or ever had, much significance. When awe of the Evil One is genuine, people fear even to pronounce his name. They refer to him obliquely as 'Old Harry', 'Old Nick', 'Old Splitfoot', 'Old Scratch', 'the old one', 'the gentleman down-stairs', or use even more indirect terms. Parish, for instance, in his *Sussex Dialect*, 1875, noted that the devil was always spoken of as 'he', with a special emphasis. ('In the Downs there's a golden calf buried.' 'Then why döant they dig it up?' 'Oh, it is not allowed; *he* would not let them.')

It is, of course, true that in London and southern England the player is usually called 'he' (often spelt 'hee', although pronounced 'ee', and some-times actually believed to be 'E'); and that in the south-west, the mid-lands, and north, the ordinary term is 'it'. (The player who is 'it' is then said to be either 'on' or 'on it'. Thus a child may say 'Who's going to be it?' Everyone cries 'No it!' and the last person to say 'no it' is 'on it'.) But although these expressions have a regional bias (see maps), it is difficult to learn much about their age. They have, it seems, always been exclusive to children, and consequently their appearances in literature are rare. The following are the terms that have been found in Britain for the player with the operative power in a game, or for the state of having this power.

Catcher. Surprisingly rare as the standard term, other than in Orkney, Shetland, Norwich, and Widecombe-in-the-Moor.

Daddy. In the district of the Lenches, near Evesham, the chaser was known as 'Daddy' in such games as 'Catch and Kill', 'Cabbage and Bacon', 'Jack Fox', and 'Daddy Daddy Touchwood' (*A Worcestershire Book*, 1932, pp. 45–50). It is not clear whether the term comes from the dialect *dad*, to hit or touch, as appears from such names as 'Off-Ground Daddy' and 'Cross Dadder'; or whether it is from *Daddy* meaning Father, see under 'Daddy Whacker'.

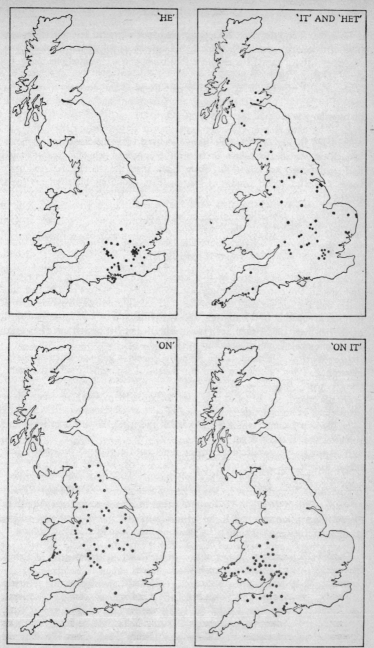

FIGS. II–V. Places where the terms 'He', 'It' and 'Het', 'On', and 'On it' predominate.

Done. Perth and district. 'We say a grace to see who is to be done' (Boy, 11). ' "You're done" is one of the many terms used at school. It simply means you are "het" ' (Girl, 11). 'This rhyme is said when you are picking one out to be done' (Girl, 11, Luncarty).

Has it on. Kirkwall, Orkney. 'One person has it on and must try to catch the rest' (Boy, 12). 'The catcher tries to touch someone and when she does the player who was touched then has it on' (Girl, 12).

He. The chaser seems to have been known as 'he' before the game was called 'He'. In *The Boy's Handy Book of Sports*, 1863, the chaser in a game of 'Touch' is specifically called *he* 'to use the approved schoolboy expression'. Earlier, in *Games and Sports for Young Boys*, 1859, the chaser is referred to as *touch* or *he*. In *Gammer Gurton's Garland*, 1810, a counting-out rhyme is given ending:

> Stick, stock, stone dead,
> Blind man can't see,
> Every knave will have a slave,
> You or I must be HE.

Het. Apparently produced by the Scots practice of broadening *i* into *e*, and prefixing an *h* to a word beginning with a vowel. 'The players race to a certain object and the last person there is "het" ' (Kinlochleven). 'The person that is heat has to run after it' (Edinburgh, also 'het', 'hit', and 'it'). 'If you are touched by the het-man you are then het' (Isle of Bute). 'The last person there has to go het' (Cumnock). Compare children in Holland, 'Ik ben het'.

Him. Whalsay, Shetland. 'I like when the person is him and when he chases me around. I don't like when I am him because I sometimes can't catch them' (Boy, *c.* 10).

Hit. Scotland, see *het*. But it is also to be heard in southern England in places such as Portsmouth.

In. Current around Axminster, Devon, in the 1920s. 'Two has to be in and run after the others.' Cf. *on.*

It. Apparently as widespread in the nineteenth century as it is today (see map). Robert Chambers, born in Peebles in 1802, says of a game such as 'Hid-ee': 'The *tig* usually catches and touches some one upon the crown, before all are in—otherwise he has to be *it* for another game' (*Popular Rhymes of Scotland*, 1842, p. 62). A Devonshire writer in 1864 reported that the player chosen for a game was said to be 'of it' (*Notes and Queries*, 3rd ser., vol. v, p. 395). Cf. the German child's 'Es', and the French child's 'Il l'est', a term of long standing (it appears in *Les Jeux des jeunes garçons*, *c.* 1810).

King. The accepted term in Orkney, the Isle of Lewis, and parts of Caithness, for as long as man can remember. Also common in Scotland for the chief player in a game such as 'You can't cross the Golden River', and formerly in the game 'England and Scotland' where, says Cromek, 'a king is chosen as leader of either party' (*Remains of Nithsdale and Galloway Song*, 1810, p. 251). In ancient Greece and Rome the leader of a game, or the player who did best, was called 'king', the duffer or player who did worst being the 'ass'. Thus in Plato's *Theaetetus*, § 10:

"He that mistakes, and as often as anyone mistakes, shall sit as an ass, as the boys say when they play at ball; but whoever shall get the better without making a mistake shall be our king, and shall order any question he pleases to be answered.'

Man, mannie, old man. Scotland, north country, Liverpool, Lincolnshire, East Anglia. 'First we decide who is to be "mannie", which just means the person who is to be out or the person who is the catcher' (Girl, *c.* 11, Aberdeen). 'When you are caught three times you are old man then it is your turn to chase the others' (Boy, 13, Attleborough, Norfolk). Hence the trick boys have in Liverpool of asking players if they are man or woman: 'If they say man they are man and have to chase you'. Doubtless the term has long been in use, but it has not been found before 1894 when a South Shields correspondent to *Notes and Queries*, 8th ser., vol. vi, p. 155, wrote: 'In this neighbourhood there used to be, and I dare say still is, a game which we called "tiggy touch wood", where if the "man" succeeded in touching a boy before he could touch wood, he in turn became the "man".'

On or *on it*. To be 'on' or 'on it' is to be in the state of being the player who has special power within the game, other than leadership. 'We play Tag. It starts by somebody touching another and saying "Touch, you're on it"' (Girl, 14, Newbridge, Monmouthshire). 'If there are a lot of people playing you will need two people on' (Girl, 9, Wolstanton). It will be seen from the maps (Figs. IV and V) that 'on' predominates in the north and midlands, 'on it' in south Wales and Wessex. Cf. the German 'dran'.

Outer. London. The common name for the player who is separated from the rest in games such as 'Kerb or Wall' or 'Peep Behind the Curtain'.

Tag, Tagger, Ticker, Tig, Tigger, Tiggy. A chaser was formerly often known as 'tag', 'tig', or sometimes 'tiggy'. 'The moment this person is touched, he or she becomes *Tig*' (*Scottish Dictionary*, 1825). 'The boy he has touched is made tiggy until he can pay the compliment to some other boy' (*Youth's Own Book*, 1845, p. 217). Today, although not often, such a player is the tigger, tagger, or ticker. 'One person is the tagger and has to count to thirty' (Girl, *c.* 11, Plympton St. Mary). 'The last one out is ticker' (Boy, 10, Oundle).

Ten. Flotta, Orkney. 'If he is caught he has to go ten with the one who was ten first' (Boy, 11). From *Ta'en*?

Touch. Formerly common, now rare. 'One volunteers to be the player who is called Touch; it is the object of the other players to run from and avoid him' (*Boy's Own Book*, 1829, p. 24). 'Last one across the road has got to be Touch' (Boy, 14, Ipswich).

Under. Liverpool and the Potteries. 'The one who was caught first is under in the next game' (Girl, 11, Stoke-on-Trent).

Up. Laverstock, Maidenhead, Portsmouth. 'We get a lot of boys and dip who's going to be up' (Boy, 9, Maidenhead).

SELECTION MADE BY CHANCE

When players exert themselves to avoid being 'on' the procedure is simple, the execution quick, but it is not always decisive, and not invariably the saving in time and temper that it ought to be. 'We all shout "no on", and we keep shouting "no on", and we go on and on until it is near the end of playtime and we don't start the game at all, like', confessed one lad. In consequence most children prefer the allotment of the disliked role to be a matter of chance. They feel that if the choice has been made by providence there is no possibility of argument, especially if the choice has fallen on someone other than themselves. Thus one boy will stand in the middle of a circle, shut his eyes, hold out his arm, and turn round and round until he has no idea which way he is facing. The player he is pointing at when he stops is 'on'. Likewise a player turns round and round with his eyes shut and throws a ball, and whoever he happens to hit is 'on'. Or a player bounces a ball in the middle of a circle, each player having his legs wide apart, and the one through whose legs the ball rolls is 'on' (see further under 'Kingy', pp. 96–7). Or a boy puts his hand behind his back, holds up a certain number of fingers, and the child who is unlucky enough to say the correct number is 'on'. Or one player goes out of hearing, the rest pick colours, including a colour for the person who is away, and when the person comes back and names a colour, the player who has chosen that colour is declared 'on'.

Such methods of determining who is to take the unpopular role are obvious enough, and might occur to anyone. Yet children also indulge in practices that are not likely to come to the minds of any but the initiated, and which, indeed, have something in common with old forms of divination now supposedly forgotten. Thus girls in South Elmsall, aged nine and ten, sometimes use mud to decide who is to be the chaser.

'We dip our fingers in the mud and the people playing choose a finger each first, and if the finger they have chosen come out with the most mud that person is "on".'

And again:

'A person wets all her fingers and wipes them on the ground. The one that is wettest, for example the fourth, well the fourth girl is "on".'[1]

Sometimes the procedures they adopt to ascertain the whim of fate on such a minor matter seem unnecessarily tortuous. One player will be appointed to make six piles of soil, or as many piles as there are players, each pile to look exactly alike, but one pile to conceal a stone. Each player

[1] Cf. Mme de Chabreul, *Jeux des jeunes filles*, 1856, p. 2, n.: 'Souvent, pour indiquer la personne qui dirigera le jeu, ou qui y remplira un certain rôle, on tire au sort, par le *doigt mouillé*.' See also Littré's *Dictionnaire* under *doigt mouillé*, where he explains that the child who picks the damp finger wins or loses what has been agreed upon.

then chooses a pile and picks it up, and 'the one who picks up the stone is "on" '. And again, one player makes a pool of spit on the back of his hand, smacks it with his index finger, and everyone notes in whose direction the spittle flies.[1] And, not uncommonly, they engage in the old and international practice of drawing lots or 'cuts'. They collect straws, or plantain stalks, or pieces of twig, or matchsticks, or ice-lolly sticks, and make one a different length from the rest. Then one person holds them in his fist so that only the ends show, and each player takes one. The player who draws the length that is different from the rest (which may be the shortest or the longest) finds himself with no option but to be the chaser.

⁎⁎ In such a manner Chaucer's Canterbury pilgrims decided who should tell the first tale (*Prologue*, 835–6):

> Now draweth cut, er that we ferrer twynne;
> He which that hath the shorteste shal bigynne.

And in our own time, the girls in Mary McCarthy's *The Group*, 1963, ch. i, arranged by drawing straws who should take on the joyless duty of inviting Kay home for the holidays.

In Japan, too, boys habitually decide their disputes by drawing straws, which they call 'kuji'. In China (Canton) the practice of drawing straws is or was prevalent, and called 'Ts'ím ts'ò' (Stewart Culin, *Korean Games*, 1895, p. 52). In Macedonia Turkish children were found choosing sides with a long and a short straw in the 1920s (*Folk-Lore Society Jubilee Congress*, 1930, p. 156). In bar rooms in South Africa, we are informed, men may be seen drawing matchsticks to decide who shall stand the next round, just as drinkers in France did in the Middle Ages with straws. The game 'Erbelette', referred to by Froissart, may have been a sport of this kind, although the more usual names in medieval France seem to have been 'Courtes-pailles', 'Longs festuz', and 'Court festu'. In fact Jehan Palsgrave, *Lesclarcissement de la langue françoyse*, 1530, is explicit: 'I drawe lottes, or drawe cuttes, as folkes do for sporte, *je joue au court festu.*'

ODD MAN OUT

Another way a player can condemn himself to the unwanted role is by the process known as 'Chinging up' or 'Odd Man Out'. For this operation, much resorted to in Greater London, the players stand in a circle facing inwards, with their hands behind their backs, and chant in unison certain

---

[1] Compare a method of testing a boy's truthfulness cited in *The Lore and Language of Schoolchildren*, p. 128.

words, which vary from district to district, but are in Walworth, for example:

'Allee in the middle, and the odd man's out!'

On the word *out* they whip their hands from behind their backs, holding them in front for all to see, either with their fists clenched, or with their fingers stretched out palms downwards, or with their hands clenched but first two fingers spread out. They then look round to see if one player is 'odd', that is to say holding his hands in one of the three positions but different from everyone else, in which case that player becomes 'ee'. If no one is odd they try again; and sometimes increase the likelihood of one player being odd by introducing a fourth finger position, known as 'grab' or 'crane', in which the fingers point downwards.

Alternatively, the players pair off and play against each other, either bringing their hands out from behind their backs, or dabbing them in the

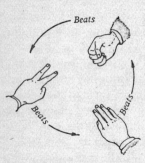

air three times in front of them, and making the finger formation at the third dab, synchronizing their movements with three vocables such as 'Ick, ack, *ock*'. The finger formations they present now have significance: the clenched fist represents 'stone', the flat hand 'paper', and the two extended fingers 'scissors'. If both players chance to produce the same sign it is a 'wash out', and they try again. But if they produce different signs one of them inevitably wins since it is held that 'stone' blunts and thus beats 'scissors'; 'scissors' cut and thus conquer 'paper'; and 'paper' wraps round and thus triumphs over 'stone'. Each winner then plays another winner until there is only one boy left. This form of elimination often goes under the name of the sounds with which they synchronize their movements, thus 'Ching, Chang, Cholly' (South London), 'Chu Chin Chow' (Enfield), 'Dib, Dob, Dab' (Camberwell), 'Ding, Dang, Dong', 'Dish, Dash, Dosh', or 'Zig, Zag, Zog' (Southwark), 'E.I.O.', or 'Ick, Ack, Ock' (Croydon), 'Hick, Hack, Hock' (East Barnet), 'Eee, Pas, Vous' (Lambeth), and 'Stink, Stank, Stoller' (Brixton).

In Manchester, where the procedure is known as 'Flee, Fly, Flo, Bank', the players divide into groups of three, and act slightly differently. They swoop their fists from side to side while chanting 'Flee, fly, flo', and then on '*bank*' they flatten out their hands, either palms up or palms down. The odd man in each group then plays the odd men of two other groups; while if there are only two players left the decision is made with what they term 'Sizz, back, or brick', which in practice is 'scissors, paper, or stone', although the players themselves do not think of their words as having any representational significance, and we admit we did not understand them ourselves at first.

As is well known 'Paper, Scissors, Stone', or 'Rock, Paper, Scissors', can also be a diversion between friends to while away the time and see who wins most often; and it is also the basis of the rather less friendly contest known as 'Stinging', which is described on p. 225 under Duels.

*** Children in London have a fixed idea that 'chinging up' is oriental, and for once a folk-theory may be correct. In Japan it is a commonplace for children to determine priorities, or settle disputes, by waving a closed hand in the air three times, while chanting the meaningless words 'Jan Ken Pon', and then making exactly the same finger formations that British children do, and with the same signification, 'hasami' (scissors), 'kami' (paper), 'ishi' (stone), or in dialect, 'choki, pä, gu:'. Indeed 'Jan Ken Pon' is so ordinary in Japan that even professors resort to it to decide, for instance, which of them shall drink the next drink at a draught, or which of them wash up afterwards. Likewise in the great ports of China, if not elsewhere, children and grown-ups alike resort to 'Chai Ken' to decide anything trivial, the decision (in Shanghai) usually being the best of three. In Hong Kong, where men commonly play for drinks, the only observable difference in the finger signs is that for 'paper' the flat hand is held vertically, not horizontally as in England. Similarly in Indonesia, a traveller has told us of her astonishment at seeing children squatting in the shade playing the game she remembered from her childhood in a north London suburb (she knew it as 'Hic Haec Hoc'), although in Indonesia the game is 'earwig, man, and elephant', the earwig overcoming the elephant by crawling through his brain. In Abyssinia it appears they compete against each other with up to eight finger formations of different value; in Choa, for instance, they have needle, sword, scissors, hammer, the Emperor's razor, sea, altar, and sky (*Jeux abyssins*, M. Griaule, 1935, p. 189). In Egypt, too, children play a finger-sign game, placing their hands in a pile, palms downwards, and then on the command 'Eat okra', pulling their hands from the pile and holding them either with palms upwards or palms downwards, they then look around to see which player is different from the rest, precisely as young Mancunians do when playing 'Flee Fly Flo Bank'. Furthermore, judging by a scene in one of the tombs at Beni Hassan in Middle Egypt, it appears that finger-flashing games have been known in Egypt since about 2000 B.C. According to Wilkinson's delineations in *Ancient Egyptians*, vol. i, 1878, p. 32, two players are shown flashing a certain number of fingers simultaneously, and it is probable that both guessed what their total would be, as in the game 'Micatia' or 'micare digitis', played in ancient Rome, and 'Morra' played in Italy today—as also 'Ch'ái múi' in China, and 'Tōjin Ken' in Japan. This well-known gambling game, although not in itself closely related to 'Odd Man Out', is yet part of the story, since in classical times it seems to have been the recognized method for deciding who should have—or avoid having—first turn in a game. Thus Calpurnius Siculus, about A.D. 60, describes two shepherds agreeing to play the best of three goes ('ter quisque manus jactate micantes'), to determine which of them should sing first in a contest (*Bucolica* ii. 26). And the poet Nonnus (fifth century A.D.) tells of Eros and Hymen deciding by the same process which should have first throw in the wine-throwing game 'Cottabe', explaining, even, that they did this with the varied movements of the fingers, holding out some, while keeping others pressed to the palm of the hand (*Dionysiaca* xxxiii. 77–80).

DIPPING

Fanciful as it would seem to somebody who had never been a child, the normal way the young decide who is to have the unpopular part in a game is to form the players up in a line or circle, and count along the line the number of counts prescribed by the accented syllables of some little rhyme, such as the following which has fifteen counts:

> Err'ie, orr'ie, round' the ta'ble,
> Eat' as much' as you' are a'ble;
> If' you're a'ble eat' the ta'ble,
> Err'ie, orr'ie, *out*!

One child gabbles the words at speed, pointing briefly at each player in turn as he does so, and if there are less than fifteen players, he continues round the circle or along the line a second time, counting himself in first. The player the last count falls on is then either made the chaser, and the game begins; or, more often, he is counted 'out' and stands aside while the rhyme is repeated and a second player eliminated, and so on, until only one player remains—on whom the count has never fallen—and that player is the unlucky one. Virtually every child in England now calls this procedure 'dipping', a term that seems to have become general only in the early 1940s. It has not been known to our older correspondents; it does not appear in the accounts of games we possess written by children in the 1920s and 1930s; and it is not current in Scotland, Canada, or the U.S.A.

In the 1920s and 1930s, however, children used to touch the ground or point at the ground when they began counting, saying 'Dip' as they did so: 'Dip—Eeny, meeny, miney, mo', 'Dip—Each, peach, pear, plum'; and it is possible that this gave rise to the term, 'dipping' being easier to say than 'counting-out'.[1] Such a derivation would be in keeping with the origin of several other juvenile terms. For instance, games sometimes acquire their names from rhymes repeated when playing them (e.g. 'Blackthorn', 'Peter Pan', and 'North, South, East, West'); and there are earlier terms for counting-out whose origins appear to be analogous. In Kettering, about 1920, the process of counting-out was known as 'imbering', from the then popular rhyme 'Imber ormer dasma dormer'; in Workington, about the same time, children spoke of 'inking it out', presumably with reference to the old rhyme 'Ink, pink, pen and ink'; and in Aberdeen where the term dipping remains unknown, a 13-year-old girl told us she 'coonted a gipsy to see who was the mannie', meaning that to find who would be the chaser she used the rhyme:

> Gipsy, gipsy, lived in a tent,
> Couldn't afford to pay the rent,
> When the rent man came next day,
> Gipsy, gipsy, ran away.

Other names there are or have been for dipping include:

Chappin' out. Mentioned by Robert Chambers, *Popular Rhymes of Scotland*, 1842, p. 62: 'He is chosen by lot, or, as the boys express it, by chappin out.'

Counting a pie. The usual term in Aberdeen. 'French Tick and Tack is played by counting a pie and the odd man out is the mannie' (Girl, 13). In the nineteenth century counting-out was often conducted by each boy putting a finger in somebody's cap, where the fingers were then counted, and the counter was not readily able to identify whose finger he counted out. This was known as 'putting in the pie'.

Counting-out. The usual term in Scotland, some parts of the north country, and in the United States. Also general amongst those who write about children. 'The operation of counting-out is a very important mystery in many puerile games', observed J. O. Halliwell in 1849.

[1] The practice of saying 'dip' before starting the rhyme apparently dates from the nineteenth century. In the *Journal of American Folklore*, vol. x, 1897, p. 319, appears the following rhyme from Penzance, Cornwall:

> Dip!
> Ickery, ahry, oary, ah,
> Biddy, barber, oary, sah,
> Peer, peer, mizter, meer,
> Pit, pat, out one.

Children seem to feel that if the first count is to the ground, the subsequent counting is more magical, or is anyway less under the control of the counter. In France when children start in a similar fashion, they say 'pouce'.

Deeming. Current about 1925 at Hurworth, County Durham.

Dishing up. In Walworth children speak of 'dipping out', 'dipping up', or 'dishing up'.

Picking. Current in the West Riding. Thus in Leeds when a player wishes to do the counting he cries 'Ferry for picking!' In Hull in the 1890s it was 'picking out' or 'knocking out'.

Telling out. The editor of *Gammer Gurton's Garland*, 1810, gives the heading 'Telling out' to the counting-out rhyme 'One-ery, two-ery, ziccary zan'. Mrs. Baker, in her *Northamptonshire Glossary*, 1854, calls such rhymes *tells*. 'One of the number repeats a *tell*, touching each play-mate in succession with the forefinger as she repeats each word, spelling the last, and the one whom the last letter falls to is to commence the game, or to preside over it.'

Titting out. Boys in the south of Scotland and in the Lake District were said to be 'titted out' when eliminated with the rhyme:

> Tit, tat, toe, here I go,
> And if I miss I pitch on this.
>
> *Mill Hill Magazine*, vol. v, 1877, p. 95

DIPS

> 'There are so many dips that I've lost count.'
> *Girl, 9, West Ham*

To the outsider it appears that any wretched doggerel will do for a dipping rhyme. A 'dip', it seems, can be of almost any length, cast in almost any mould, be either sense or nonsense (usually the latter), and need not even be a rhyme.[1] On occasion the dipper may merely count to twenty-one, and whoever the number twenty-one falls on becomes the catcher. He may recite the alphabet to the twenty-first letter, 'A, B, C, D, E, F, G, H, I, J, K, L, M, N, O, P, Q, R, S, T, *You*', or just repeat the vowels, 'A, E, I, O, *You*', or say 'One, two, sky blue, all out but you', or 'Red, white, and blue, all out but you', or occasionally 'A, B, C, D, E, F, G, H, I, J, K, L, M, N, O, P, Q, R, S, T, U—are—He'; or in areas where the chaser is known as 'it'—A, B, C, D, E, F, G, H, I for It' (Pendeen), or 'I – T spells it, thou art it' (Blackburn), or 'Tom Tit you are it' (Cleethorpes).

More often, as we have already said, the process of dipping is more anxious and time-taking: the dip has to be repeated as many times, less one, as there are players, since the player the dip ends on is not selected but eliminated:

[1] Several collections have been made of dipping rhymes, and references to the following in the next few pages are given with only the editor's name or short-title: Henry Carrington Bolton, *The Counting-out Rhymes of Children*, 1888; W. Gregor, *Counting-out Rhymes of Children*, 1891; Patricia Evans, *Who's It?*, 1956; Jean Baucomont, Roger Pinon, and others, *Les Comptines de langue française*, 1961; Matizia Maroni Lumbroso, *Conte, cantilene e fila-strocche*, 1965.

> Ip, dip, sky blue,
> Who's it? Not you.
> God's words are true,
> It must not be *you*.
> *East Dulwich*

The person pointed to stands aside, and the rest face the ordeal of being counted again. Thus certain phrases are thought appropriate simply because they end with the word *out*, as 'Pig's snout, walk out', and 'Boy Scout walk out' which sometimes gets extended to:

Boy Scout walk out,   *Or*,   Boy Scout walk out
Girl Guide step aside.   With your breeches inside out.

(In Glasgow, 'Oot Scoot you're oot'.) Likewise: 'Ice cream sold out', 'Wring the dirty dishcloth out', 'Egg shells inside out', 'Wee jelly biscuit is out', 'In pin, safety pin, in pin out', 'Little Minnie washed her pinnie in-side-out', 'I-N spells in, O-U-T spells out'. Sometimes they have nothing more original to offer than 'Horsie cartie rumble oot' (Aberdeen), 'Eggs and ham, out you scram' (Lydney), 'Elephant trunk, out you bunk' (Market Rasen), 'Smack, wallop, thump, you're knocked out' (Wilmslow), 'A car went up the hill and conked out' (Bristol), and 'Black shoe, brown shoe, black shoe, out' (Barrow-in-Furness). Yet, despite appearances, it is clearly not true that any phrase will do; for unless the children being counted are already familiar with the words the dipper is using, they cannot be sure that he is being fair. In consequence we find the same formulas used over and over again not only in one place by one group of children, but by many children over wide areas, and even around the English-speaking world. Further it appears that this has long been the case; and many of the dips familiar today have been doing playground duty for longer than the oldest teacher can remember. The words 'One, two, sky blue, all out but you', may be nonsensical, but their eight counts have been found to rule playgrounds alike in Manchester and Maryland, and appear to have ordained which child should be chaser at least since Victorian days (see Bolton, 1888, p. 92, and Gregor, 1891, p. 30). Likewise 'Red, white, and blue, all out but you', is known to have been recited in America as well as Britain for eighty years or more, and it was not necessarily new when Bolton came across it in Pennsylvania in 1888. The ubiquitous 'Pig's snout, walk out', has been recollected by correspondents who were at school in 1910; and the equally improbable 'Egg shells inside out' was already on the lips of Lancastrian children in the 1920s. The Scottish 'Wee jelly biscuit is out', collected today in Luncarty,

Helensburgh, and Bute, has been current at least since 1911 (*Rymour Club Miscellanea*, vol. ii, p. 70). And children in Alton were still commemorating in the 1960s the long defunct Board of Education:

> London County Council, L.C.C.,
> Board of Education, you are he.

The durable nature of these dips becomes still more evident when versions of a single rhyme, or rhyme-form, are set side by side. For instance, children have delighted in the following spell-like jingle for the past eighty years; and the recordings show both how the number of counts in a dip can remain constant, despite differences in wording and even meaning; and how variations can in themselves be traditional and, as it were, 'correct' for particular areas.

Iggy oggy,
Black froggy,
Iggy oggy out.
Girl, 13, Dulwich

Iddy oddy,
Dog's body,
Iddy oddy out.
Girl, 13, Ruthin

Iddy oddy,
Dog's body,
Inside out.
Lancashire, 1920s

Iddy oddy,
Cock's body,
Inside out.
Northumberland, c. 1890

Iddle oddle,
Black poddle,
Iddle oddle out.
Somerset, 1922

Ibble obble,
Black bobble,
Ibble obble out.
Gloucester, Forest of Dean, Lydney, Newport, South Molton, Welshpool, and Croydon

Ibble ubble,
Black bubble,
Ibble ubble out.
Children, Hampstead

Ickle ockle,
Black bottle,
Ickle ockle out.
Girls, Swansea

Ickle ockle,
Ink bottle,
Out goes she.
New York, Lincolnshire, c. 1930

Ickle ockle,
Chockle bockle,
Ickle ockle out.
9-year-olds, Titchmarsh, Northamptonshire

Ickle ockle,
Chockle chockle,
Ickle ockle out.
Witherslack, Westmorland, 1937

Ickle ockle,
Chocolate bottle,
Ickle ockle out.
Manchester and Radcliffe; also Oldham, c. 1933

Iggle oggle,
Blue bottle,
Iggle oggle out.
Girl, 14, Ruthin

Iggle oggle,
Black bottle,
Iggle oggle out.
Girl, Truro

Eettle ottle,
Black bottle,
Eettle ottle out.
> Girl, 12, Aberdeen; also from Helens-
> burgh and Orkney, 1961, Berwick,
> c. 1935, Forfar, c. 1910, Stromness,
> 1909

Eettle ottle,
Black bottle,
You are out.
> Girl, 12, Edinburgh; and Gregor, 1891,
> Edinburgh

Eettle ottle,
Black bottle,
My dog's deid.
> Forfar, c. 1910; and Gregor, 1891,
> Fraserburgh

Eetter otter,
Potter bottle,
Out jumps the cork.
> Boy, 10, Dublin

Eetie ottie,
Horses are naughty,
Eetie ottie out.
> Girl, 14, Aberdeen

Ingle angle,
Silver bangle,
Ingle angle out.
> Children, c. 10, Newcastle upon Tyne,
> North Shields, Birmingham, Welshpool

Ingle angle,
Golden bangle,
Ingle angle out.
> Boy, 12, Golspie

From these examples it will be seen that the 'Eettle ottle' versions belong to Scotland, that the ones beginning 'Iddy oddy' are traditional in the north-west, and that 'Ibble obble' appears to centre on the Severn. In addition, extended forms of these jingles have long coexisted, for example:

Eettle, ottle, black bottle,
Eettle ottle out;
If you want a piece and jam,
Please step out.
> Aberdeen, Ballingry, Edinburgh, Flotta,
> Forfar, Golspie, Kirkcaldy, Luncarty,
> and Stromness

Eettle ottle, black bottle,
Fishes in the sea,
If you want a pretty one,
Just catch—me!
> Young children, Ballingry

Ickle, ockle, black bockle,
Fishes in the sea,
If you want a pretty maid,
Please choose me.
> James Kirkup, 'The Only Child',
> p. 147, referring to South Shields,
> c. 1926

Ickle ockle, blue bottle,
Fishes in the sea,
If you want a pretty maid,
Please choose me.
> Girl, 13, Spennymoor

Ickle ockle, black bottle,
Ickle ockle out,
If you want a lump of jelly,
Please walk out.
> Knighton, and similar Ruthin

Ickle ockle, blue bottle,
Ickle ockle out,
If you see a policeman,
Punch him on the snout.
> Girl, c. 9, Ipswich

Ingle angle, silver bangle,
Ingle angle out,
If you want another bangle,
Please walk out.
> Girls, c. 11, Swansea

Ingle angle, silver bangle,
Ingle angle out,
Turn the dirty dish cloth
In – side – out.

Forest of Dean

Hibble hobble, black bobble,
Hibble hobble out,
Turn the dirty dish cloth
In – side – out.

Forest of Dean

Eettle ottle, black bottle,
Eettle ottle out.
If you had been where I had been,
You would not have been out.

Kirkcaldy; also Glasgow, c. 1925

Eettle ottle, black bottle,
Eettle ottle out,
Tea and sugar is my delight,
And o-u-t spells out.

Edinburgh; also Midlothian, c. 1905

Eatle autle, blue bottle,
Eatle autle out,
Tea and sugar's my delight,
Tea and sugar's out.

Children, Kirkcudbright, 1904

Eettle ottle, black bottle,
Eettle ottle out,
Shines on the mantelpiece,
Just like a threepenny piece,
o-u-t spells out.

Girl, c. 11, Aberdeen

Eetle ottle, black bottle,
Eetle ottle out,
Standing on the mantelpiece,
Like a shining threepenny piece,
Eetle ottle, black bottle,
Eetle ottle out.

Edinburgh, c. 1900, and Caithness, c. 1915[1]

Needless to say these rhymes with additional lines are ill received if the players do not expect them. They know the count will now fall on a different person, and feel certain they are being imposed upon if the person the count falls upon is themselves. In many places a special ending is customary, and there can be no cause for complaint. The dipper regularly ends a rhyme with the words 'o'-u'-t' spells' out', and' out' you' must' go'', or 'o'-u'-t' spells' out', please' walk' out'', or 'One', two', three', out' goes' she''. In parts of Aberdeen the counterman may add:

> Out goes a bonny lass, out goes she,
> Out goes a bonny lass, one, two, three.

In Harrogate:

> Raggle taggle dish cloth torn in two,
> Out goes you.

In Gloucester:

> o-u-t spells out, so out you must go,
> With a dirty dish cloth on your head because I said so.

---

[1] Compare:
> Look upon the mantelpiece,
> There you'll find a ball of grease
> Shining like a threepenny piece,
> Out—goes—she.
>
> Stratford-le-Bow, 1890s, and 'London Street Games', 1916, p. 56

And in Market Rasen:

> O-U-T spells out, so out you must go,
> With a dirty wet dish cloth wrapped round your big toe.[1]

In Liverpool the authority of the dip is felt to be enhanced when the dipper adds:

> O-U-T spells out, and out you must go,
> Because the king and queen says so.

In Birmingham the dipper may add:

> And out you go with a jolly good clout
> Upon your ear-hole spout;

and he tries to put his words into practice; as he also does in Blackburn, when he says:

> If you do not want to play
> Go away with a jolly good slap across your face like that.

And in Manchester, as in many other places, the dipper ends:

> And if you do not want to play
> Just take your hoop and run away.[2]

But should a girl from Portsmouth be on holiday, and extend a rhyme with lines her country cousins have not heard before, they are not likely to be impressed with their fairness, even if the words she says are:

> As fair as fair as it can be,
> The king of Egypt said to me,
> The one that comes to number three
> Must be he. One – two – three.

THE MOST-USED DIPS

In general the best-known dips today were also the favourites fifty and sixty years ago; and the following verses though not the most beautiful compositions in the English language, must be some of the most useful.

[1] Dish cloths or clouts (usually old, dirty, or torn in two) have for years been verbally awarded to the player who is counted out. In *The American Boy's Book of Sports and Games*, 1864, p. 32, one of the 'best known' rhymes ends:

> O-U-T—spells out,
> With the old dish-clout—
> Out, boys, out!

[2] At the beginning of the century this was, in the slang of the time,

> If you do not want to play
> You can 'sling your hook' away.

(*Rymour Club*, vol. i, 1911, p. 237, and *London Street Games*, 1916, p. 63.)

Eeny, meeny, miney, mo,
Catch a —— by the toe,
If he squeals let him go,
Eeny, meeny, miney, mo.

> *Current for at least eighty years. See
> 'Oxford Dictionary of Nursery Rhymes',
> pp. 156-7*

Eeny, meeny, miney, mo,
Sit the baby on the po,
When he's done
Wipe his bum,
Tell his mummy what he's done.

> *Current since the nineteenth century*

Each, peach, pear, plum,
Out goes Tom Thumb;
Tom Thumb won't do,
Out goes Betty Blue;
Betty Blue won't go,
So out goes you.

> *The first couplet, at least, has been
> current since c. 1915*

Inky, pinky, ponky,
My daddy bought a donkey,
The donkey died,
Daddy cried,
Inky, pinky, ponky.

> *Current since c. 1900*

If you had been where I'd been
You'd have seen the fairy queen;
If you'd been where I've been
You'd have been out.

> *Possibly a descendant of the Jacobite
> song 'Killiecrankie':*
>
> > *An' ye had been where I ha'e been
> > Ye wadna been sae cantie, O;
> > An' ye had seen what I ha'e seen,
> > I' the braes o' Killiecrankie, O.*

Ink, pink, pen and ink,
Who made that dirty stink?
My mother said it was you.

> *Current for at least fifty years[1]*

A pennorth of chips
To grease your lips.
Out goes one,
Out goes two,
Out goes the little boy
Dressed in blue.

> *Current since the 1920s*

Dip, dip, dip,
My blue ship,
Sailing on the water
Like a cup and saucer
Dip, dip, dip,
You are not it.

> *Current for fifty years, and possibly the
> most popular dip today amongst small
> girls. Alternatively:*
>
> > *Dash, dash, dash,
> > My blue sash, etc.*

Red, white, and blue,
The cat's got the flu,
The baby has the whooping cough
And out goes you.

> *Cf. 'Lore and Language of School-
> children', p. 106*

Round and round the butter dish
One, two, three,
If you like a nice girl,
Please pick me.

> *A version, 'Round about the punch
> bowl', appears in 'Traditional Games',
> vol. ii, 1898, p. 84*

---

[1] In Lancashire:

> Ip, dip, pen and ink,
> Who made that great big stink?
> I believe it was you.
> You shall have it for your supper
> On a piece of bread and butter.
> O-U-T spells out.

Dipping rhymes are not often rude, although small facetiae such as 'Ip, dip, bull's shit, you are not it', and 'Blib, blob, hoss tod, blib, blob, out', are repeated with open humour.

Oh deary me,
Mother caught a flea,
Put it in the kettle
To make a cup of tea.
The flea jumped out,
And bit mother's snout,
In come daddy
With his shirt hanging out.

> *As well known in late nineteenth century
> as today. See also 'Lore and Language
> of Schoolchildren', p. 19*

I know a washerwoman,
She knows me.
She invited me to tea.
Have a cup of tea, ma'm?
No, ma'm.
Why, ma'm?
Because I have a cold, ma'm.
Out goes she.

> *Current since before the First World
> War. Part quoted in 'London Street
> Games', 1916, p. 60*

Paddy on the railway
Picking up stones;
Along came an engine
And broke Paddy's bones.
Oh, said Paddy,
That's not fair.
Pooh, said the engine-driver,
I don't care.

> *Current since the beginning of the cen-
> tury[1]*

Have a cigarette, sir?
No, sir.
Why, sir?
Because I've got a cold, sir.
Let me hear you cough, sir.
Very bad indeed, sir.
You ought to be in bed, sir.
O-U-T spells out.

> *Robert Graves printed a version, 'I have
> a little cough, sir', in his 'Less Familiar
> Nursery Rhymes', 1926*

Old Father Christmas,
What do you think he did?
He upset the cradle,
And out fell the kid.
The kid began to bubble,
He hit it with a shovel,
O-U-T spells out.

> *All recordings from north of the Wash,
> other than one from Cape Town. Known
> in Halifax, c. 1900*

One, two, three, four, five, six, seven,
All good children go to heaven.
Penny on the water,
Tuppence on the sea,
Threepence on the railway,
Out goes she.

> *Current since the 1880s*

Two, four, six, eight,
Mary's at the cottage gate,
Eating cherries off a plate,
Two, four, six, eight.

> *Cf. 'Oxford Dictionary of Nursery
> Rhymes', p. 334, where the rhyme is
> traced back to Regency days*

Hickety pickety i sillickety
Pompalorum jig,
Every man who has no hair
Generally wears a wig.
One, two, three,
Out goes he.

> *General since nineteenth century[2]*

---

[1] In other versions of this encounter the aggrieved party is said to be 'Piggy', 'Peggy', 'Polly', 'Tommy', or 'Teddy'. The rhyme apparently stems from an old Scots ditty 'Pussy at the fireside suppin' up brose'. See our *Puffin Book of Nursery Rhymes*, 1963, p. 140.

[2] The saying 'Every man who has no hair may lawfully wear a wig' appears in *Peter Prim's Pride, or Proverbs, that will suit the Young or the Old*, 1810.

Oranges, Oranges, four a penny,
All went down the donkey's belly;
The donkey's belly was full of jelly,
Out goes you.

The first line introduced several other puerile rhymes in the nineteenth century

Three white horses
In a stable,
Pick one out
And call it Mabel.

Best known in the north country

The following are favourites in Scotland:

Skinty, tinty, my black hen,
Lays an egg for gentlemen,
Sometimes nine, and sometimes ten,
Skinty, tinty, my black hen.

Already current in 1853. See 'Oxford Dictionary of Nursery Rhymes', pp. 201–2

Eachie, peachie, pearie, plum,
Throw the tatties up the lum.
Santa Claus got one on the bum,
Eachie, peachie, pearie, plum.

Ayr, Cumnock, Flotta, Forfar, and Golspie

Engine, engine, number nine,
Runs along the bogey line,
Pea scoot, you're oot,
Engine, engine, number nine.

This version has been repeated since the 1920s. Others date back to nineteenth century[1]

Eenty, teenty, orry, ram, tam, toosh,
Ging in alow the bed, an catch a wee
  fat moose.
Cut it up in slices, and fry it in a pan,
Mind and keep the gravy for the wee
  fat man.

Aberdeen and Edinburgh

Three wee tatties in a pot,
Tak ain oot and see if it's hot,
If it's hot cut its throat,
Three wee tatties in a pot.

The first couplet was known in Philadelphia by 1888. Bolton no. 716

Oh dear me,
Ma grannie catcht a flea,
She roastit it an toastit it,
An' took it till her tea.

Popular throughout this century, and recited by J. J. Bell's 'Wee Macgreegor', 1902. 'Sich vulgarity!' exclaimed Aunt Purdie

⁎⁎* Presumably there was once a time when children were so eager to begin their games that they could do so without the benefit of rhyme, but it is difficult now to imagine. There were, or so it seems, special formulas for counting-out in the seventeenth century, anyway in France. Cotgrave in 1611 described the game 'Defendo' as:

'A play with bits of bread (ranked one by another) which the player counts with certaine words, and the last his words end on, he takes, whither it be little or great.'

[1] See *Midwest Folklore*, Winter 1951, p. 255. Compare:

Engine number nine;
Ring the bell when it's time.
O-U-T spells out goes he,
Into the middle of the dark blue sea.

Pennsylvania, Bolton, 1888, no. 709

Engine, engine, number nine,
Running on Chicago line;
When she's polished, she will shine.
Engine, engine, number nine.

North Carolina Folklore, vol. i, 1952, p. 168, from 1925

Some rhymes which are known to be old, such as 'Who comes here? A Grenadier' (part-quoted 1725), and 'Hickory, Dickory, Dock' (recorded 1744), are or have been at one time used for dipping. But no English counting-out rhyme, as such, is known to us earlier than the one the antiquary Francis Douce collected from his 'pretty little Sister Emily Corry' in 1795 (Bodley, Douce Adds. R 227):

> Doctor Foster was a good man,
> He whipped his scholars now and then,
> And when he had done he took a dance
> Out of England into France;
> He had a brave beaver with a fine snout,
>> Stand you there out.

It is possible, however, that the 'certaine words' which Cotgrave knew were not words merely discounting sense, such as we have recorded so far, but were an example of the outright gibberish that some children still favour for counting-out.

CHINESE COUNTING

'This is one you won't know,' said a 10-year-old, 'because it has only just been made up:

> Addi, addi, chickari, chickarı,
> Oonie, poonie, om pom alarie,
> Ala wala whiskey,
> Chinese chunk.'

It was not for us to tell her that children knew it in other parts of the country, and said it was 'Chinese counting'; that children knew this or similar nonsense in the United States, and thought it was 'Indian counting'; that its progenitor had been known in Britain and America when her great-grandmother was a child;[1] and that people had often wondered about such gibberish and were not certain where it came from, but tended to honour it with an ancestry more remarkable than the imaginings of any child.

Possessing now, as we do, above a thousand recordings of gibberish counting-out rhymes, the possibility is alluring that something will be

---

[1] Compare:

Ra, ra, chuckeree, chuckeree,	Rye, chy, chookereye, chookereye,
Ony, pony,	Choo, choo, ronee, ponee,
Ningy, ningy, na,	Icky, picky, nigh,
Addy, caddy, westce,	Caddy, paddy, vester,
Anty, poo,	Canlee, poo.
Chutipan, chutipan,	Itty pau, jutty pau,
China, chu.	Chinee Jew.
Fraserburgh, Gregor, 1891, p. 30	*Pennsylvania, 'Journal of American Folklore', vol. x, 1897, p. 321*

discovered if the rhymes are ordered geographically, or chronologically, or like-with-like, or indeed all these ways at once. It has been found, for instance, that there is one piece of gibberish that reigns supreme in British playgrounds, and that it is everywhere recited with extraordinarily little variation. Thus in Scotland:

Eeny, meeny, macca, racca,
Rae, rye, doma, anca,
Chicca, racca,
Old Tom Thumb.

Girl, c. 11, Aberdeen

Eeny, meeny, maca, racar,
Er, I, domeraca,
Ali baba, sugaraca,
Om, tom, toosh.

Girl, c. 11, Helensburgh

Eeny, meeny, macker, acker,
Ere, o, dominacker,
Ala packa, pucker acker,
Um, pum, push.

Girl, c. 12, Kirkcaldy

Eany, meany, maca, raca,
Red rose, doma naca,
Ali Baba, suva naca,
Rum, tum, toosh.

Boy, 12, Langholm

In England, proceeding southwards:

Eeny, meeny, mack-a, rack-a,
Rare, o, domino,
Ala-balla, jooba-lalla,
Hom, pom, flesh.

Boy, 13, Bishop Auckland

Ena, mena, macka, racka,
Rai, ri, domi nacka,
Chika lolla, lolla poppa,
Wiz, bang, push.

Girl, 13, Cleethorpes

Eeny, meeny, mackeracka,
Rari, jackeracka,
Rari, sackeracka,
Pon, pom, puss.

Girl, 11, South Elmsall

Eenie, meenie, macca, racca,
Ie, rie, dumma racca,
Ticka racca, lollipop,
Rum, pum, push.

Children, c. 10, Birmingham

Eeni, meeni, macaraca,
Rare, ri, domeraca,
Chiceraca,
Rom, pom, push.

Girl, 9, Ipswich

Eenimeenimackeracka,
Airidominacka,
Chickawallalollipopa,
Ompompush.

Boy, 12, Croydon

Iney, meney, macker, acker,
Air, I, donnal macker,
Chica chica, wolla wolla,
Om, pom, push.

Boy, 11, Oxford

Eani, meani, macker, racker,
I, o, domin acker,
Chicka pocka, lolipoppa,
Om, pom, push.

Boy, c. 12, Portsmouth

In Wales:

Ina, mina, maca, raca,
Re a, rom, domi naca,
Chica pica, lollie popa,
Om, pom, push.

Children, Amlwch, Anglesey

Eny, meeny, maca, racka,
Rare ri, dom er racka,
Chic a packa, lollipap,
You are out.

Girl, c. 12, Welshpool

Ina, mina, macaraca,
Ri a rie, a dominacer,
Chica boccer, lollipoper,
Rum, tum, tush.
Girl, 13, Ruthin

Eeny, meeny, macca, racca,
Ere, ree, dominacca,
Icaracca, omaracca,
Om, pom, push.
Girls, c. 12, Swansea

In Australia:

Eena, meena, micka, macka,
Eyre, eye, domma nacka,
Icky chicky,
Om, pom, puss.
Children, Melbourne, 1922

In New Zealand:

Eeny, meeny, macka, racka,
Rare, rye, domma nacka,
Chicka pocka, ellie focka,
Om, pom, puss.
Girl, 11, Wellington

Each day of the year hundreds of children must initiate their chosen game with a recital of this count, sometimes leading into the gibberish with the words:

> I went to a Chinese laundry
> To buy a loaf of bread;
> They wrapped it up in a tablecloth
> And this is what they said:
> *Eenie, meenie, macca, racca, etc.*

There are many children who know no other gibberish rhyme like 'Eenie, meenie, macca, racca'. They are amused by it, intrigued by it, and treasure it as an old and special possession. Indeed the meaningless words would appear to have been carefully passed down to them by previous generations as a talisman from the long ago, when it might be presumed to have been understood, and to have had wonderful significance, were it not that its recorded history goes back no further than our own day. 'Eenie, meenie, macca, racca' was not known to Bolton in 1888, nor to Gregor in 1891, nor to the alert members of the Rymour Club collecting before the First World War. We have no records of it ourselves before the 1920s. Earlier than this it is found only in embryo:

Eener, deener, abber, dasher,
Ooner, eye-sher,
Om, pom, tosh.
Iggery-eye, iggery-eye,
Pop the vinegar in the pie,
Harum scarum, pop canarum,
    Skin it.
London, c. 1910 (two recordings)

Haberdasher, isher asher,
Om, pom, tosh.
London Street Games, 1916, p. 95

Ena, dena, dasha, doma,
Hong, pong, toss.
Stromness, 1909. 'Notes and Queries', 10th ser., vol. xi, p 446.

Further, 'Eenie, meenie, macca, racca' seems to be unknown, or little known, in the United States, where the nearest equivalents popular today have some soda added:

Icka backa, icka backa,
Icka backa boo;
Icka backa, soda cracka,
Out goes you.

Acker backer, soda cracker,
Acker backer boo;
If your father chews tobacco
He's a dirty Jew.

The seeds of 'Eenie, meenie, macca, racca' are nowhere to be discovered, unless they are present in a rhyme which now has only local currency:

Inty, minty, tipedy, fig,
Delia, doilia, all munig,
Eicha, peicha, dol muneicha,
Om, pom, tush.
                        Girl, 11, Ruthin

Inty, minty, tipsy, tee,
I-la, dila, dominee,
Occapusha, dominusha,
Hi, tom, tush.
                        Yorkshire, 3 recordings

Yet 'Inty minty' is a count that has always had a more lively existence in America than in Britain. Not only did Bolton collect thirty versions in the United States in 1888, but the rhyme appears to have been as well known there in the mid nineteenth century as in the mid twentieth century:

Eeny, meeny, tipty, te,
Teena, dinah, domine,
Hocca, proach, domma, noach,
Hi, pon, tus.
        *Philadelphia, 'Notes and Queries', 1st
        ser., vol. xi, 1855, p. 113*

Aila, maila, tip-tee tee,
Dila, dila, dominee,
Oka, poka, dominoka,
High prong tusk . . .
        The American Boy's Book, 1864, p. 32

Inty, minty, tippety, fig,
Delia, dilia, dominig;
Otcha, potcha, dominotcha,
Hi, pon, tusk.
Huldy, guldy, boo,
Out goes you.
        *Hartford, Connecticut. Bolton, 1888,
        no. 624*

Eenie, meenie, tipsy, toe,
Olla, bolla, domino;
Okka, pocha, dominocha,
Hy, pon, tush.
O-U-T spells out goes he,
Right in the middle of the dark
    blue sea.
        Washington, D.C. Bolton, 1888, no. 621

Inty, minty, tibbity, fee,
Delia, doma, domini;
Eenchi, peenchi, domineechi,
Alm, palm, pus,
Alicka, balicka, boo,
Out goes Y-O-U.
        *New York City, 1938. Dorothy
        Howard MSS.*

Impty, dimpty, tibbity fig,
Delia, dauma, nauma, nig,
Heitcha, peitcha, dauma neitcha,
Ein, pine, pug,
Ullaga, bullaga, boo,
Out goes YOU.
        *San Francisco. Patricia Evans, 'Who's
        It?' 1956, p. 17*

Where then did such a formula come from? If every known version was fed into a computer would its original home be found on the continent of Europe? Compare:

Eenie, meenie, tipsey, tee,
Alabama, dominee,
Hocus, pocus, deminocus,
I pon tust.
> *Pasadena, California, 1938. Dorothy Howard MSS.*

Anemane, mikkelemee,
Hobbel, den dobbel, den dominee,
Flik, flak, floot, eik en lood,
Jij bent dood.
> *'Nederlandsche Baker en Kinderrijmen', 1874, cited by Bolton*

Engete pengete zukate me
Abri fabri domine
Enx penx
Du bist drauss.
> *Hungary. 'Zeitschrift für deutsche Mythologie', vol. ii, 1855, p. 218*

Eckati peckati zuckati me,
Awi schwavi domine,
Quitum quitum habine
Nuss puff kern
Du bist drauss.
> *Austro-Czechoslovakian border. 'Ztschr. d. Myth.', ii, p. 218*

However, the opening phrases that have been predominant—'Inty minty' and 'Eenie meenie'—recall other rhymes of English-speaking children that seem to have been well known to earlier generations: the Scots 'Inty tinty tethery methery' and 'Eenty teenty figerty feg'; the American charm that was Eugene Field's song of a far-off year:

> Intry-mintry, cutrey-corn,
> Apple-seed and apple-thorn . . .

and 'Eenie, Meenee, Mainee, and Mo', that Kipling declared were the First Big Four of the Long Ago.

Kipling was well aware, nevertheless, that 'the terrible rune', whose edict no boy or girl dare disobey, was not the verse of comparatively modern evolution concerning a hapless African but its forerunner of mysterious meaning, or no meaning, that was documented before his day:

Any, many, mony, my,
Barcelony, stony, sty,
Harum, scarum, frownum ack,
Harricum, barricum, wee, wi, wo, wack.
> *'Northamptonshire Glossary', vol. ii, 1854, p. 333*

Eeny, meeny, moany, mite,
Butter, lather, boney, strike,
Hair, bit, frost, neck,
Harrico, barrico, we, wo, wack.
> *Philadelphia. 'Notes and Queries', 1st ser., vol. xi, 1855, p. 113*

Hana, mana, mona, mike,
Barcelona, bona, strike,
Hare, ware, frown, venac,
Harrico, warrico, we, wo, wac.
> *New York, 1815. 'Notes and Queries', 1st ser., vol. xi, p. 352*

Eena, meena, mona, my,
Pasca, lara, bona, by,
Elke, belke, boh,
Eggs, butter, cheese, bread,
Stick, stock, stone dead.
> *F. W. P. Jago, 'Dialect of Cornwall', 1882*

Elimination rhymes which are sound-related, to say the least, are found in France, Germany, Austria, Romania, Holland, and Poland, for example:

Une, mine, mane, mo,
Une, fine, fane, fo,
Maticaire et matico,
Mets la main derrière ton dos.
>   *Dauphiné, Savoie. 'Les Comptines', 1961, p. 126*

Ene, tene, mone, mei,
Paster, lone, bone, strei,
Ene, fune, herke, berke,
Wer? Wie? Wo? Was?
>   *E. Fiedler, 'Volksreime in Anhalt-Dessau', 1847, p. 53*

Additional interest attaches to the following rhyme current in Norway in which the second line starts 'Katta', in the way it starts 'Cat a' or 'Cat'll-a' in some English versions, the presumed starting-point of 'Catch a' when the lines were anglicized. Compare:

Ina mina maina mau
Katta lita bobbi sau
Di va nokså gau.
Ina mina maina mau.
>   *Bergen, Norway. Helge Børseth, 'Min mann Mass', 1959, p. 12*

Ena deena dinah doe,
Cat aweazel awhile awoe
Spit, spot, must be done,
Twiddlum, twaddlum, twenty-one.
>   *Children in Somerset, 1922. Macmillan MSS.*

Zaina, daina, dina, disk,
Kittla, faila, fila, fisk,
Each, peach, must be done,
Tweedlum, twadlum, twenty-one.
>   *North-east Scotland. Gregor, 1891, p. 31*

Eena, meena, mina, mona,
Jack the jeena, jina, jona,
Ah me, count 'em along.
You shall be the soldier's man
To ride the horse, to beat the drum,
To tell the soldiers when to come,
One, two, three,
Out goes thee.
>   *S. O. Addy, 'Household Tales', 1895, p. 148*

Een-a, deen-a, dine-a, dust,
Cat'll-a, ween-a, wine-a, wust,
Spin, spon, must be done,
Twiddlum, twaddlum, twenty-one.
O-U-T, spells out,
A nasty dirty dish-clout;
Out boys out!
>   *'Games and Sports for Young Boys', 1859, p. 68*

It appears that the theory that these rhymes are centuries old is not to be lightly dismissed. It will ever be a wonder that children who cannot remember their eight-times table for half-an-hour, can nevertheless carry in their heads assemblages of rhythmical sounds, and do so with such constancy that gibberish remains recognizable although repeated in different centuries, in different countries, and by children speaking different languages.

Eeny, meeny, mink, monk,
Chink, chonk, charla,
Isa visa varla,
Vick.

> *Girl, 11, Lydney*

Eeny, meeny, mink, monk,
Chink, chonk, chow,
Oozy boozy, vacadooza
Vay, vie, vo,—vanish.

> *Children, Canberra, Australia*

Ene, mene, mink, mank,
Klink, Klank,
Ose, Pose, Packedich,
Eia, weia, weh.

> *Karl Simrock, 'Das deutsche Kinder-*
> *buch', 1857; also two similar versions*
> *from Berlin correspondents*

Ene, mene, ming, mang,
Kling klang,
Osse bosse bakke disse,
Eje, veje, vaek.

> *Glostrup, Denmark. E. K. Nielsen,*
> *'Det lille Folk', 1965, p. 96. Also*
> *Swedish version, p. 118*

Eena, meena, ming, mong,
Ting, tay, tong,
Ooza, vooza, voka, tooza,
Vis, vos, vay.

> *White children, Rhodesia*

Ala mala ming mong,
Ming mong mosey,
Oosey, oosey, ackedy,
I, vi, vack.

> *Boys, Enfield*

Yet, as we have seen, there is no certainty that a formula that is known in one country, or even in several, will be repeated in a neighbouring country. The following which is current in Germany and Denmark, which is part of the American child's repertory and has been so since Bolton's day (1888), and which has been familiar in north-east Scotland for the past fifty or sixty years, has for some reason never permeated England:

Ipetty, sipetty, ippetty sap,
Ipetty, sipetty, kinella kinack,
Kinella up, kinella down,
Kinella round the monkey o' town.

> *Girls, Aberdeen. Well known north-east*
> *Scotland*

Ellerli, Sellerli, Sigerli, Sa,
Ribedi, Rabedi, Knoll.

> *Maria Kühn, 'Alte deutsche Kinder-*
> *lieder', 1950, p. 170*

Ibbity, bibbity, sibbity, sa,
Ibbity, bibbity, vanilla.
Dictionary, down the ferry.
Fun, fun, American gum
Eighteen hundred ninety-one.

> *Chicago. Margaret Taylor, 'Did You*
> *Feed My Cow?', 1956, p. 78*

Iberdi, biberdi, ziberdi, zab,
Iberdi, biberdi, kanalie.

> *Denmark. E. K. Nielsen, 'Det lille*
> *Folk', 1965, p. 97*

This count is recognizable both by its rhythm and by the ending of the second line where Danish children say *kanalie*, German *knoll* (pronounced *kenoll*), Scots *kinella*, and American, consulting their pleasure, say *vanilla*. The only rhyme having any resemblance in England is one that is also

known in Scotland independently of 'Ippetty, sippetty', and thus is
probably of separate origin:

Ibsy, bibsy, ibsy, I,   Ibse, ibse, ibse, ah,
Ibsy, bibsy, sago.   Ibse, ibse, zebo.
    Girl, 12, Ford, Shropshire   *Boy, 15, Forfar*

Mishearings may, for instance, easily have made this concoction from the
words Mrs. Baker knew in 1854 as 'a Tell, to decide who is to commence
a game':

        Izzard, izzard, izzard, I,
        Izzard, izzard, izzard, I.
        'Northamptonshire Glossary', vol. i, p. 352

    The rhyme 'Inty tinty tethery methery', also known as 'Eenty teenty
ithery bithery' and 'Zeenty teenty tether a mether', etc., and first recorded
in 1820, was remarkably popular in Scotland in the nineteenth century (it
has not been found outside Scotland except amongst emigrants), and remains moderately well known, the following two specimens being separated
by at least one and a half centuries:

Zinti, tinti,   Zeenty teenty
Tethera, methera,   Heathery bethery
Bumfa, litera,   Bumful oorie
Hover, dover,   Over Dover
Dicket, dicket,   Saw the King of easel diesel
As I sat on my sooty kin   Jumping over Jerusalem wall.
I saw the king of Irel pirel   Black fish, white trout,
Playing upon Jerusalem pipes.   Eerie, oarie, you are out.
    Charles Taylor, 'The Chatterings of the   *Girl, c. 13, Cumnock*
    Pica', 1820, p. 31, where described as
    being old

This basic rhyme has a companion that is remarkable as much for its
survival amongst children in Glasgow for a hundred years without wear
and tear, as for the traces it retains of its ancestry:

Eenty teenty haligalum,   Eenty teenty tuppenny bun.
The cat went out to get some fun,   The cat went oot to hae a fun.
It got some fun on Toddys grond,   Hae a fun, play the drum,
Eenty teenty haligalum.   Eenty teenty tuppenny bun.
    Girl, c. 9, Glasgow   *Girl, 12, Aberdeen*

Eenty-peenty, halligo lum,
The cat gaed oot to get some fun;
It got some fun on Toddy's grun—
Eenty-peenty, halligo lum.

Calder Ironworks, nr. Glasgow, c. 1855.
'Rymour Club', vol. i, 1906, p. 5

Cf. Indi tindi alego Mary,
Ax toe, alligo slum.
Orgy porgy, peel-a-gum;
Itty gritty, francis itty,
Ordellum joodlum pipes.

Victoria, Australia, c. 1895[1]

The starting-point, or inspiration, or source of occasional words in 'Eenty teenty' and its associates, would appear to be versions of the 'shepherd's score', so called, the numerals reputedly employed in past times by shepherds counting their sheep, by fishermen assessing their catch, and by old knitting women minding their stitches. These scores have been found principally, but not exclusively, in Yorkshire, Lancashire, Westmorland, and Cumberland. In the neighbourhood of Keswick they remain so familiar they are known not only to the old folk but to children, both boys and girls, who acknowledge 'When we are at school we count in Cumbrian when we are dippin'.' According to a 12-year-old girl in Borrowdale, tape-recorded for us by Father Damian Webb, their count proceeds in fives:

1 yan	6 sethera	11 yan-dick	16 yaner-bunfit
2 tan	7 lethera	12 tan-dick	17 taner-bunfit
3 tethera	8 hothera	13 tether-dick	18 tethera-bunfit
4 methera	9 dothera	14 mether-dick	19 methera-bunfit
5 pimp	10 dick	15 bunfit (or bumfit)	20 gigert

A boy, who did not state where he came from, gave the numerals slightly differently: 'Yan, tan, tethera, methera, pimp, othera, sothera, hin, twin, dick.' A girl from Braithwaite, however, confirmed the first girl's version, other than 'overa' for 'hothera', and this version is, in fact, virtually the same as the Borrowdale version collected in 1877, which was said to have been obtained from shepherds in the vale sixty years previously, in 1818 (*Transactions of the Philological Society*, 1878, p. 354).

Since these scores have an unfamiliar sound, appear to be old, and are still something of a philological mystery, theories about them abound. It has been suggested they were brought from Wales by drovers in late medieval times; that they are a relic of the language spoken by the ancient rulers of Strathclyde, or more particularly of Cumbria; that they are

---

[1] A reader's contribution to *John O'London's*, 14 April 1950. An almost identical rhyme from New Town, Tasmania, appears in the *Journal of American Folklore*, vol. x, 1897, p. 319; and another similar has been sent us by Dr. Ian Turner, emanating from a Tasmanian born 1853.

a survival of the language preserved by the Celts when they retreated into the hills from the Anglo-Saxon invaders; and, more popularly, that they were the charms of white witches or the incantations of the ancient Druids. Certainly the scores bear relationship to the Welsh numerals: the words for 4, 5, 10, and 15, in particular, have resemblance; the counting proceeds 1+10, 2+10, 3+10, 4+10, 15, 1+15, 2+15, 3+15, 4+15, while in other languages, even in other Celtic languages, the count from 15 is 6+10, 7+10, 8+10, 9+10. On the other hand, no records of the score have been found in Wales other than the one here printed for the first time, and this is reputedly an importation. It has always been in the north-west and in Yorkshire that the tradition of old men counting with curious numerals has been strongest.[1] And the word for twenty suggests the Gaelic *fichead*, the equivalent of the Latin *viginti*. Remembering that Froissart in his journey south from Scotland, about 1364, noted that the common people in Westmorland still spoke the ancient British tongue, there may be weight in Henry Bradley's assertion that the numerals 'are entitled to be regarded as a genuine remnant of the British dialect of the north-west of England, and as proving that the dialect was nearly identical with the oldest known Welsh' (*The Academy*, vol. xv, 1879, p. 438). Comparison can best be made if the modern Welsh numerals are set in column beside some of the scores.

Welsh numerals	Score from Borrowdale	Score from Bishopdale	Score from Litchfield, Connecticut	Score from Wales	A Roxburghshire count
1 Un	Yàn	Een	Rene	Ainy	Zeendi
2 Dau	Tan	Teen	Tene	Bainy	Teendi
3 Tri	Tethera	Peever	Tother	Banny	Taedheri
4 Pedwar	Methera	Pepperer	Feather	Batry	Muudheri
5 Pump	Pimp	Pence	Fib	Bin	Baombe
6 Chwech	Sethera	Sather	Solter	Aithy	Heeturi
7 Saith	Lethera	Lather	Lolter	Karthy	Zeeturi
8 Wyth	Hothera	Luther	Poler	Kary	Aover
9 Naw	Dothera	Nogger-go-lence	Deborah	Katry	Daover
10 Deg	Dick	Hine-er-giggle	Dit	Kin	Dek
11 Un-ar-ddeg	Yan-dick	Tine-er-giggle	Rene-dit	Ainy kin	Mu
12 Deuddeg	Tan-dick	Pear-er-giggle	Tene-dit	Bainy kin	Daonul
13 Tri-ar-ddeg	Tether-dick	Pepper-er-giggle	Tother-dit	Banny kin	Rahn
14 Pedwar-ar-ddeg	Mether-dick	Pomfit	Feather-dit	Batry kin	Tahn
15 Pymtheg	Bunfit (or Bumfit)	Heen-er-bun	Bumpum	Bwmfa	Toosh
16 Un-ar-bymtheg	Yaner-bunfit	Teen-er-bun	Rene-bumpum	Ainy bwmfa	
17 Dau-ar-bymtheg	Taner-bunfit	Pear-er-bun	Tene-bumpum	Bainy bwmfa	
18 Deunaw	Tethera-bunfit	Mepper-er-bun	Tother-bumpum	Banny bwmfa	
19 Pedwar-ar-bymtheg	Methera-bunfit	Pepper-er-bun	Feather-bumpum	Batry bwmfa	
20 Ugain	Gigert	Figgit	Giggit	Icka	

Very many recordings of sheep-scoring numerals have been made over the past hundred years. The above examples, other than the Borrowdale

[1] The score was still being used by boys in Yorkshire in the 1940s, when a 12-year-old reported that when he counted out he said: 'Ya, ta, tethera, pethera, pip, slata, lata, covera, dovera, dick' (*Daily Mirror*, 22 May 1948, p. 7).

score still known today, have been selected not for their conformity but their variety.[1]

The relationship between the children's rhymes and the shepherds' scores is not close. Yet it may be felt that the opening of the rhyme 'Zinti, tinti, tethera, methera' is closer to the Roxburghshire count 'Zeendi, teendi, taedheri, muudheri', and indeed to the Cumberland score 'Yan, tan, tethera, methera', than are any of the scores to the Welsh numerals. The similarity of the fifth and sixth counts 'bumfa, litera' and 'baombe heeturi' is not remarkable in itself, but becomes increasingly apparent the greater are the number of children's rhymes collected, as 'bamfy, leetery', 'bamfaleerie', 'bamber oozer', 'bumpanary', and 'bump and airy'. Further 'bumfa, litera' and 'bamfy, leetery, heetery' (Bolton, no. 873), bear comparison with 'mimph, hithher, lithher', which are five, six, and seven in a Wensleydale knitter's score (*Notes and Queries*, 3rd ser., vol. iv, 1863, p. 205); while 'hover, dover, dicket' of the rhyme would seem to be related to 'hothera, dothera, dick', eight, nine, and ten, of the Cumberland score, as also, perhaps, to the nursery rhyme 'Hickory, dickory, dock'.

In the companion counting-out rhyme, the first line 'Eenty, teenty, tuppenny bun' may be no more than a 'sound' coincidence with 'Een, teen . . . teen-er-bun' of the Bishopdale score. Likewise only an unpoetic ear, perhaps, will hear an echo of the Welsh numerals 'Un, dau, tri, pedwar, pump' (pronounced *een, daay, tree, pai'dwaar, pimp*) in the Scots rhyme 'Eetum, peetum, penny pump'; but it is a rhyme which takes a diversity of forms:

Eetum, peetum, penny pump,
A' the laadies in a lump,
Sax or saiven in a clew,
A' made wi' candy glue.
    W. Gregor, 'Folk-Lore in North-east Scotland', 1881, p. 169

Eetum, peetum, penny pie
Popaloorum chicken chie,
Black pudding, white troot,
I choose the first one oot.
    Boy, 10, Stromness, 1961

[1] Respectively from A. E. Pease, *Dictionary of North Riding Dialect*, 1928, p. 164 (a particularly eccentric score); Bolton, 1888, p. 121; a correspondent who had it from her Welsh grandmother, who said it was learnt from soldiers disabled in the Napoleonic wars; A. J. Ellis, *Transactions of the Philological Society*, 1878, p. 357, 'a boy's school method of "counting out"'. For further examples and references, see A. L. J. Gosset, *Shepherds of Britain*, 1911 (article by Skeat); Melius de Villiers, *The Numeral Words: Their Origin, Meaning, History and Lesson*, 1923; *Notes and Queries*, vol. clxii, 1932, pp. 332, 373, and 411–12; and Michael V. Barry, *Transactions of the Yorkshire Dialect Society*, 1967, pp. 21–31. Some shepherds' counts appear to belong to a different tradition. The following is quoted in *Word-Lore*, vol. i, 1926, p. 148, as usual among Sussex shepherds, the sheep being counted in couples: 'Onetherum, twotherum, cocktherum, qutherum, setherum, shatherum, wineberry, wigtail, tarrydiddle, den.' The formula, at Lewes, was said to be 'Egdum, pigdum, fifer, sizer, cockerum, corum, withecum, taddle, teedle, ten.'

Eatum, peatum, penny, pie,
Pop a lorum, jettum I,
Ease, oze, ease ink,
Pease porridge, man's drink.
> *Argyllshire. 'Folk-Lore', vol. xvi, 1905, p. 450*

Heetum peetum penny pie,
Populorum gingum gie,
East, West, North, South,
Kirby, Kendal, Cock him out.
> *J. O. Halliwell, 'Nursery Rhymes', 1853, p. 188*

Zeetum, peetum, penny, pie,
Poppy-lorry, jinkum, jie,
Fish guts, caller troot,
Gibbie, gabbie, ye're oot.
> *Edinburgh. 'Rymour Club', vol. i, 1911, p. 90*

Eetem, peetem, penny pie,
Pop-a-lorie jinkie jye,
Stan' ee oot bye
For a bonny aipple pie,
Black fish, fite troot,
Eery airy ee're oot.
> *Huntly district, Aberdeenshire, c. 1920*

Ikey pikey penny pie,
Popalorum jiggum jye,
Stand thee oot lug.
> *W. Dickinson, 'Cumberland Glossary', 1881*

Eetem, peetem, penny pie,
Cock a lorie, jenky jye,
Ah, day, doot,
Staan ye there oot bye.
> *New Deer. Gregor, 1891, p. 15*

The objection that the shepherds' numerals, like the Welsh numerals, are based on the digital system, reckoning in groups of five, while the counting-out rhymes are in verse form, usually with four beats to the line, and that this precludes kinship, does not seem to us insurmountable. When children count out before a game they do not count on their fingers, as they do in school, for no sums are involved. They are not trying to seek a total as men ordinarily do when they are counting sheep, or otherwise wrestling with numerals. Children are merely marking-off; their requirement is rhyme and regular rhythm to help the memory, and four beats seem to come more naturally than five (compare 'Tinker, tailor, soldier, sailor' and 'Silk, satin, cotton, rags'); their pleasure is in assonance, and reduplication ('eenie meenie', 'zinty tinty', 'heetum peetum', 'om pom'). These factors can easily affect the measure. There is, in addition, as in all forms of oral transmission, a tendency to rationalize, to substitute known words for unknown. Thus a score learnt by a lady in Bransdale, near Scarborough, went:

> Yan, teean, tethera, methera, pip,
> Seeaza, leeaza, catra, coan, dick.

But the words a shepherd in Weardale is said to have used were:

> Yen, tane, tether me, leather me, dick,
> Caesar, lazy cat, or a horn, or a tick.

What is noticeable is that the counting-out rhymes that are native to

Scotland are the ones which bear most resemblance to the scores. The following rhyme was common enough in the late nineteenth century, and remains so today:

Zeenty, teenty, figery, fell,
Ell, dell, dominel,
Urky, purky, taury rope,
An, tan, tousy, joke,
You are oot.

Boy, 12, Helensburgh

Eenty, teeny, figury, fell,
Ell, dell, dominell,
Irky, pirky, tarry rope,
An, tan, tousy, Jock.

Girl, 14, Kirkcaldy

Inty, tinty, figgery, fell,
Ell, dell, rumble dell,
Ucky, pucky, toosie row,
An, tan, toosie row.

Girl, 14, Flotta, Orkney

Zeeny, meeny, feeny, fig,
El del dominy ig,
Zanty panty hithery mithery,
Bafaleery over dover
Nicky divy den.

Girl, 15, Musselburgh

No record of it has been found before 1880, so it may derive from another rhyme, rather than from a score. But the thirteenth, fourteenth, and fifteenth counts, 'an, tan, toosie', are reminiscent of the thirteenth, fourteenth, and fifteenth counts of the Roxburghshire numerals 'rahn, tahn, toosh', which were communicated to A. J. Ellis by the editor of *The Oxford English Dictionary*, J. A. H. Murray, born at Hawick in 1837, and may be presumed to be a childhood memory. And if the third count is dropped from the first five of a score known near Penrith, about 1840, 'Iny, tiny ... fethery, phips' (Ellis, F. 1), or from a Rhode Island 'Indian' count of about the same date 'Ene, tene . . . fether, pip' (Ellis, F. 5), we are not much removed from 'Inty, tinty, figgery, fell' and 'Zeeny, meeny, feeny, fig'. No relationship is apparent, however, between the scores and the gibberish rhymes which circulate in southern Britain. The following, for instance, of which two strains are discernible (the 'X, Q' strain and the 'Dutch cheese' strain), is particularly popular around London, and has not been found north of Manchester.

Inkey pinkey ellakamar,
X, Q, santa mar,
Santa mar, ellacafa,
Sham.

Student, Luton

Ip, dip, dalabadi,
Dutch cheese, santami,
Santa mi, dalabadi,
Sham.

Children, Enfield

Inky pinky ellakama,
X, Q, santa fa,
Santa fa, ellakama,
Trot, trot, trot.

Student, Plymouth

Ip, dip, alaba da,
Dutch cheese, chentie ma,
Chentie ma, alaba da,
Dutch cheese, Scram.

Girls, c. 12, Wembley

Inka vinka vinegar,
X ma, polinimar,
Polinimar, franc, franc.
Girl, c. 11, Lydney

Dip, dip, alla ber da,
Dutch cheese, sentima,
Sentima, alla ber da,
Dutch cheese, scram.
Girl, 11, Aberystwyth

Inka ponka pinka pa,
Tish U, allah-ma-gah,
Inka ponka pinka pa.
*10-year-olds, Headington, who be-
lieved the rhyme was exclusive to
themselves*

Dip, dip, allabedar,
Duck shee, shantamar,
Shantamar, allebedar,
Duck shee, shantamar.
Boys, c. 12, Sale

These rhymes have been conveyed moderately carefully by oral tradition from late Victorian times, when they seem to have been particularly well known in the Portsmouth area.

Ickledee, pickledee, elleka-mah,
Dex Q elleka-fah,
Awnty Sawnty, elleka-see,
Trance.
*Portsmouth, 1895–1900; Gosport, c.
1900*

Ecklie, picklie, eleka fa,
Fix, Q, salty fa,
Sonti fonti, eleka fee,
Trons.
Portsmouth, c. 1915

But their pedigree can be traced back no further, unless the family came from America where counts such as the following seem to have flourished:

Ickama, dickama, aliga, mo,
Dixue, aliga, sum,
Hulka, pulka, Peter's gun,
Francis.
San Francisco. Bolton, 1888, no. 675

Ikkamy, dukkamy, alligar mole,
Dick slew alligar slum,
Hukka, pukka, Peter's gum,
Francis.
*Massachusetts and Baltimore, 1848–
58. Bolton, no. 674*[1]

Very different, however, is the recorded life-span, now over, of rhymes beginning 'Onery, twoery' or 'Anery, twaery'. Early in the nineteenth century when antiquarians first became interested in children's gibberish rhymes, nearly every example they collected or recollected began 'Onery, twoery'.

---

[1] In these earlier versions there is a word at each line-end giving a natural four beats that has often been dropped in the modern English rhyme, so that the third measure has to be stretched to make two beats by accenting the terminal *ar*. This possibly illustrates the ease with which a numeral can be dropped if a digital score is becoming accommodated to counting-out.

One-ery, two-ery, ziccary, zan;
Hollow bone, crackabone ninery ten:
Spittery spot, it must be done;
Twiddleum twaddleum Twenty one.
Hink spink, the puddings stink,
  The fat begins to fry,
Nobody's at home, but jumping Joan,
  Father, mother and I.
Stick, stock, stone dead,
  Blind men can't see,
Every knave will have a slave,
  You or I must be HE.
  'Gammer Gurton's Garland', *1810*, *p. 31*

One-ery, oo-ry, ick-ry, an,
Bipsy, bopsy, little Sir Jan,
Queery, quaury,
Virgin Mary,
Nick, tick, toloman tick,
O-U-T, out,
Rotten, totten, dish-clout,
Out jumps—He.
  Philip Gosse, Dorsetshire schooldays,
  c. 1820, 'Longman's Magazine', 1889

Anery, twaery, tickery, seven,
Aliby, crackiby, ten or eleven;
Pin-pan, muskidan,
Tweedlum, twodlum, twenty-one.
  Current in Edinburgh, 1821, and also a
  generation earlier. 'Blackwood's Maga-
  zine', August (Part II), 1821, p. 36

One-erie, two-erie, tickerie, seven,
Allabone, crackabone, ten or eleven;
Pot, pan, must be done;
Tweedle-come, tweedle-come,
  twenty-one.
  Surrey, England. Jamieson's 'Scottish
  Dictionary, Supplement', vol. ii, 1825,
  p. 169

Onery, uery, ickory, Ann,
Filisy, folasy, Nicholas John,
Queevy, quavy, Irish Mary,
Stingalum, stangalum, buck.
  New England, c. 1820. Newell,
  'Games of American Children', 1883,
  p. 197

No explanation is readily forthcoming why the most popular gibberish rhyme of the nineteenth century (Bolton gives eighty versions) should be unknown to children today; while the gibberish that is now most common, 'Eenie, meenie, macca, racca', was unknown in the nineteenth century. Children certainly enjoy 'Chinese' counting as much today as they ever did; and they retain the ability, as we have seen from 'Eenie, meenie, macca, racca', to transmit gibberish with little variation. Further, they are not keen on novelty for the sake of novelty since, as far as the children are aware—and as far as anyone else was aware—'Eenie, meenie, macca, racca' is ancient. We know only that dips, even magical-sounding dips in which not a word is understandable, are not necessarily old, and need not originally have been for elimination. The well-known nonsense refrain of the song 'The Frog and the Mouse' has been pressed into ser- vice for counting-out; specimens of secret language, 'Ifficky Ikey hadikey mikey gunnicky', have been used for counting-out; so have scraps of mock Latin, 'Orcum, porcum, unicorcum, herricum, merricum, buzz'. The dipping rhyme 'Addi, addi, chickari, chickari' probably started life as the chorus of a music-hall song. And in recent times children have adopted the flight of nonsense 'Teenie, weenie, yellow polka-dot bikini'.

Rhymes, sayings, and beliefs do not have to be old to become traditional, nor do they have to have had any special significance. We have shown elsewhere how an item can become traditional merely because it fits the requirements of a particular channel of communication. Children obviously have a disposition for the authority that attaches to a rhyme which seems to be in a foreign tongue. As Southey remarked, if such rhymes are 'not in a known tongue, they may by possibility be in an unknown one'. Children saying 'Eenie, meenie, macca, racca' or 'Icketty, picketty, eye selicketty umpelaira jig' can have the pleasurable feeling that although the words sound like nonsense, if a Chinaman chanced along *he* would understand what they were saying. We suggest that the counts beginning 'Onery, twoery' lost their attraction simply because they were too ordinary; the words 'Onery, twoery, ickery, Ann' were coming to sound too much like straightforward English.

COUNTING FISTS OR FEET

Not infrequently children feel that the process of dipping is fairer—which means less likely to go against them—if each player counts as two, and his hands are counted rather than his person. The dipper (or 'spudder') cries 'Spuds up'. The players hold out their clenched fists, thumbs uppermost, and the dipper taps each fist in turn, counting as he does so,

> One' potato (pronounced *bertater*), two' potato,
> Three' potato, four',
> Five' potato, six' potato,
> Seven' potato, more'.

He includes himself by banging his right fist on his left, and his left fist on his right, and the eighth fist he comes to, with the word 'more', he bangs harder than the rest, and the player puts the fist behind his back. The dipper goes on counting round eliminating further fists (in some places his words are 'One spud, two spud, three spud, four . . .' and in the north '. . . five potato, six potato, seven potato, *raw*'), and when both of a player's fists have been knocked down that player is out, and the count continues among the rest until only one player is left holding up a spud.

This system increases the suspense, makes more of a game of finding who is to be on, leaves less time for the playing of the chosen game, and—such love has the younger part of mankind for complicating a ritual—is resorted to almost more often than ordinary dipping. 'One potato, two potato' has been in constant use throughout the twentieth century; it is much employed in America; and counting fists is also standard practice in many European countries, where, as in Britain, the operation is generally associated with a particular rhyme.

In France:

Sancta fémina goda,
Caracas et Quito,
Villes principales Cayenne
Et Paramaribo.
Cache ton poing derrière ton dos.
   *'Les Comptines', 1961, p. 102; also
   oral collection, Vichy, 1965*

In Holland:

Olleke, bolleke,
   Rubisolleke,
Olleke, bolleke,
   Knol.
                              Current since nineteenth century

And since for the past fifty years children in Britain have had a second rhyme which they regularly associate with counting fists, a gibberish count (recordings from places as far apart as Aberdeen, Amlwch, and East Barnet), which goes—

Olicka bolicka,
Susan solicka,
Olicka bolicka,
Nob

—it appears that the young in Britain and Holland have been associating the same actions with the same meaningless words for the past two or three generations, and thus unwittingly giving further evidence of the internationality of the juvenile community.

When fists are counted the number of participants in the dip, as we have seen, is effectively doubled; when feet are counted the process is slightly less time-consuming, since each player in the circle puts forward only one foot at a time. Each player puts out his right foot, and the dipper crouches on the ground and touches each toe-cap in turn, saying:

'Your' shoes' are' dirty', please' change' them'.'

The player whose shoe is touched at 'them' changes his feet, putting forward his left foot. The dipper continues round with 'Your shoes are dirty, please change them', and when his words end on someone's left foot that person is out. Counting feet is now as popular, or more so, than counting fists (children may feel that the dipper, doing his work at speed amongst their shoes, does not always know whose feet he is counting out); and slight variations to the formula are current, as 'My mother says your shoes are dirty, please change your feet', and 'Your shoes need cleaning with Cherry Blossom Black Boot Polish', and at Accrington:

Your shoes are dirty, your shoes are clean,
Your shoes are not fit to be seen by the Queen,
   Please change them.

⁂ Elimination by counting-out players' feet has not been long recorded in England, but it is well known elsewhere. At Vichy in France the formula is compounded of nonsense syllables:

> Di pi tic,
> Di pi toc,
> Carabou azinel,
> Vire, vire, forekel.

In Italy a nonsense rhyme is used beginning 'Pe u, pe do, pe tre', and the counting-out becomes a game in itself (*Conte, cantilene e filastrocche*, 1965, p. 24). In Abyssinia the players pray to be first to withdraw their feet, 'I pray, Mother Marie, that mine shall be first', for such a player is acclaimed master or 'fortunate one'; while the child whose foot remains out longest is punished, his only choice being whether he will be punished 'in the sky or on the earth'. If he replies 'in the sky' his foot is held high, but is allowed to hit the ground lightly; if he says 'earth', his foot is lifted less high but brought down with force (Marcel Cohen, *Jeux abyssins*, 1912, pp. 13–17). Games have also been played on this principle in Scotland. In the Highlands one of the company counted out the players' feet using a Gaelic gibberish rhyme, beginning:

> Ladhar-pocan,
> Ladhar-pocan,
> Pocan seipinn
> Seipinn Seonaid
> Da mheur mheadhon.

When only one player remained his foot was placed in the hook used for hanging the cooking-pot over the fire. A further rhyme was recited to determine whether he should be punished, and if the player was again unlucky he was blindfolded and made to kneel down. The leader would then hold something over his head asking 'Ciod e so os-cionn am bodach?' If the player who was blindfolded guessed correctly what was being held over his head he was freed; but if his guess was incorrect, the article (perhaps a peat) was laid on his back, and he had to attempt to guess some other object (Maclagan, *Games of Argyleshire*, 1901, pp. 92–3). In Ireland, too, where the sport was known as 'Trom, trom, cad tá os do chionn?' ('Heavy, heavy, what is on your back?'), the men at a wake would sit on their haunches and extend one leg; and one of the company would count out their feet with a doggerel, either in English or Irish, and he whose foot was not counted out had to bend down and be loaded with objects until he guessed one of them (Sean O Súilleabháin, *Irish Wake Amusements*,

1967, p. 120). It is remarkable that this amusement is almost exactly duplicated at the furthermost point of Europe, in Armenia. In 1888, or thereabouts, one of Bolton's correspondents stationed at Harput reported that the local children sat in a circle putting their feet forward, and one of the party repeated a jingle, touching a foot at each word. The jingle ended,

> Alághĕná,
> Chalághĕná,
> Akh dedí,
> Chekh dedí,

and the foot upon which the last word was pronounced was withdrawn. The player who was left at the end with a foot not counted out was compelled to stoop over, while the rest of the players stacked their hands on his back. Then, as in Scotland and Ireland, the player was given a chance to escape punishment. He was asked whose hand was topmost. If his guess was correct he was freed, and there was a new counting-out of the feet. But if he guessed incorrectly the hands were lifted in a body and brought down with a thump upon his back.

PARTICIPATION DIPS

Part of the schoolchild's genius (as also of others whose minds have not grown with their bodies) is to be perpetually troubled by possibilities of unfairness, and ever to be contriving methods to overcome them. The player who frequently takes the part of dipper is soon suspected of knowing how the dip will work out, and a safeguard is felt to be necessary. The dip takes the form of a question:

> My mother made a nice seedy cake:
> Guess how many seeds were in the *cake*?

The player reached with the word *cake* gives any number he likes, and the dipper has then to continue dipping for that number of counts. Participation dips are thought rather jolly, which in fact they are:

Old Mother Ink	Charlie Chaplin
Fell down the sink,	Sat on a pin,
How many miles	How many inches
Did she fall?	Did it go in?
—*Three*.	—*Four*.
One, two, *three*.	One, two, three, *four*.

Dic – dic – tation,   Dic-a dic-a dation,
Cor – por – ation,   My operation,
How many buses   How many stitches
Are in the station?   Did I have?
—*Five*.   —*Six*.
One, two, three, four, *five*.   One, two, three, four, five, *six*.

However, the suspicious-minded become aware that a new danger has replaced the old one. The child who gives the number can, with a little preparatory mathematics, make the dip fall on whomever he wishes, even himself. It is therefore felt more satisfactory if the player's answer is spelt round.

> Engine, engine, on the line,
> Wasting petrol all the time.
> How many gallons does it take,
> Five, six, seven, or eight?
> —*Eight*.
> E-I-G-H-T spells *eight*.

Even this is not manipulation-proof, and it is noticeable that in most of the older dips the spelling of the number or of a colour does not immediately conclude the dip.

As I went down the Icky Picky lane   As I went up the Piccadilly hill
I met some Icky Picky people,   I met some Piccadilly children,
What colour were they dressed in—   They asked me this, they asked
Red, white, or blue?   me that,
—*Red*.   They asked me the colour of my
R-E-D spells red.   best hat.
And that's as fair as fair can be   —*Green*.
That you are not to be *it*.   G-R-E-E-N spells green, and O-U-T
> *Manchester, Welshpool, Trowbridge,*   spells *out*.
> *Enfield, Shrewsbury, Ruthin. Known in*   > *Swansea, Golspie, St. Peter Port.*
> *Gloucestershire in 1898*   > *Known in Somerset, 1922*

My mother bought me a nice new   Father Christmas
  dress.   Grew some whiskers,
What colour do you guess?   How many inches long?
—*Green*.   —*Four*.
G-R-E-E-N was the colour of the   One, two, three, four.
  dress.   And if you do not want to play
> *Manchester. Current since Edwardian*   Just take your joy and run away
> *days. 'Rymour Club', vol. i, 1906–11,*   With a jolly good smack across
> *p. 105*   your face
                                            Just like *this*.
                                            > *St. Peter Port and Wakefield*

I know a doctor,
He knows me.
He invited me to tea.
Have a cigarette, sir?
No, sir.
Why, sir?
Because I've got a cold, sir.
How many blankets do you need?
—*Three.*
One, two, three, and out you must *go.*

> *Well known in the Home Counties.*
> *Alternatively 'How many tablets do you*
> *need?' and 'How many weeks did you*
> *stay in bed?' Part quoted 'London*
> *Street Games', 1916, p. 91*

Engine, engine, number nine,
Running on Chicago line.
If the train should jump the track
Do you want your money back?
—*Yes.*
Y-E-S spells Yes,
So if you do not want to play
Please take your hoop and run *away.*

> *Wilmslow. Also Harrogate and Dublin;*
> *Melbourne and Philadelphia. First*
> *couplet goes back to 1890 in the*
> *United States*

Fear of fraud may be further allayed if the respondent is made to shut his eyes.

Up the ladder, down the ladder,
See the monkeys chew tobacco,
How many ounces did they chew?
Shut your eyes and think.
—*Six.*
One, two, three, four, five, six,
And out you must go for saying so.

> *Ubiquitous since 1920s*

Mickey Mouse bought a house,
What colour did he paint it?
Shut your eyes and think.
—*Red.*
R-E-D spells red,
And out you must go for saying so
With a clip across your ear-hole.

> *Berry Hill version. Widely known.*
> *Earliest recording 1936*

My mother and your mother
Were hanging out the clothes,
My mother gave your mother
A punch on the nose.
What colour was the blood?
Shut your eyes and think.
—*Blue.*
B-L-U-E spells blue, and out you go
With a jolly good clout upon your big nose.

> *Birmingham version. Apparently known*
> *everywhere, but a particular favourite in Scot-*
> *land. An Edinburgh version appears in*
> *'Rymour Club Miscellanea', vol. i, 1906–11,*
> *p. 107*

Sometimes the condition is made that the player on whom the count ends must be wearing the colour named, otherwise he—or more likely she, for this is a feminine diversion—is not allowed to be out.

My mother and your mother
Were chopping up sticks.
My mother cut her finger tips.
What colour was the blood?
—*Pink.*
P-I-N-K spells pink,
So pink you must have on.
                    Swansea

My wee Jeanie
Had a nice clean peenie,
And guess what colour it was.
—*Blue.*
B-L-U-E spells blue,
That bonny bonny colour of blue,
And if you have it on you are out.
            Edinburgh. Versions throughout Scotland[1]

Any deceit that may still be supposed possible, is finally frustrated if the question asked in the dipping rhyme is one to which a player can give only one answer, and which yet varies with the player asked, for instance his birth date.

Eachie, peachie, pear, plum
When does your birthday come?
—*Fourteenth of December.*
1, 2, 3, 4, 5, 6, 7, 8, 9, 10, 11, 12, 13, 14,
D-E-C-E-M-B-E-R. You are out.

The most ardent disciplinarian should be satisfied with a dip such as this, even if he finds after carrying it out that no time remains for playing the game to which it was intended to be a preliminary. But this, he will find, troubles only the serious-minded. It is evident that even participation dips are repeated as much for fun as for fairness. The two participation dips that follow are almost the most popular of all, yet the responses they elicit have no effect on the count whatever.

There's a party on the hill, will you come?
Bring your own cup and saucer and a bun.
    Dipper aside: What's your sweetheart's name?
    Player: Mary.
Mary will be there with a ribbon in her hair,
Will you come to the party will you come?
                Versions current throughout Britain. Based on a nineteenth-century song 'Will you come to my wedding, will you come?'

[1] The same stipulation is made in Austria, where Carinthian children say:

Mein Vata geiht ins Wirtshaus,
Was for House hat er oaun?

The player pointed at names a colour.

Hast du eine solche Farbe an dir,
So zoag s' mir.

If the girl can show she is wearing that colour she is out.

As I climbed up the apple tree
All the apples fell on me.
Bake an apple, bake a pie,
Have you ever told a lie?
—No.
Yes you did, you know you did,
You broke your mother's teapot-lid.
What colour was it?
—Blue.
No it wasn't, it was gold,
That's another lie you've told.

Well known in England, Scotland, Wales, and
the United States, since nineteenth century

To the majority of young players dipping is not so much a means of getting a game started as part of the game itself. When children describe a game they may spend as much time giving details of how they decide who is to be on as they do in describing the game; and these details they will very properly repeat with renewed earnestness when they describe further games. Preliminaries such as we have given here precede most games in which one player has a part different from the rest. It will however be appreciated that in the descriptions of games that follow we have—with the impatience of juvenile affairs which is a well-known adult characteristic—not felt it necessary to recount these preliminaries each time.

2

Chasing Games

'The rules are very simple, if you are ticked on any part of the body, you are man. But that's when the trouble starts. Some players deny that they were took, and a fight starts.'

Boy, 13, Liverpool

IN chasing games a touch with the tip of the finger is enough to transform a player's part in the game. It is as if the chaser was evil, or magic, or diseased, and his touch was contagious. His touch can immobilize a player, or make him clutch his body as if hurt, or put him out of the game. Simultaneously it can free the chaser of his task, and enable him to be an ordinary player again. It is not even necessary that the touch be given fairly for it to be effective ('Sometimes when I can't catch anyone I pretend to be giving up in a huff,' said an 8-year-old, 'then I turn round quickly'). Nor is it necessary that a player be aware who is the chaser (in one game, 'Tig No Tell', this is deliberately kept obscure). Rather the touch seems to have power in itself; and chasing games could well be termed 'contaminating games' were it not that the children themselves do not, on the whole, think of the chaser's touch as being strange or contagious. Their pleasure in chasing games seems to lie simply in the exercise and excitement of chasing and being chased; and the contagious element, which possibly had significance in the past, is today uppermost in their minds only in some unpleasant aberrations, which are here relegated to a subsection.

TOUCH

The basic game, in which one person chases the rest, can start almost spontaneously, be played virtually anywhere, and once started, is self-perpetuating. 'It is an endless game', as one child observed. No sooner has the chaser succeeded in touching someone (and perhaps said 'it' or 'tick' to emphasize the touch) than that person becomes the new chaser. Sometimes the unceasingness of the game is stressed by a name such as 'No Barlies Tick' (Welshpool), making it clear that players are not permitted to drop out or claim respite, even after using the truce word 'Barley'. Indeed girls sometimes complain that the game goes on until they are puffed out: 'The only thing you can do if you want to stop is run into the toilets.' The game is usually played within a defined area. The only other restrictions are that the chaser must not keep after the same person all the time; and that when a person is touched or 'tigged' he cannot 'tig back', or, as they express it in many places 'You can't tig your butcher', or, at Hemsworth, 'No sharps'. The player must first chase someone else.[1]

If, however, the chaser is slow, or seems uncertain whom to follow, the game is liable to be enlivened by some confident fellow flaunting himself in front of the chaser, or by the introduction of vocal stimuli. The chaser is goaded with little rhymes that are no less traditional for being witless, the usual couplet (recited in places as far apart as Swansea and Golspie) being:

> Ha, ha, ha, hee, hee, hee,
> Can't catch me for a bumble bee.

Or, with little difference, in Helensburgh and Forfar:

> Ha, ha, ha, hee, hee, hee,
> You canna catch me for a wee bawbee.

Or, in Birmingham, 'You can't catch me for a penny cup of tea'; or in Leamington Spa, Banbury, Oxford, and Alton, even more senselessly, 'You can't catch me for a toffee flea'. In Nottingham they upbraid the half-hearted player with the Midlands term 'mardy':

> Mardy, mardy mustard,
> You can't eat custard;
> Hee, hee, hee,
> You can't catch me.

In Harrogate they chant:

> Look at 'im, look at 'im,
> Chuck a bit of muck at 'im.

And in Bristol, according to a 10-year-old:

> Hurry up, hurry up, step on the pace,
> You silly old, silly old squashed tomato face.

[1] This rule keeps being expounded in the playground, as if it had just been invented. Yet the necessity for it has been recognized for generations. In *Games and Sports for Young Boys*, 1859, p.1, it was laid down: 'When Touch succeeds in touching another, he cries "Feign double-touch!" which signifies that the player so touched must not touch the player who touched him, until he has chased somebody else.' That is to say (in contemporary parlance): 'The one he tiggies tries to tig one of the others but he can't tiggy the one what tug him' (Boy, 14, Dovenby). The same rule holds on the Continent. In France the chaser 'ne peut reprendre son père'. In Berlin children cry 'Widerschlach gildet nich' or 'Widerschlach is Katzendreck!'

But if, states our informant, the chaser succeeds in touching his tormentor he exclaims, 'You're the squashed tomato now. If ever there was a nit it was you.'

THE LANGUAGE OF THE CHASE

In juvenile speech the word for the significant touch that affects another player is not synonymous with either of the standard English words *touch* or *catch*. Children in the north country, for instance, 'tig' each other, but they do not say they 'tig' wood, even when they touch it significantly for protection in 'Tiggy Tiggy Touchwood'. Again, 'tig' cannot be equated to *catch* when, in a game such as 'Tig and Relevo', a person has to be 'tigged' to be released. The word 'tig', which is now used only by the young, subsists throughout the greater part of northern Britain, the past tense very often being 'tug' or 'tugged' or even, in New Cumnock, 'tuggen'. However, in the far north of Scotland (Golspie, Inverness, Stornoway), and in parts of Wales (e.g. Fishguard and Ruthin), they 'tip' a person. In the west midlands they 'tick' him, and he is then said to have been 'took', 'tuck', or sometimes 'tucked'.[1] In Monmouthshire, Gloucestershire, and Oxfordshire, they speak of 'tagging' each other, and the person who 'tags' may be called 'tag' or 'the tagger', as he also is in the United States, not only in chasing games but in baseball when a runner is put out by being 'tagged' with the ball.[2] In Nottingham children 'dob' each other, in Romsey they 'dab', in the Forest of Dean they 'dap', at Hereford and Crickhowell they always 'tap', and in Jersey they 'take' ('First of all you dip to find out who takes'). In London, and fairly generally in the south-eastern counties, the 'He' or 'Ee' who chases strives to 'have' someone, a term which gives rise to such verbal infelicities as:

'He has to "have" another boy, if he "has" one the boy he "has" has to hold the place where he was "had".'

Boy, 13, Croydon, describing 'French He'

And:

'We played He and I was had, so I had to be he.'

Girl, 8, Dulwich, describing 'He'

[1] It is worth noting that the juvenile preterites 'tug' and 'tuck', which are sometimes cited as modern examples of 'lazy' speech are at least as old as their oldest critic. 'Any player "tug" before his return "home" becomes the toucher' was reported from Holderness, Yorkshire, in 1884 (*Notes and Queries*, 6th ser., vol. x, p. 266). 'I was the fust in the gaime to be tuck', was recorded in south Staffordshire in 1905 (*EDD*, vol. vi, p. 134). And the reinforced preterites 'tugged', 'tuggen', and 'tucked', are also of long standing. Robert Chambers describing 'King and Queen of Cantelon' in 1842 said that when a boy was caught 'the runner places his hand upon their heads, when they are said to be *taned*'—not *ta'en*, for taken, as would be expected (*Popular Rhymes of Scotland*, 2nd ed., p. 63).

[2] This has long been so in North America. The 'tagger' and the game of 'tag' were reported from Philadelphia in *Notes and Queries*, 1st ser., vol. xi, 1855, p. 113.

Occasionally the past tense 'had' itself becomes the verb and the English language is placed under considerable strain:

'If she hads a person when she is he the person she hads becomes he.'
Girl, 11, West Ham, on 'Ball He'

Further confusion may arise, at least in the adult mind, in Devon, Lincolnshire, and parts of Scotland where, no matter how lightly a person is touched, he is said to be 'hit'. 'When you get 'it on the shoulder you hold the place where you got 'it' (Girl, 9, Plympton St. Mary, describing 'Fleabite Its'). 'If the man does hit him he must stand with his legs parted' (Girl, 13, Whalsay, where the chaser—whether boy or girl—is always 'man'). And confusion may turn to alarm in Orkney, where the word for tigging is 'stoning', if a child is overheard saying, 'When somebody stones you, you have got to go and stone somebody else.'

The following are the names for the ordinary game of Touch:

Catch, Catching, Catchings, Catchy. Probably of no dialect significance but all our recordings are from Norfolk, e.g. 'We dip to find the catcher in catchings' (Norwich). In the 1930s a common name for the game in Norwich was 'Adjins' (Had-yous).

Catchers. Widecombe-in-the-Moor, Devon.

Chase, Chasers, Chasey, Chasing, Chasings. Current here and there throughout Britain, for instance, the above names were found respectively in Cleethorpes, Lossiemouth, Alton, Runcorn, and Plympton St. Mary. Sometimes they are alternative names, or present only in compounds, such as 'One-Chase-All' (Dulwich) and 'Chase me Charlie' (a name for 'Touch' in Ipswich, when played in the water).

Dets or *Detter.* Recently current in West Sussex and the Avon Valley. *EDD* gives 'Ditter' in Wiltshire and Dorset. William Barnes, the Dorset poet, knew it as 'Datter'.

Dipanonit. Contraction of 'Dip-and-on-it'. The only name known to boys at Henstridge, in south-east Somerset.

Dobby. From *dob*, to touch or tig, Nottingham. Hence such games as 'Dobby Off-Ground'. (In the southernmost parts of Yorkshire the *dobby* is the den or home.)

He. The predominant name in London and the Home Counties. Norman Douglas found his urchins knew only the names 'Touch' and 'He'—'Called Ee: all touch-games are "he" games, and this is the grandfather of the whole family'. Understandably a writer in *The Yorkshire Post*, 18 April 1961, who had always thought of the game as being *Tig*, found the name incomprehensible when he came to teach at a south London school. But it is, or appears to be, a relatively modern name (late nineteenth century?), the chaser being known as 'he' long before the game was so called (see p. 22).

It. Common in the west country, and relatively uncommon elsewhere, except in Cambridgeshire and Huntingdonshire.

Kip. The local name in Cardigan and nearby Blaenanerch. Cf. *Tip*.

Nag. Given by one juvenile informant as a name in Birmingham.

Picka. Stromness, Orkney. Hence 'High Picka', 'Funny Picka' (French He), and such-like names. 'Picka' was listed amongst games played at Stromness in 1909 (*Notes and Queries*, 10th ser., vol. xi, p. 445). Cf. *Stony Picko*.

Picky. Scalloway, Shetland. To *pick* is to strike cr touch.

Runabout-Tig. Name to distinguish ordinary *Tig* from the types with 'dens'. Isle of Bute.

Skibbie. Caithness, nineteenth century (*EDD*). A correspondent says 'Skibbie Lickie'. In Golspie it was 'Skeby'. Thus a 13-year-old girl: 'The girl who is skeby . . . tries to give them skeby. If she gives them skeby then they are skeby' (*Golspie*, E. W. B. Nicholson, 1897, p. 120).

Stony Picko. The usual name at Kirkwall, Orkney. The term for touching is 'stoning', so 'Stony Picko' is equivalent to saying 'Touch Touch'.

Tackie. 'A game in which one is appointed to pursue and catch the others. Often played in the stack-yard, and it is then commonly called "tackie amo' the rucks"' (Gregor, *Dialect of Banffshire*, 1866). In Aberdeen, today, it is generally 'Tick an' Tack'.

Tag. Predominant name in Monmouthshire, south Herefordshire, south Worcestershire, Gloucestershire, and north Wiltshire. Hungerford seems to be on the border between Tag and He. It was formerly much more widespread, even a generation ago; and at one time may have been the usual name in southern England, as it is today in Canada and the United States. However it has not been found earlier than in *The Craftsman*, 4 February 1738; and in Henry Brooke's *The Fool of Quality*, vol. i, 1766, p. 177, where 'they all played Tagg till they were well warmed'. (Brooke was brought up in Ireland.)

Takesit. St. Helier, Jersey, where to *take* is to chase and touch, and the one who does this is the *taker*.

Tapped You Last. Talgarth, Breconshire.

Tick. The usual name in Northamptonshire, Warwickshire, north Worcestershire (in Kidderminster it is 'Tick-on-tag'), Shropshire, Staffordshire, Cheshire, and south Lancashire. In some places, as Wigan and Manchester, it is often or usually 'Ticky'. In 1905 *EDD* gave *Tick* much the same area. Michael Drayton, born 1563 at Hartshill, near Atherstone in north Warwickshire, seems to have known the game as 'Tick':

> The Mountaine Nymphs . . . doe giue each other chase,
> At Hoode-winke, Barley-breake, at Tick, or Prison-base
> '*Poly-Olbion*', *1622, xxx, p. 144*

'Tick' is also general in north Wales (alternative 'Tip') including Anglesey, and as far south as Radnorshire.

Tig. Prevailing name from the Wash to the Hebrides; also in Birmingham, around Aberystwyth, in Cornwall, and in Guernsey.[1] The recordings in *EDD* indicate that *Tig* was similarly the standard name in north Britain and in

---

[1] We kept being struck, when we were in Guernsey, by the affinity the Guernsey lore had with that of Yorkshire and Lancashire. Children not only spoke of 'tigging' each other, and played 'Tig', 'Ball Tig', 'Chain Tig', 'Four Den Tig', and other games with northern names,

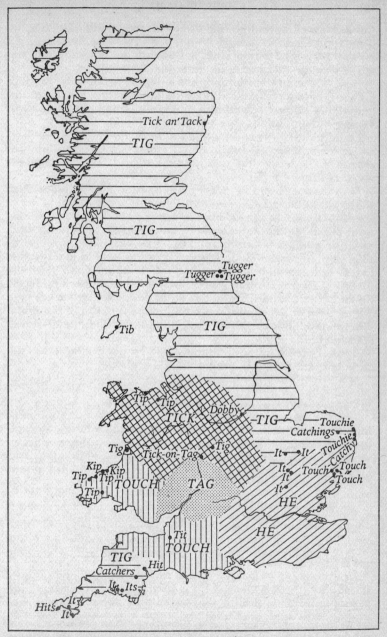

Fig. VI. Predominant names for touch-chasing in Great Britain.

Cornwall in the nineteenth century. Compounds such as 'High Tig', 'Chainy Tig', 'Tig on Lines', are common. A writer in *Blackwoods*, no. liv, 1821, p. 38, mentioned 'Tig me if you can' played in Edinburgh. Jamieson (1825) said that in Fife when one child touched another he said 'Ye bear my tig'.

Tiggy. The usual name in York, and an occasional alternative to 'Tig' elsewhere, especially in the north country. In Manchester it is sometimes 'Stiggy'; in Hexham, Northumberland, it becomes 'Tuggy' (the u as in *soot*). Common in Australia.

Tip. Alternative name in parts of Wales, hence 'French Tip', 'Ball Tip', etc. In Lewis, Orkney, and the north of Scotland (e.g. Golspie), where the children 'tip' each other, the game is sometimes called 'Tippy' or 'Tippies' instead of 'Chasers' or 'Tig'.

Tit. The name for ordinary chasing at Bridgwater, Somerset, where they also play 'Off Ground Tit', 'Tree Tit', 'Cabbage Tit' and so on.

Touch. This was the standard name in juvenile books of games in the nineteenth century, e.g. *The Boy's Own Book*, 4th ed., 1829, p. 24. Sometimes the name was extended to 'Touch and Run' (Lady Granville, *Letters*, vol. i, 1894, p. 80, referring to 1815), 'Touchlast' (given by Jamieson in 1825 as the English equivalent to the Scottish 'Tig'), and 'Touch and Catch' (Morris's *Glossary of Furness*, 1869). In Brighton today the game is sometimes 'Touch Chasing'. The chief areas for 'Touch' in the present day are south Wales, Bristol, Somerset, Dorset, the New Forest, and parts of East Anglia, hence 'Release Touch' (Ipswich), 'Sticky Touch' (Rushmere), 'Off Ground Touchie' (Yarmouth).

Tugger. Prevailing name in Gateshead and Newcastle upon Tyne (the chaser being said to be 'on'); sometimes it is 'Tigger'.

PLAYERS RESTRICTED TO PARTICULAR WAY OF MOVING

In some chasing games the ordinary rules of Touch are maintained, but all players, including the chaser, are restricted in the way they may move. They play 'Walking He' in which no player may run, 'Hopping He' in which no player may put both feet on the ground at once, and 'Bob He' in which the players squat, and progress is made by bunny jumps ('it is quite a hot game'). At Timberscombe in Somerset the game of the moment when we paid a visit in 1964 was 'Spider Touch', the players progressing on all fours, but with their fronts facing the sky. In Lincolnshire, where a popular game is 'Crab Tiggy', the players attain the 'crab' position by

they even celebrated Mischief Night on 4 November as do their northern contemporaries (see *The Lore and Language of Schoolchildren*, pp. 276–80). It transpired that when in 1940 about 5,000 children, virtually the whole schoolchild population, were evacuated, it was to the north of England that the majority were sent, and remained for the rest of the war. Afterwards, when they returned home, they naturally played the games and used the language they had become familiar with in Britain. Today, unconsciously, children in Guernsey are bearing witness to the northern hospitality given their predecessors years before they were born.

doing a handstand and dropping backwards on to their feet, so that their bodies are *arched* inside out. (A headmaster commented: 'If we tried to *make* them do this, parents would write to their M.P.s'.) In Camberwell, where one of the games is 'Butterfly He', they run in pairs holding hands, but with one player back to front. And another form of Touch in which the players move in pairs is 'Piggyback He' (otherwise known as 'Donkey Touch' or 'Horsie Tig'), in which the important rule is that the chaser must be properly mounted, or his touch does not count. The same rule applies in 'Bike He' (in Wigan known as 'Ticky on Bikes'), but in this game the chaser's art is to steer his bike across the front of the person he is chasing, thus forcing the rider to put a foot on the ground, which makes him the chaser as effectively as if he had been touched. Each of these games is best played in a small area (even 'Bike He'), and with plenty of players, although there must not be too many players in 'Bike He', or the result is hair-raising, or, more precisely, knee-grazing.

CHASES IN A DIFFICULT ENVIRONMENT

In some chasing games the players are free to move as they will, but running is difficult or impossible because of the environment in which they are played. In 'Tree He', for instance, otherwise known as 'Tree Touch', or 'Tree Tiggy':

'All the people but the person who is on climbs up a tree, and the person who is on gives them a minute and then climbs up after them. If he touches someone that person is on, but they cross over to some more trees while he is up the first tree.'   *Boy, 12, Market Rasen*

Likewise in 'Monkey Tig', played in Orkney 'in a place where there are plenty of walls', the children clamber from one elevation to another and 'if the catcher catches one of the "offers" (those who have gone off) that person is "king" '. But as in every game of chase played above ground level, 'if an offer goes off the wall, he or she is king' (Girl, 10, Stromness), that is to say, if a player puts a foot on the ground he becomes the chaser just as if he had been touched. The domestic form of 'Tree Touch' or 'Monkey Tig' is generally known as 'Pirates' or 'Shipwrecks', and is sometimes played in gymnasiums and sometimes, but not always happily, attempted at home. Every chair, table, cupboard, and stool that can be found is spread about the room, so that the pirate who chases, and the rest of the players who attempt to scramble from him, can clamber from one object to the next without putting a foot on the ground, for the floor is the sea, and anyone who steps in the sea has to become the pirate.

In the playground the younger children commonly play 'Line Touch' on the painted lines of the netball courts. All players have to stay on the lines and keep moving forward. 'You mayn't turn back even if the boy is coming towards you.' Those being chased may only change direction when they come to the end of a transverse line, while the semicircle in front of each goal is generally taken to be a place of safety. (Other names: 'Ticky Line', 'Tiggy on Lines', 'White Line Touch'.) Likewise in the street they play 'Policeman, Policeman, on the Bridge' (Birmingham), drawing chalk lines on the road, one line leading to another, and saying there is water between the lines, and anyone who steps on the road will get wet.

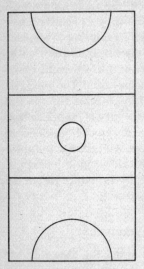

Netball court. The lines are much used for games other than netball.

In the park they play 'Swing Tig' (called 'Yo-ho' in Glasgow), in which they have one more player than they have swings, and this player is chaser:

'The person "out" runs round about the swings and tries to tig any of the people who are either swinging or have their feet on the concrete slabs. When the runner approaches the person on the swing she swings to the slab at the other side. If a person is tigged or puts her feet down in the middle the runner shouts 'Yo-ho' and takes her place in the swing.'   *Girl, 13, Glasgow*

And they play 'Roundabout Tiggy' on the little roundabouts in the park playgrounds. The chaser stands on the ground inside the roundabout with his eyes shut, and tries to touch someone as the ring of the roundabout goes round. Anybody who is touched, or who falls off the roundabout while avoiding being touched, takes the chaser's place in the middle.

§ *Touch Conveyed by Substitute for Hand*

In some chasing games the touch is made not with the hand but with an object carried in the hand, or with the hands carried in such a way that they represent something else.

BULL

In West Ham young children play a game in which the chaser, known as 'the bull', has to keep his hands clasped together in front of him, with two fingers pointing forwards for horns. Although he runs after the others and tries to touch one of them with his hand, as in ordinary chase, his touch is ineffective if his hands are not held properly, and if the touch is not made with one of the two outstretched fingers. 'The bull has to touch you with his horn, and if he touches you, you are the bull, and then you have to catch us', explained an 8-year-old boy. And the new chaser, too, must hold a proper pair of horns in front of him the whole time he chases, or his touch will not count.

This game has not been reported from elsewhere in England, but it is played in Kirkcaldy, in exactly the same way, although there known as 'Hornie'.

✲✲ The requirement that the chaser keeps his hands clasped in front of him as he chases is clearly not a rule that has been made up by the players. It not only occurs in several other games (see under 'Widdy' and 'Bully Horn') but appears to be the game described in 1813 in *A Nosegay, for the Trouble of Culling*, under the name 'Staggy Warner':

'The boy chosen for the stag, clasps his hands together, and holding them out threatens his companions as though pursuing them with horns . . . he who is struck by the stag's horns (or more properly speaking *hands*) becomes stag in his turn.'

Horned animals, other than bulls, are more often chased than avoided; but horned creatures of little earthly form, horned men, or horned gods, seem to be as old as the caveman's drawing at Ariège; and like 'Auld Hornie' himself have constantly brought consternation to men's gatherings. It may be pertinent that in *Notes and Queries*, 11th ser., vol. viii, 1913, p. 34, a Lincolnshire game is described in which the pursuer strikes his captive on the back, calling 'Horney, Horney, Horney'; and that a further writer (vol. viii, p. 115), recalled that when he was a lad a game was in vogue called 'Hunt the Devil to Highgate' in which 'He', as he ran, was flicked with the ends of moistened pocket handkerchiefs. At Painswick in Gloucestershire children used formerly to rush down the street crying 'Highgates', after taking part in the ancient ceremony of 'Clipping the Church', conceivably a survival of the Roman festival of *Lupercalia*, when youths struck every woman they met with thongs made of goat skin. It is probably no more than a coincidence that in the earliest description of Touch known to us, in Ammann's *26 nichtigen Kinderspiele*, 1657, the children are chased by one of their number, who tries to slap a player with his hand and then run away; and the name of the game was 'Zicken', the chaser being said to be a goat.

WHACKO

In Hanley, and other districts in the Potteries, children play the game 'Whacko', in which the power of touch rests in a scarf. Two boys twist the scarf as tight as they can, and then bring the ends together so that it 'twists itself', and makes a serviceable weapon that can be held in one hand. The boy who is 'on' carries the scarf and has to chase and hit some-one with it, who then takes the scarf and chases someone else.

In Senghenydd, Glamorganshire, the game is known as 'Slopper'.

* To twist a piece of material so that it can be used for hitting someone seems to be traditional also in France. In the game 'Mère Garuche', a form of 'Warney', each player arms himself with a handkerchief 'en garuche, en le pliant et le tressant en anguille' (*200 Jeux d'enfants, c.* 1892, p. 132).

DADDY WHACKER

In some chasing games in which a stick or whip is used to make the effec-tive touch, there remains an atmosphere of the days when working men and apprentices took part in these sports, and rough play was tolerated. In Wigan, where children play a game called 'Stick', the pursuer has the right to chastise whom he touches. 'The person who is "it",' says a 10-year-old, 'finds a small stick about the size of a clothes peg and a long one about the size of a poker.' He 'tries to hit one of the others with the small stick or by throwing or by any other way. When the chaser has succeeded in hitting somebody with the small stick, he whacks them with the long one.' The player has to stand submissively to receive this punishment, after which the chaser 'hands the long stick to the person he has just hit and throws the short one away. The boy he has just caught is then "It" and has to find the small stick before he can begin chasing'.

In Annesley, on the edge of Sherwood Forest, where the children speak of 'running' somebody, rather than of chasing them ('The girls run us, then the boys run the girls'), the game is called 'Running' and the chastise-ment is known as giving a person 'his dubbins'. Here, however, the chaser does not forfeit his wand of office: 'When he catches you he hits you with the stick and when he has finished hitting the person he lets them go and he counts ten and starts running you again.'

In Monmouthshire where the game is known as 'Daddy' or 'Daddy Whacker', the play is further ritualized: 'When they are caught once the boy who is chasing them gives them three whackings, then lets them go, but when they are caught the third time they have to take the place of the one who is "on it"' (Caerleon).

Other names: 'Stick Tig' (Kinlochleven), 'Stick Touch' (Newbridge, Monmouthshire), and 'Whips' (Golspie).

✲✲ This game appears to be depicted in a rhyming-alphabet chapbook *The First Step to Learning*, printed by J. Catnach about 1820. The woodcut illustration to the letter 'O' shows a boy running with a stick in his hand, and two other children who are perhaps trying to avoid him. The inscription reads 'O was Old Daddy, who shall he be?' It thus appears that 'Daddy' is a traditional name for the chaser, and this seems to be confirmed by a version of 'Ticky, Ticky Touchwood' reported in *Notes and Queries*, 11th ser., vol. viii, 1913, p. 34, in which the children taunted their pursuer 'Daddy! Daddy! I don't touch wood'. Whether or not coincidental, this challenge used also to be made in Silesia and Switzerland, where in a game of touch-iron the cry was 'Father, I have no iron, hit me' (H. Handelmann, *Volks- und Kinder-Spiele aus Schleswig-Holstein*, 1874, p. 66, quoted by Newell); and the saying in France that a player who has just been touched 'ne peut reprendre son père' will already have been noted. D. Parry-Jones, *Welsh Children's Games*, 1964, pp. 110–11, says that in Pembrokeshire the game was known as 'Data Meddw' (Drunken Daddy), and in Glamorgan it was 'Daddy Gransher' and 'Padi-racsyn' (Paddy the ragged). See also under *Daddy*, p. 20, and 'Daddy Grandshire', p. 174.

BALL HE

In 'Ball He' the ball becomes an extension of the chaser; he may run with it, throw it when he likes at whom he likes (or dislikes), and if he hits someone that person becomes the chaser; but if he misses he has the wearisome task of fetching the ball himself while the others run off to the other end of the playground. It is a game that is liable to exhaust both body and temper. It seems to be played only when boys are cooped up together with nothing better to do; and it is a game which an unusual number of children speak of as being rough and breeding disputes. 'Sometimes someone throws the ball too hard and a fight starts or the person gets mad and goes after the person to get his own back' (Boy, 13, Liverpool). 'Usually there are some black eyes or bleeding noses. Also, if the ball just scrapes you or somebody and "they" say it did and you say it didn't, there is a "free for all" fight, fighting "all in"' (Boy, 15, Guernsey). Sometimes it is the rule that players must be hit below the knee for it to count; and generally anyone who touches the ball, even accidentally, automatically becomes the chaser, although in some places players are allowed to fist the ball away as in 'Kingy' (q.v.), and in Bristol they have a version called

'Scoop' in which the ball may be scooped up and thrown away with the cupped hands.

Names for the game are mostly as would be expected: 'Ball Tag', 'Ball Tig', 'Ball Touch', 'Ball Chase', 'Tiggy Ball', and 'He with the Ball'. In Welshpool the game is known as 'Poison Ball'; in Selborne it is 'Stingy'; and in some places it is 'Dodge Ball', although 'Dodge Ball' is usually the game in which one boy is in the middle of a circle of players, trying to dodge the ball.

It is not necessary, of course, that the object thrown be a ball. 'The most exciting game of tig that there is, is "Slipper Tig",' says a Lincolnshire lad, 'where one boy gets a slipper and chases the others and tries to hit them by throwing the slipper at them.' And at one school in Greater London 'Coke He' is the popular game, partly, it seems, because of the availability of large supplies of boiler fuel, and partly because 'it is against the rule of the school to throw coke about in the playground'.

THREE LIVES

In 'Three Lives', which is a more civilized game than 'Ball He' and played principally in Scotland, no player is appointed chaser, and it is as if the ball itself was 'het'. Whoever is nearest the ball throws it at someone else, and whoever is hit by the ball below the thigh loses a 'life'. The game usually begins with the players standing in a small circle with their feet apart touching the feet of their neighbours on either side. One player bounces the ball in the middle of the circle, and as soon as the ball rolls under somebody's legs, that player picks it up while the rest run. Staying where he is he throws the ball at someone, trying to hit him. Thereafter anyone near the ball can pick it up and fling it at any other player, but a person must not move with the ball in his hand. The other players can run where they like within bounds. Each person has three lives, and may continue in the game until he has been hit a third time, when he usually has to stand aside until only one player is left, who starts the next game. But at Stornoway in Lewis a player who loses three lives becomes the centre of attention. 'When the ball has hit you three times below the knee you go through the mill. Everyone puts one hand against a wall and the person runs through and everyone gives him a smack on the back.'

Names: In Aberdeen, where the game is tremendously popular, it is called 'Sappy Soldiers'; in Perth it is 'Wounded, Dying, Dead' (the first time a person is hit he is wounded, the second time dying, and the third time he is dead and out of the game); in Langholm it is called 'Deadies'; in

New Cumnock it is 'Chipped, Cracked, and Broken' ('When you are broken you have a last throw and if you hit anyone you have to be broken again, but if you miss you are out of the game'); and in Glastonbury it is 'Multiplication Touch' ('If you are touched once you are "double one", if you are touched twice you are "treble two", if touched three times you are "treble three", but if you are touched four times you are out').

§ *The Touch having Noxious Effect*

FRENCH TOUCH

The fancy that a chaser's touch is contagious or even wounding is here given logical expression. The game starts as ordinary 'Touch', but when the chaser succeeds in touching someone, the new chaser has to keep one hand covering the spot where he was touched until he, in his turn, manages to touch someone else. The sport, therefore, is for the chaser not merely to touch another player, but to touch him on a part of his anatomy that will be an embarrassment to him to keep a hand on while he is chasing, for instance, the top of his head, his nose, backside, a shoulder-blade, elbow, knee, or foot. 'It looks so funny when someone is tuck on the foot and one sees him hopping about', remarks a 12-year-old. In fact one of the joys of the game seems to be envisaging the awkward places where a person might be touched. 'He might be tug on the eyes then he won't be able to see. Then he can't tig us at all', suggested a Birmingham 8-year-old. It is, of course, a strict rule of the game that the chaser has no power of touch unless one of his hands is covering the affected spot.

Names: The majority are evocative of the game, for instance: 'Hospital Touch', 'Poison Touch', 'Poisonous Tiggy', 'Body Tick', 'Flea-Bite Its', 'Ticky Wounded', 'Wounded Tiggy', 'Doctor Touch', 'Lame Tig', 'Ticky Lame Horse', 'Sticky Touch', 'Sticky Glue Touch', and, in Kirkwall, 'Funny Picko'. The name 'French Touch' probably reflects the feeling that a touch which is so pernicious must be an importation. This is also implicit in the names 'London Touch' (in Bristol) and 'Chinese Touch' (in London). And it is consistent that children in Berlin think of it as foreign too. They call it 'Englisch Zeck' or 'Englisch Einkriege' (Reinhard Peesch, *Das Berliner Kinderspiel der Gegenwart*, 1957, p. 34).

THE DREADED LURGI

There has long been and still remains a sinister side to the game of Touch. Generation after generation of children have, at certain times, felt that something evil or sickening was transmitted by the chaser when he made his touch, and this especially when he touched a person on the skin. The headmistress of a village school in Wiltshire writes: 'For several years we have occasionally seen "P.A.L." written in ink on arms and legs. Our children are usually embroidered with something or other, and "Wash it off" was the nearest we had to conversation on the subject. If we had thought about it we should have assumed it was merely a gang sign or a hero's initials. In any event it was so small and blurred that we did not always recognize it as being "P.A.L.". Recently, however, a small girl was crying because everybody said she had "it". I could not find out what "it" was, and was really exasperated before a boy volunteered that "it" was "Lurgy", and, thereafter, that "Lurgy" was "it". This was deadlock, until, rolling up the leg of his shorts, the boy said, "She should have had this—P.A.L.—Protection Against Lurgy".'

Following on this report, inquiries have shown that a number of schools, large and small, have been troubled with the complaint. 'The dreaded Lurgi' is passed on by touching a person's skin, or, sometimes, by throwing a small contaminated parcel at him (as in 'Ball Tig'). Clearly the immediate source of infection has been 'The Goon Show'. On 9 November 1954 an edition of 'The Goon Show' (a B.B.C. sound programme featuring Peter Sellers, Harry Secombe, and Spike Milligan) was entitled 'The Lurgi strikes Britain', and 'the dreaded Lurgi' became a by-word in subsequent programmes.[1] It seems, however, that diseases, even imaginary ones, do not occur spontaneously. In Norwich we found the word 'lurgy' in everyday use amongst children—'You're lurgy'—but with the restricted inference that the person was 'stupid, goofy, looney, nuts, a nit' ('nit' was a favourite word in the early sixties). Boys in Norwich also said to each other, 'You've got the touch', 'the flea touch', 'the silly touch', or 'the lurgy touch', and this touch was one that could be passed on. Almost certainly they were not referring to the 'dreaded lurgi', but sustaining the dialect *lurgy* and *lurgy fever*, which have long been in use in East Anglia to distinguish the idle, and the ailment from which they suffer.

In Hampshire we found children who had the feeling that a person's characteristics could be transmitted by touch. Thus in one school, there was a girl whom we shall call Jill Benney, with the reputation of a cry-baby, it was felt any child could be contaminated who was given the

[1] Lurgi came to Britain in reality, on 24 October 1958, in the shape of the Lurgi High-Pressure Coal Gasification Plant at Westfield in Fife. This name had been derived from the middle of the first word of Metallurgische Gesellschaft AG, a company established in 1897, the predecessor of the present-day owner of the Lurgi companies, Metallgesellschaft AG of Frankfurt-am-Main.

'Benney touch'. In Liss children transmit something, which only they can understand, when they make the 'aggie touch'; and the touch may even be passed on by the deceit of shaking hands. In Wolstanton children play a game called 'Germ':

'Germ is a kind of tick you dip and the person that is out has the germ but if he ticks somebody that person is on and it goes on like that.' *Boy, 9*

In Glastonbury they play 'Minge', a game in which players can obtain no protection from the touch, as they ordinarily can, by saying their truce-term 'fens':

'In the game Minge someone is on-it, and the person that is on-it has a disease. But if the person who's on-it touches somebody else they've got the disease and they chase. There is no fens in the game.' *Boy, 11*

In Swansea girls obtain a morbid thrill playing 'The Plague'; in Cranford, Middlesex, the game is called 'Fever'; in Sale, Manchester, it is 'The Poo' ('the one that has it at the end of the day smells'); at Castel in Guernsey it is 'Poisonous Fungi'; and at St. Peter Port it is 'Lodgers':

'A bundle of paper is wrapped up tightly in a ball and one person is on. He throws the ball of paper at somebody while the others shout out "How much do they pay a week?" or "How many have you got?" The person who is on is supposed to have fleas.' *Boy, 11*

And while these games are being played, and even afterwards, the suspension of disbelief in the game's pretence can be absolute: the feeling is unfeigned that the chaser's touch is unhealthy.

⁂ Such games seem to be played around the world. In Auckland, New Zealand, when a boy is tagged by a girl, the others deride him shouting 'You've got girl fleas'. In Valencia the ordinary game of chase is 'Tu portes la pusa' (You carry the flea). At Massa in the Bay of Naples, the game is 'Peste'. And in Madagascar, according to *The Folk-Lore Journal*, vol. i, 1883, p. 102, the child who did the chasing was *bôka*, a leper, and when he touched someone his leprosy was conveyed to the one he touched, who in turn tried to rid himself of the disease on someone else. At the end of the game all the children spat, saying '*Poà*, for it is not I who am a leper'.

Indeed the importance to a child of not being the one who receives the 'last touch' has frequently been noticed, particularly in the nineteenth century. (See, for example, 'Last Bat' in Brockett's *North Country Words*, 1825, and 'Tig' in Patterson's *Antrim Glossary*, 1880.) It has been observed that a child who received the 'last touch' on the way home from school, when he has no way of passing it on, was liable to feel genuinely ill at ease.

But as Blakeborough pointed out in his *Wit of the North Riding*, 1898, 'this last tig had to be given on the skin, not on the jacket, or the boy would call out, "I wasn't born with my clothes on" '.

§ *Immunity from the Touch*

In the following games those being chased can make themselves immune to the 'Touch' merely by themselves touching a particular substance, or by staying in a specific place, or by adopting a certain posture. It is, however, generally expected of a player that he will relinquish his security when the chaser has moved away; and in some games he has no option, as when a second fugitive arrives at his sanctuary and his departure is obligatory.

TOUCHWOOD

In 'Touchwood' security from the chaser is obtained by touching any object which can be termed 'wood', as a door, window-sill, fence, or trunk of a tree. Players run at will from one piece of wood to another, and even back to the piece of wood they have just left. The attraction of the game, as in all games of this type, is the chance it gives the less-good runner to feign boldness. He can taunt the chaser with the words 'Tiggy tiggy touchwood, I don't touch wood', and the moment the chaser turns on him scurry to a nearby sanctum which his pursuer has not observed. However, as an 11-year-old remarks, 'It is not fair if you carry wood about with you when you are playing this game. And it is not fair to climb up trees.'

Names: 'Touchwood' (general), 'Ticky Touchwood' (Ipswich), 'Tig-on-Wood' (Rossendale), 'Tiggy Tiggy Touchwood' (Thirsk), 'Tuggy Tuggy Touchwood' (Etton), 'Wood Tick' (Knighton), 'Wood Touch' (Timberscombe).

** 'Touchwood' is less a favourite with children today than with antiquarian impresarios who wish to exhibit the game as evidence that wood anciently possessed religious or magical properties. Not that there is any evidence that the game is old. The earliest known reference to it is in Robert Anderson's *Ballads in the Cumberland Dialect*, 1805, p. 35, 'Tig-touch-wood' (*EDD*). Thereafter mention of it is made in *Blackwood's Magazine*, August 1821, p. 33, 'tig touch timmer'; *The Boy's Own Book*, 4th ed., 1829, p. 24; *Exercises for the Senses*, 1835, p. 123; and *The Book of Games*, *c.* 1837, p. 119. A joke about a boy carrying a pencil with him, which he produces from his pocket when touched, occurs in *The Youth's Own Book of Healthful Amusements*, 1845, pp. 217–18. All indications are that 'Touch Iron' is the older form of the game.

TOUCH IRON

This is the same game as 'Touchwood' except that, as a Knottingley boy says, 'it is played in a good spot which has gratings, fences with nails, or any other iron objects', and 'the players are chased about until they see a metal object and grasp it'. However, in Knottingley, according to one informant, the game tends to become more virile than elsewhere, for 'it is up to the players to knock each other off the piece of iron to try to make them be on' (cf. 'Budge He' hereafter). 'Touch Iron' is not as popular as it used to be. It seems to have fallen out of favour in the latter part of the nineteenth century, possibly because by then every boy had begun to carry iron with him in the nails of his boots, and recourse to iron had become too easy.

Other names: 'Iron Tig' (Blackburn), 'Metal Touch' (Bristol), 'Tig on Metal' (Bacup), 'Tiggy on Iron' (South Elmsall).

** References to 'Touch Iron' considerably antedate 'Touchwood'. A humorous writer in *The Craftsman*, 4 February 1738, claiming access to an old manuscript, asserted 'In Queen Mary's Reign, Tag was all the Play; where the Lad saves himself by touching of cold iron.' The engraver of *Les Trente-six figures contenant tous les jeux*, 1587, shows boys saving themselves by holding on to the iron bars of windows while a chaser waits for them to let go; and the caption states it to be common for boys to play 'qui retiendra fer'. Florio gives 'Ferri' as 'a kind of play so called' in his Italian–English *Dictionarie*, 1611; and in Italy today, where iron is touched to give protection from the plague in the game 'Peste' (already noted), and where the touching of iron is common to counteract ill luck, the game is 'Tocca ferro'. In Britain it is of course notorious, as Robert Kirk wrote in 1691, that 'all uncouth, unknown wights are terrified by nothing earthly so much as cold iron'; and an amount of superstitious practice has developed in the light of this knowledge, specifically in the hanging of horseshoes over doorways. Peesch, in *Das Berliner Kinderspiel der Gegenwart*, 1957, p. 34, says that in the elementary form of 'Einkriege' players are safe who hold on to either wood or iron.

TOUCH COLOUR

If 'Touch Iron' was the game of the eighteenth century, and 'Touchwood' of the nineteenth, 'Touch Colour' is the game that prevails today. Possibly because the architecture of contemporary school-playgrounds efficiently eliminates trees and other signs of nature, the children are forced to play

'Touch Colour', or, depressingly often, 'Green Touch'—the green being not grass but paint.

> 'At our school we have lots of green doors and posts and spots of green paint, so we run backwards and forwards to different spots of green.'   *Boy, 10*

At other schools, according to the surroundings, they play 'Green Drainpipe Touch', 'Touch Stone Wall', and 'Door-knobby Tig', and such-like perversions of the traditional amusement, for although children will remain faithful to a traditional game for as long as they can, they are not slow to adapt if the environment makes an old game impracticable.

Other names: 'Tiggy Touch Colour' (Birmingham), 'Dobby Colours' (Nottingham), 'Ticky Off-Green' (Wilmslow), and, less common than 'Green Touch'—'Black Touch', 'Brown Touch', 'Red Touch', etc.

** A writer in *The Boy's Own Paper*, 5 November 1887, p. 88, describing his schooldays in the middle of the nineteenth century, recalled playing 'Touch-wood', 'Touch-iron', and 'many absurd Touches' including 'Touch-paint'. An 8-year-old Italian girl told us that she and her friends decide before beginning to play whether they will touch wood, touch iron, or touch a colour. 'Touch Colour' is also much played in Sweden; and, as in Britain, it is now played more often than either 'Touchwood' or 'Touch Iron'.

COLOURS

This is a variation of 'Touch Colour' in which the monotony of green is dispelled. Whoever is appointed chaser names the colour that will give immunity while he is chasing; and those players who are not wearing the colour, and cannot see any of it, have no alternative but to keep running (cf. 'Farmer, Farmer, May We Cross Your Golden River?'). When someone is caught and becomes the new chaser he chooses a new colour; and at Arncliffe in the West Riding the naming of the colour becomes a small ceremony. The children form a circle round the person who is 'on' and sing:

> Charlie over the water, Charlie over the sea,
> Charlie caught a blackbird, and can't catch me.

The person in the middle calls out the colour (or it may be a type of object), and the players have to find it without being caught.

Names: 'Colours' (Bristol), 'Colour Tag' (Cwmbran), 'Colour Touch' (Swansea), and 'Charlie over the Water' (Arncliffe).

OFF-GROUND HE

'Off-Ground He' is such a favourite that in some places children give the impression it is the only chasing game they know. The motif is simple: a player cannot be caught if he is above the ordinary level of the ground. He is safe as long as he can balance on a brick, hang from the branch of a tree, straddle a fence, or climb on to a dustbin. Of course, should a player be half-falling from his perch the chaser will 'guard' him closely, and touch him the moment he falls; but, as several children remark, 'You must not pull someone off of the object they are on, and then touch them, that is cheating.'

Names: 'Dobby Off-Ground' (Nottingham), 'Feet Off Ground' (London and environs), 'High Picka' (Orkney), 'High Tig' (general in Scotland), 'Last Off Ground' (Aberystwyth), 'Jack Above' (Barrow-in-Furness, Barlby and Market Rasen), 'London Town' and 'No Feet' (alternative names in Norwich), 'Off-Ground Catch' (Gorleston), 'Off-Ground Daddy' (Poole), 'Off-Ground He' (general north and west Home Counties), 'Off-Ground Its' (Lizard), 'Off-Ground Tag' (Pontypool and Forest of Dean), 'Off-Ground Tick' (Frodsham, Wolstanton, Welshpool, Knighton), 'Off-Ground Tig' (Accrington and Halifax), 'Off-Ground Touch' (west country and East Anglia), 'Tick Off-Ground' (Upton Magna), 'Tig Off-Ground' (Bishop Auckland and St. Peter Port), 'Tig on High' (Bacup), 'Tiggy Off-Ground' (Lincolnshire), 'Tiggy on High' (Cumberland and Durham), 'Tuggy, Tuggy, Off-Ground' (Etton).[1]

Occasionally the nature of the 'off-ground' sanctuary is specified, as at South Elmsall where they play 'Stone Tiggy' and, says a 10-year-old, 'they cannot tig us if we are on stone and we shout "Lick-Lock I'm in my den" '.

** 'Feet Off-Ground' was one of the games listed in *Notes and Queries*, 11th ser., vol. i, p. 483, as being played by children in London elementary schools in 1910. It is also given in *200 Jeux d'enfants*, c. 1892, pp. 100–1, under the name 'Le Chat Perché'. It is known to be popular today in Sweden, Italy, and Spain, and is probably played throughout Europe; but it does not seem to have been common in the United States up to the early 1950s. Brewster mentions only 'Hang Tag', in which players obtain immunity by hanging from the branch of a tree.

---

[1] In recent years, possibly under the influence of this game, children on the way to school have become victims of an absurd game or superstition that they must be 'off-ground' when a car passes. The consequent dodging on to doorsteps and hedgebanks when a car is seen approaching assists neither their own progress nor the serenity of the driver.

BUDGE HE

This is an elaboration of 'Off-Ground He', and—for the energetic—
a more satisfactory game. Only one player may take refuge on any one
elevation. If a second player comes to the safety-place, the first player is
obliged to leave. Hence the game is played with a certain assertiveness:

> 'A person decides he wants to move, so he runs over to another person who
> is off-ground and says "Bunk you skunk". The other person has to move and
> the person who is on has to try and tig him.'   *Boy, 13, Market Rasen*

It is while the expelled player is considering whom, in his turn, he shall
discompose with a visit, that the chaser commonly makes his touch.

The game can also be played with posts for sanctuaries (e.g. netball
posts), or with the players standing on places which are distinguishable,
although not elevated, such as the covers of drains and water hydrants.
'There can only be one person on a drain', explains a Fulham girl, 'but say
a girl wanted to change drains she would run on to the next one and say
"Buzz" and the person who was on that drain would have to get off, and
most probably would be had.' 'I don't like playing "Buzzing Bee",' commented an 11-year-old Walworth girl, 'because in the summer the drains
smell terribly.' In Kirkwall, Orkney, the game is known as 'Syre Buzz',
a syre being a drain.

Other names, some of them aptly imperious, include: 'Scoot' (Widecombe-in-the-Moor), 'Scram' (Timberscombe), 'Shift' (Ipswich), 'Shoo'
(St. Martins), 'Buzz' (Bristol and parts of London), 'Bunk Tiggy' (Market
Rasen), 'Tiggy Budge' (York), 'Off-Ground Budge' (Croydon), 'Hoppit'
and 'Hook' (Norwich), 'Clear-Off He' (Yarmouth), 'No Two Birds in One
Nest' (Shoreham-by-Sea), and 'Feet-Off-Ground-He-Scram' (Broadbridge Heath).

TWOS AND THREES

'Twos and Threes' is the orderly adult-approved form of 'Budge He'.
Children stand in pairs, one child behind the other, in a circle with all
the players facing the centre, and an equal distance between each
pair. One pair are appointed chaser and fugitive. The player being
chased runs where he likes outside the circle or across the circle or
dodging around the pairs, and he obtains safety by placing himself in
front of any one of the pairs, whereon the person at the back of that
pair becomes the new person to be chased. There are thus never more
than two people running at once, but the game is non-stop, for should a

player be caught he becomes the chaser and turns round and chases the one who caught him.

'Twos and Threes' is not a game children readily play on their own. Unlike 'Budge He' there must be not less than six players, their number must be even, the formation of the game does not occur spontaneously, and there is none of the subsidiary sport of moving at will from safety to evict somebody else. Indeed, when the game is played in the street, as in Putney, they do not bother about the circle, they simply 'cling together in pairs' and shout 'Help', and the game is called 'Help'.[1]

⁑ The name 'Twos and Threes' has predominated only in the twentieth century; earlier it was 'Round Tag' or, when the players stood in a line rather than a circle, 'Long Tag' (*Cassell's Sports and Pastimes*, 1888, p. 272). In polite society it was 'Fox and Goose', 'Faggots', 'Tertia', or 'Touch-Third' (Henry Dalton, *Drawing-Room Plays*, 1861, p. 329). Francis Kilvert seems to have known it as 'Thirds' (*Diary*, 31 August 1871). *The Cumberland Glossary*, 1881, has 'Hinmost o' Three', probably the same game, 'played on village greens'; *The Modern Playmate*, c. 1870, has 'Tierce'; and the usual name in the north country was apparently 'Tersy' or 'Tarsy' (*EDD*). This suggests that the game came from France where *le Tierce* or *le Tiers* has long been popular. The young Gargantua played 'au tiers' (1534); and, even earlier, the Ménagier de Paris, c. 1393, described young wives 'en la rue avecques leurs voisines jouans au *tiers*' (vol. i, 1847, p. 72). Moreover *le tiers* seems to have been a popular sport in medieval France, for Martial D'Auvergne makes it the occasion of some horseplay between young men and maids in *Les Arrêts d'Amour*, written about 1460-5, where a gallant complains of a girl shoving a handful of grass down the back of his neck; and the girl replies that she did so merely in the spirit of the game:

'Jouant au tiers en ung beau grant preau vert, et par joyeusete en courant par derriere elle mist audict gallant ung tantinet d'herbe entre la chemise et le dos, ce gallant se despita si terriblement que il lui vint incontinent bailler deux grans soufletz.'   *Edition c. 1520, 51st judgement, ll. 19-26*

Descriptions of the game in the seventeenth and eighteenth centuries show it to have been played in much the same way as today, but with the pursued player keeping outside the circle until he moved in front of a pair, a rule still observed in some places (e.g. at Ardingly College), and particularly in the United States. In Germany the game was 'Das

---

[1] Compare the fancy names given to the game by those in charge of children: 'Cat and Sparrow', 'Dog and Rabbit', 'Fox and Rabbit'. Children on their own are not so childish.

Drittenabschlagen' (J. C. F. Gutsmuths, *Spiele für die Jugend*, 1796, p. 276), and the chaser usually carried a *Plumpsack* or knotted handkerchief with which to hit the runner. Kampmüller says that in Austria the game has been superseded by a version 'Der dritte schlägt', in which the third in the row, instead of running away, chases the one who had been chasing (*Oberösterreichische Kinderspiele*, 1965, pp. 142–4).

In the United States the game is generally known as 'Three Deep', sometimes 'Third Man'. Mrs. Child in *The Girl's Own Book*, Boston, 1832, called it 'Tiercé, or Touch the Third'.

TOM TIDDLER'S GROUND

'Tom Tiddler's Ground' varies from other sanctuary-site games less in form than in emphasis, which is on the peril of entering the chaser's territory, rather than the security of being off it. The game has in fact an element of make-believe or exaggeration, and is mostly played by younger children. A straight line or a large circle is drawn on the ground, and beyond the line or within the circle is 'Tom Tiddler's Ground'. The children gather along the edge of the forbidden territory, daring each other to dash in for a few moments when they think it safe, and calling attention to themselves when they do so, the traditional words being:

> I'm on Tom Tiddler's ground,
> Picking up gold and silver.

Tom has to stay within his ground, and touch someone who is upon it. This player then takes his place, or, very occasionally, is made prisoner and has to be rescued.

The name 'Tom Tiddler' is, however, sometimes forgotten. The game becomes more dramatic, and the ground more dangerous to step on, when the landowner is a vampire, a dragon, a ghost, or other bogie. In a Hampshire village young girls were seen playing 'Gorillas' beside the school toilets. There was a division in the wall between the boys' toilets and the girls', and when a player went beyond this mark she was on the gorilla's ground, and the gorilla tried to 'have' her. They put a foot into the gorilla's territory and withdrew it sharply when the gorilla lunged at them. One girl ran a semicircle through the gorilla's territory. Another invaded it by climbing along a ledge of the toilets. Several ran round the back of the little building and came out the other side. All the while they jeered, 'Silly old gorilla', 'Stupid old gorilla'. Almost the greatest part of the fun seemed to be thinking of rude things to say to the gorilla.

⁎⁎⁎ In a recollection of the game, played about 1803, one boy was 'Tom Tidler', and his ground was marked off with a boundary line:

'He had heaps of sticks, stones, &c., supposed to be his treasures. The game consisted of a lot of boys invading his ground, and attempting to carry off his treasures, each calling out, "Here I'm on Tom Tidler's ground, picking up gold and silver". Meanwhile Tom was by no means a sluggard, but briskly defended his property, and drove off the thieves with a whip or switch.'

'Notes and Queries', 3rd ser., vol. iv, pp. 480–1

Yet the name of the groundlord was not always Tom Tiddler in the past, any more than it is today. In *The Craftsman*, 4 February 1738, and in Henry Carey's *Namby Pamby*, 1725, he was a friar:

> Now my Namby Pamby's found
> Sitting on the Friar's Ground,
> Picking Silver, picking Gold,
> Namby Pamby's never Old.

In Edward Moor's recollection of about 1780 he was 'Tom Tickler' (*Suffolk Words*, 1823); in Charlotte M. Yonge's *The Stokesley Secret*, 1861, ch. 2, he was 'Tommy Tittler'; in Mrs. Child's *Girl's Own Book*, 1832, p. 44, he was 'Old Man in his Castle'; in the *Alphabet of Sports*, 1866, an 'Old Man in Orchard'.

Jamieson in his *Scottish Dictionary*, *Supplement*, 1825, describes the game under the names 'Canlie' (' a very common game in Aberdeenshire') and 'Willie Wastell'. Newell, *Games of American Children*, 1883, gives the alternative names 'Dixie's Land' (New York), 'Golden Pavement' (Philadelphia), 'Van Diemen's Land' (Connecticut), and 'Judge Jeffrey's Land' (Devonshire, England). Gomme, *Traditional Games*, vol. ii, 1898, adds 'Old Daddy Bunchey' (Liverpool), and 'Pussy's Ground' (Norfolk). At Chirbury the game was called 'Boney' (i.e. Bonaparte), 'I am on Boney's ground' (*Shropshire Folk-Lore*, 1883, pp. 523–4). In Cornwall it was 'Mollish's Land' (*Folk-Lore Journal*, vol. v, 1887, p. 57). Correspondents have supplied the names 'I set my foot on Airlie's Green' (Strathmore, c. 1920), and 'Willy, Willy Wausey, I'm on your causey' (Carlton, c. 1898, Nottingham, c. 1925).[1] Perhaps it was Dickens who popularized the name 'Tom Tiddler'. The game seems to have been a favourite with him. In addition to the Christmas story 'Tom Tiddler's Ground', he refers to it in *Dombey and Son* and in *David Copperfield*.

---

[1] *Causey*, causeway or pavement. Similar games are 'Hopping on my Granny's Causey', *Games of Argyleshire*, 1901, p. 134; 'Mannie on the Pavement' in Aberdeen, *Traditional Games*, vol. ii, 1898, p. 443; and 'Padrone marciapiede' played by children in Rome (*Giochi*, 1967, p. 305).

On the Continent versions of the game seem to have a long history. It is apparently depicted in *Les Trente-six figures*, 1587, plate ix, under the name 'Je suis sur ta terre vilain'. It seems to be the game 'Man, man, ik ben op je blokhuys' listed in a Dutch translation of Rabelais, 1682. And in Swabia, where the protected territory was a 'kingdom', the invaders' cry was remarkably similar to the English:

> König, ich bin in deinem Land,
> Ich stehl dir Gold und Silbersand.
> *E. Meier, 'Deutsche Kinder-Reime', 1851, p. 121*

The variety of European parallels extant is well shown by Roger Pinon in *Arts et Traditions Populaires* (Strasburg), vol. ix, 1961, no. 1, pp. 16–23.

SHADOW TOUCH

Of all the methods of obtaining safety in a game, that in 'Shadow Touch' is the strangest, for it is a player's shadow that has to be touched, not his body; and to make himself safe a player has to make his shadow disappear. 'This is a tig for sunny days only, where a person tigs another person's shadow with his foot.' So it is that children are to be seen flattening themselves against the sides of houses, crouching beside cars, and crawling into bushes, for 'when a boy is in a shadowy spot the chaser cannot chase him no longer'. It is therefore not surprising, as one informant confided, that 'the person who is it may be it for a very long time'.

Despite the game's obvious defects (a shadow does not feel when it is trodden upon, nor remain under the chaser's foot for identification), 'Shadow Touch' is undoubtedly popular. Country children may complain that 'it is rather inconvenient because it can only be played when the sun is out'; but no such disadvantage mars the pleasure of the city child. 'When it is dark and the street lamps come on it is then we try to catch each others shadders', explained a 12-year-old.

Other names: 'Shadows' and 'Tig on Shadow'.

** The game is also played in Canada and the United States ('Shadow Tag'), Australia ('Shadows'), and New Zealand ('Shadow Tick').

Norman Douglas noted it in *London Street Games*, 1916, p. 139.

THREE STOOPS AND RUN FOR EVER

Security in this game is attained by posture. It starts as ordinary 'Touch' with one person chasing the rest, but any runner can make himself safe when he thinks he is about to be touched simply by crouching on the ground, and (usually) uttering some special word. This way of obtaining immunity is so easy, and so amusing to put into practice, that the attraction the game has for children is readily understandable. But it is, of course, too easy a way of becoming safe to make a good game. The amusement is infinitely frustrating for the chaser; and the number of times each person may stoop has to be strictly limited. A player who stoops more than the permitted number of times (usually three) is immediately penalized. He is made what he most dreads being—the new chaser; and he knows that he is likely to be the chaser for a long time, for once there is a new chaser the rest of the players are all allowed the pleasure of having three more stoops.

The great popularity of this game is reflected in the variety of names and local rules it possesses. The comments of the players on the spot, given in the following list, show, however, that the variations in the way the game is played are in manner rather than matter.

Bob Down Bunny. West Ham (a P.T. name). 'If you are had before you bob down you are still hee even if you haven't had one bob' (Girl, 11).

Bob Down Tick. Frodsham, Cheshire.

Bob Tig. Scarborough.

Bobbin. Berry Hill, Forest of Dean.

Bobs and Excuses. West Ham and Golders Green. 'When the person comes very close, and the person running away is out of breath he is allowed to bob down or make an excuse something like "Ah, my knee hurts" or "Oh, my side hurts". You are allowed to bob down three times and make three excuses. If you are had before you have said an excuse or made a bob you must be he' (Girl, 11).

Bops. Attleborough, Norfolk. 'Once you have bopped six times you are the old man' (Boy, 13). Cf. *Bop*, to dip, or duck suddenly, in Forby's *Vocabulary of East Anglia*, 1830.

Cabbage. Griffithstown, Monmouthshire. 'You get down on the floor and say "Cabbage" ' (Girl, 12).

Dippsy. Middleton Cheney.

Excuse me Touch. Ponders End, Enfield. 'Players say "Excuse me" when they duck.'

Fish and Chips. Swansea.

Ground Tiggy. South Elmsall. 'You bend and touch the ground with your hands' (Girl, 10).

Low Picko. Kirkwall, Orkney.

Low Tig. The usual name in Scotland. 'One sits on the ground if one does not want to be tug' (Girl, 14, Forfar). Cf. 'High Tig' p. 81.

Mercy Touch. Bristol, Truro, and Helston. 'You kneel on the ground and say "Mercy" ' (Girl, 10).

Pounds, Shillings, and Pence. Ramsey, Huntingdonshire. 'If you want a rest you shout out "Pounds" and sit on the ground. The second time . . . "Shillings".

The third time you say "Pence". Then you must run because you can't have any more rests' (Girl, 12).

Six Stoops and Run for Ever. Broadbridge Heath.

Squashed Tomatoes. Euston. 'Bend down and say "Squashed Tomatoes" ' (Girl, 11).

Squat. Hounslow.

Stoop Tick or *Stoops*. Banbury.

Stooping Tick. Wolstanton, Staffordshire.

Teddybear's Touch. Glastonbury. 'You say "Teddybear" and sit down' (Boy, 10).

Ten Stoops and Run for Your Life. Alton. 'In some games there are no squibs [truce term] or homes, but if you stoop you're safe' (Boy, 9).

Three Bobs and Run for Your Life. Blackburn.

Three Bobs and Three Excuses. York.

Three Bobs and Two Pokers. Yarmouth. 'Stand stiffly at attention for pokers' (Girl, 12).

Three Bops and Run for Your Life. Ipswich. *See* 'Bops' above.

Three Squads and Three Excuses. Camberwell, Dulwich, and Peckham. 'When a new person is hee you start again with your squads and excuses' (Girl, 10).

Three Squats and Run for Your Life. Peckham Rye and Walworth.

Three Stoops and Run for Ever. Welwyn. 'We shout "Stoop" when we get into crouch position' (Boy, 11).

Three Stoops, Three Pokers, and Run for Your Life. Petersfield.

Three Stoops, Three Statues, and Run for Your Life. Banbury. 'You either stoop or stand like a statue or the person chasing you can have you' (Girl, 11).

Tick Cuckoo. Welshpool. 'The one that's on it comes after you and you duck down and call Cuckoo' (Girl, 12).

Ticky Bob-Down. Wilmslow.

Ticky Little Man, *Ticky Stoop Down*, *Ticky Toadstool*. Alternative names at Sale, Manchester.

Tiggy Bob-Down. South Elmsall. 'If you crouch down and shout "Cabbage", "Carrots", or "Turnip", you can't be tug' (Girl, 10).

Tomato. Vale, Guernsey. 'When you have bob down three times you must run for your life' (Girl, 10).

Touch Ground It. Helston.

Tuggy Little Man. Hexham, Northumberland.

** Newell noted 'Squat Tag' being played in the United States in the nineteenth century (*Games of American Children*, 1883, p. 159), and 'Squat Tag' or 'Stoop Tag' seem to be the usual names there today. In Edmonton, Alberta, it is called 'Animal Squat Tag'. It seems that squatting gives protection in one chasing game or another in many parts of the world. Newell says that it does in Spain, and cites Maspons y Labrós, *Jochs de la Infancia*, 1874, p. 81. Lumbroso says it does in Lombardy in the game 'Cuciù' (*Giochi*, 1967, p. 306). Kampmüller says it does in Austria in the game 'Hockerlbot' (*Oberösterreichische Kinderspiele*, 1965, p. 136). Brew-

ster says it does in India in the game 'Uthali' (1953, p. 64). Culin says it
does in Korea in the game 'Syoun-ra' (*Korean Games*, 1895, p. 51).

Northall, in *A Warwickshire Word-Book*, 1896, gives the name 'Tick-
and-tumbledown'.

§ *Proliferation of Chasers*

In the following games the chaser is not relieved of his task when he
touches someone. He continues chasing throughout the game, and those
he touches become chasers with him. Thus at the climax of the game all
the players are chasers except one.

HELP CHASE

'Help Chase' is the straightforward game in this category: those who are
touched by the chaser help him to chase the rest, or, as the children put it:
'If the man who is hee has another man, he is hee with him and so on.'
'Eventually all are had and the time comes to start again. The first one
had in the last game is hee in the next.'

Names and variations: 'All Man He' (West Ham), 'Help Chase' (St.
Leonards-on-Sea and elsewhere), 'Tig and Help Chase' (Oxford), 'American Tig' (New Cumnock), 'Gorilla' (Fulham, 'Immediately one is court he
becomes a goriller with the original one'), and 'Devil's Den' (Swansea,
'Whoever is caught becomes another devil').

In Stoke-on-Trent, where the game is called 'Scatter', a fair start is
ensured by having the players stand round 'the one that is under' (i.e. the
chaser) and each take hold of a piece of his clothing. They may not begin
running until the one who is under says 'Scatter, scatter, scatter, one, two,
three'. In Bishop Auckland the game often starts with two chasers, and is
known as 'Two On' or 'Skip Jack'. To ensure that everybody starts at once
the chasers cross their arms in front of them, and each player takes hold of
a finger. To start the game the chaser then says 'Nash',[1] or 'Skip Jack, run
and never come back'. When a person is caught the catcher says 'Skip
Jack, help to catch, by one, two, three'.

CHAIN HE

'Chain He' is a game which once, perhaps, possessed a sinister or other-
worldly significance. When the chaser manages to touch somebody that

[1] Cf. J. H. Vaux, *Vocabulary of the Flash Language*, 1812, 'Nash, to go away from, or quit,
any place or company; speaking of a person who is gone, they say, he is nash'd or Mr. Nash is
concerned'. See also *Lore and Language of Schoolchildren*, p. 373, and, purely for fun, Ogden
Nash, *Everyone but Thee and Me*, 1963, p. 155.

person has to join hands with him, and from then onwards they run together. Each person touched joins the chasers, taking the hand of the person who touched him, so that there is an ever-lengthening 'chain' of chasers; although in one form of the game, now popular, when there are four chasers they split into pairs, and when these pairs catch two more players they split into further pairs. Both forms of the game have been described as 'the game of the moment', and both have the same rules: that a player can be 'tigged' or 'caught' only by a chaser's disengaged hand (i.e. the outside hand of a player at one or the other end of the chain); that a player cannot be caught if the chain is broken (the chasers must first join up again); and that the first person caught becomes the chaser in the next game.[1]

Of the two forms, the game in which the chasers split into pairs is the fastest; but the game with the long chain is the one that is memorable. When twelve or more players are linked together they make a formidable-looking chaser, and it is fortunate for those who are still free that the longer the chain grows the more awkward are its movements, and the more likely it is to break when the two ends strive to touch different players. Some children describe the wonder of playing the game at night under the street lamps, when a chain of children spreads out across the road and moves forward in the half-light to hem in those who are still free. The free players charge the chain, aiming at the weakest link, and time after time manage to break through before the ends can curl round to touch them. Other children tell how experienced players learn to close their ranks so that a charge from even the largest boy can be contained, and his chance of escape becomes slight. 'I do not like this game', complained a 10-year-old (probably a weak link in the chain), 'because when you keep 'old of 'ands the person who is on keeps dragging you and you easily get tired of it and if you leave loose of 'ands and tig somebody the person you tig is not caught.' But a 13-year-old Liverpool lad affirmed: 'It is an exciting game, and the last person to be caught is quite pleased with himself for being the last one.'

Names: The commonest names are descriptive, 'Chains', 'Chainy', 'Chain He', 'Chain Had', 'Chain Tig', 'Chain Touch', 'Ticky Chain', 'Ticky Join Up', 'Linky', 'Strings', 'String Tiggy', 'String Touch', 'Stringing Up' (Helston), and 'Altogether Tiggy' (Northampton). In some places more than one name is current, for instance, in Liverpool, 'Chain

---

[1] Occasionally the last person caught is the chaser in the next game, but this means that the one who has played best in the last game gets the least popular part in the next one, and this is not usually considered fair.

Tick', 'Stag-Eye', and 'Pea-Wak-Fly'. Around London the smart name is 'Sticky Toffee'.[1] At Chudleigh, Welshpool, Thirsk, and Stenness in Orkney, the game is 'Fishes in the Net', and when there are three or four people in the chain they must completely encircle a person to catch him. In Stromness it is 'Fisherman's Net' (cf. the French 'L'épervier' in which the first two chasers are termed 'les pêcheurs'). Most names are for either form of the game; but 'Couple Tag' (St. Ives, Cornwall), and 'Pairs' (Scalloway, Shetland), can only refer to one form, while 'Long Ticker' (Spaldwick), 'Long Chain Tag' (Sleaford), and 'Dragon's Tail' (east Montgomeryshire), distinguish the other. At Annesley the game is called 'Fly' from the chaser's command to the players to start running (cf. 'Scatter'). And in Dulwich, where the game is called 'Sheep Dog', at Blaenavon, where it is 'Shepherds', and in Scarborough, where it is 'Shepherd's Crook', they chase only in pairs, and each captive is taken back to base to make an additional pair.

Other names current include: 'Aller Beyroot' (Neath), 'Buckle' (East Meon), 'Bully' (Caister-on-Sea), 'Clawer' (Tewkesbury), 'Corner to Corner' (Snelland), 'Cree' (Knighton), 'Cree-cree' (Llanfair Waterdine), 'Crackum' (Stamford, Lincolnshire), 'Doctor' (Cwmbran), 'Join up Bucket' (St. Ives), 'Knocky-knole' (Llangunllo, Radnorshire), 'Stag' or 'Stags' (fairly common), and 'Tommy Early' (Presteigne). Also 'Hawks and Doves' and 'Fox and Rabbits', but pretty-pretty names usually come from teachers' manuals.

** In the United States 'Chain Tag' or 'Link Tag'. In Austria 'Kettenfangen' (Kampmüller, 1965, p. 151). Further names and antecedents appear in the next section.

VESTIGIAL FEATURES IN 'CHAIN HE' VARIANTS

In widely separated areas of Britain hand-linking touch continues to be played under strange names and fanciful-seeming rules that are possibly the consequence of custom and pastime in former days.

Chain a Wedding. In the neighbourhood of Pontypool the game of 'Chain a Wedding' begins with two children holding hands and running after the rest. They make a chain until all have been caught except one, whereupon they break up to catch the last player. First and last caught then become the chasing pair for the next game. It seems probable that the name 'Chain a Wedding' is an echo of the old practice, prevalent

---

[1] This name seems to have been current for some years. A speaker, quoted in *English Dance and Song*, summer 1967, p. 40, knew it in 'barrel organ days'; and Gomme, 1898, gives 'Sticky Toffey' as the name of a game (undescribed) played by schoolchildren in Hoxton, London, N.1.

particularly in south Wales, of obstructing the wedding procession by a rope tied across the road, so that a toll could be exacted for allowing the otherwise happy pair to continue on their way (see Kilvert's *Diary*, 15 May 1875, and *The Lore and Language of Schoolchildren*, 1959, p. 304).

Warning. The special emphasis on the player who remains uncaught also occurs in Swansea, where the game is called 'Cock Warren' and the last child is similarly chased by all the others running freely. 'Cock Warren' appears to be a corruption of 'Cock Warning', a name for 'Chain He' in the Rhondda Valley; and the name of a 'favourite game' (undescribed) played at Sedgley Park School, Staffordshire, about 1805.[1] The game of 'Warning', otherwise known as 'Widdy', was popular throughout the nineteenth century. In *The Boy's Own Book*, 4th ed., 1829, pp. 23–4, it is described as a game in which one player, standing behind a line, delivers the following challenge or caution:

> Warning once, warning twice, warning three times over;
> A bushel of wheat, a bushel of rye,
> When the cock crows, out jump I!
> Cock-a-doodle-doo!—Warning!

'He then runs out, and touches the first he can overtake, who must return to bounds with him. These two then (first crying "Warning" only) join hands, and each of them endeavours to touch another; he also returns to bounds, and at the next sally joins hands with the other two. Every player who is afterward touched by either of the outside ones, does the like, until the whole be thus touched and taken. It is not lawful to touch an out-player after the line is broken, either accidentally, or by the out-players attacking it, which they are permitted to do. Immediately a player is touched, the line separates, and the out-players endeavour to catch those belonging to it, who are compelled to carry those who capture them, on their backs, to bounds.'

Basically this is 'Chain He', but clearly a slower game since the chasers return to base each time they have touched someone (as in 'Shepherd's Crook' played today in Scarborough); and the quaint practice of those who have been chased retaliating by chasing their chasers in the hope of riding to the boundary on their backs, seems to have been vestigial even in George IV's time. When boys in ancient Greece played the game 'Ostrakinda' any fugitive they caught was called the '*ass*', this signifying that he had to carry them home (Pollux, ix. 111; and see further under 'Crust and Crumbs'). Indeed, a vestige of this practice survives to the present day in Grimsby, where the last one to be caught when playing 'Chains' is given an enforced ride on the shoulders of his companions.

[1] F. C. Husenbeth, 1856, p. 107. In Whitland, Carmarthenshire, the name is 'Cocks-a-morning'.

Further, at the Lizard, in the heel of Cornwall, an 11-year-old boy (seemingly posted there for visiting folklorists) made our journey worth while with the information that 'the game where you all join up in one long string, we call "Warney". The last person who's caught, he's Warney. He says, "Warney one, warney two, warney three", like that, up to ten, then he starts chasing'.

Staggy. In the Manchester borough of Sale a version of the game is known as 'Staggy in the Button Hole', and contains the singular feature that 'the one who was tuck first and is on next gets beaten on the back while the others chant:

> Staggy in the button hole, one, two, three,
> Staggy in the button hole, one, two, three,
> Staggy in the button hole, one, two, three,
> Staggy in the button hole, can't catch me.

Then the game starts again' (Boy, 12). This is almost identical to the treatment that awaits the newly appointed chaser at Accrington, who has first to allow himself to be hit ('but not very hard') ten times on the head before he begins chasing. And at Rossendale, also in east Lancashire, where the game is known as 'Stagger-Ragger-Roaney', the boy has to suffer being 'shaken up and down' while the rest of the players chant:

> Stagger-Ragger-Roaney, my fat pony,
> One, two, three, four, five, six, seven.

'They then let him go and run away, and when he has recovered he chases them.'[1] The names 'Stag', 'Staggy', 'Ticky-Stag' (and at Knighton 'Staga-lonia') are not uncommon denominations for 'Chain He', although more often they refer to the catching game in which players run back and forth across some open space (pp. 128–30). It is curious that in the games 'Staggy in the Button Hole' and 'Stagger-Ragger-Roaney' there should be cere-monial discomfiture for the player who acquits himself poorly, when it is recollected that in the west country deceiving husbands or scolding wives used to be made the subject of a rough ceremony termed a 'Stag Hunt'.[2]

Widdy. At Rushmere St. Andrew, near Ipswich, one of the places where the game is called 'Stag', the person who has been appointed 'it' clasps his

---

[1] The pacifist Joshua Rowntree is on record as saying that when he was a boy at the Friends' School, Bootham, York, about 1854, 'Stag-a-rag was one of the best playground games'— S. E. Robson, *Joshua Rowntree*, 1916, p. 25. Northall in his *Warwickshire Word-Book*, 1896, has the game 'Stag-alone-y' with the signature verse 'Stag-alone-y, My long pony, Kick the bucket over'.

[2] See S. Baring-Gould, *Red Spider*, 1887, ch. xxiv; *Transactions of the Devonshire Association*, vol. lxxxiii, 1951, p. 77; *Folk-Lore*, vol. lxiii, 1952, pp. 102–9.

hands together in front of him, counts slowly up to ten, and then announces his animation with the shout 'Squirrel'. Similarly, 600 miles away in Whalsay, one of the Shetland Islands, when the children play 'Humpty Dumpty'—said to be a corruption of 'Hunty Bunty'—the first chaser clasps his hands together in front of him and issues the innocuous-seeming warning or invitation 'Who's coming in my fun Humpty Dumpty?' and keeps his hands clasped, as he runs, until he tigs somebody, and joins up with him. In Peckham Rye the game is known as 'Chain Widdy', in Thirsk it is occasionally 'Whitee', in Helston it is 'Willie Willie Way'. The game 'Widdy', as set out in *Games and Sports for Young Boys*, 1859, pp. 1–2, is identical in all respects to 'Warning' except that the chaser is described as first clasping his hands together, calling out 'Widdy, widdy, way—cock warning!' and striving 'to overtake and touch one of the others without dividing his hands'.[1] It seems possible, as has been suggested under 'Bull', that this pursuit and 'touching' with clasped hands formerly had a greater significance than the mere hampering of the chaser. Jamieson gives the earliest account we have of touch-linking in the *Supplement* to his *Scottish Dictionary*, 1825, when he defines the Lothian game of 'HORNIE' as:

'A game among children, in which one of the company runs after the rest, having his hands clasped, and his thumbs pushed out before him in resemblance of horns. The first person whom he touches with his thumbs, becomes his property, joins hands with him, and aids in attempting to catch the rest; and so on till they are all made captives. Those who are at liberty, still cry out, Hornie, Hornie!'

Jamieson wondered 'whether this play be a vestige of the very ancient custom of assuming the appearance and skins of brute animals, especially in the sports of Yule' (as described by Strutt), or whether it might 'symbolize the exertions made by the devil, often called *Hornie*, in making sinful man his prey, and employing fellow-men as his coadjutors in this work'. Whatever its antecedents may be, a children's game called 'Hornes Hornes' was listed by Randle Holme in 1688, and the name 'Hornie' for a linking game subsists to this day in Kirkcaldy; while in Stromness, where in 1909 'Long Horny' and 'Short Horny' were played, the name for 'Chain He' continues to be 'Horny'.

[1] This warning cry recalls the chant of the hideous boy 'Deputy' in *Edwin Drood*, 1870:
'Widdy widdy wen!
I – ket – ches – Im – out – ar – ter – ten,
Widdy widdy wy!
Then – E – don't – go – then – I – shy,
Widdy Widdy Wake – cock warning!'

Bully Horn. In *The Book of Sports*, *c.* 1837, the editor, William Martin, who was born in 1801 at Woodbridge in Suffolk, recalls the game 'Stag Out'.[1] The chaser is again likened to a stag. He 'clasps his hands before him and rushes at the other boys'; and the additional point is made that the other boys seem to 'taunt and bay him', yet, if the stag succeeds in touching one of the 'bayers', he can make him submit to the ordeal of bearing him (the stag) back to bounds, before the two of them rush out together with linked hands to touch another if they can. The game of 'Bully Horn', played to this day in Ballingry, Forfar, and Golspie, is remarkable in that it embodies not only the return to bounds with each new person 'tigged', but the tormenting of those caught, and the systematic goading of him—or it may be her—who chases. The players are thus preserving—unconsciously, but in no irresolute manner—all the rudimentary features of play in former times. The following is a description of 'Bully Horn' by a 12-year-old Forfar schoolgirl:

'In the winter about eight o'clock a gang of us get together the likes of Isabella, Sandra, Margaret, Mary, another Margaret, David, Gus or Angus, Harold, and Duncan. We all stand on the pavement and run to a lamp-post, the last one is out. The one that is out stands on the pavement and we shout names to her to annoy her, then she comes and chases us. When she gets someone they have to run to the pavement and we run after them and thump them, pull their hair, before they reach the pavement. We run up to the pavement and catch the one that is tigged. We twist her fingers and do everything we like till she says "Bully Horn". Then they join hands and try to get someone else out. If they let go hands we shout out aloud "Chainy Broken—Get them". We run and hit them again before they reach the pavement. This goes on till everybody is out. The one that was last out is first out next time.'

KINGY

This fast-moving game has all the qualifications for being considered the national game of British schoolboys: it is indigenous, it is sporting, it has fully evolved rules, it is immensely popular (almost every boy in England, Scotland, and Wales plays it), and no native of Britain appears to have troubled to record it.[2]

'Kingy' is a ball game in which those who are not He have the ball hurled at them, without means of retaliation, and against ever-increasing

[1] Cf. 'Stag' still played at Rushmere St. Andrew, only some five miles from Woodbridge
[2] Children and teachers alike often regard the game as being special to their school, as well they might from the absence of literature on the subject. Norman Douglas in *London Street Games*, 1916, p. 5, mentions a ball game called 'King' which may or may not be this game; and Sutton-Smith in *The Games of New Zealand Children*, 1959, p. 150, gives a brief account of 'Kingy' as played in Wellington South.

odds, an element that obviously appeals to the national character. Anyone who is hit by the ball straightway joins the He in trying to hit the rest of the players. Those who are throwing may not run with the ball in their hands, but pursue their quarry by passing the ball to each other. Those being thrown at may run and dodge as they like, and may also punch the ball away from them with their fists. For this purpose players sometimes wrap a handkerchief round their hand, as 'fisting' the ball can be painful. The game continues until all but one have been hit and are 'out', and this player is declared 'King'. When the contestants are skilled (and boys of fifteen and sixteen readily play the game), the ball gets thrown with considerable force: it shoots back and forth across the street or playground, and the game can be as exciting to watch as a tennis match.

As befits a sport in which so much energy is expended, the preliminaries are sometimes wonderfully ritualistic. At Bishop Auckland, for instance, one person shouts 'King' to start the proceedings, and two others follow up by crying 'Sidey'. The players then form a circle round the King, with the two who shouted 'Sidey' standing on either side of him like heirs-apparent. The players making the circle stand with legs apart, each foot touching the foot of their neighbour on either side. The King picks up the ball and bounces it—or, as they say in Bishop Auckland, 'stounces' it—three times in the ring, and then lets it roll. Everyone watches to see whose legs it will go through. If it does not roll through anybody's legs the King picks it up and bounces it again, and if his second turn fails he has a third try. If the ball still has not passed between anyone's legs, he hands it to the first sidey (the 'foggy-sidey') who, as necessary, repeats the performance—for the moment the ball does pass between someone's legs that person is 'on', and everyone runs. At the end of the game whoever becomes King takes the place in the centre of the ring to start the next game, and the first two people to shout 'Sidey' stand beside him.

In Grimsby, where they also start with the circle, they select the person who is first to bounce the ball in the centre by counting round the players with the words:

> Double circle's not complete
> Till it goes through someone's feet.

The person pointed to at 'feet' goes into the centre. In some places, however, it is the person who provides the ball who first goes in the centre.

In Scarborough, especially among younger children, the circle-start to the game becomes virtually a game in itself. If the ball is about to roll between a person's legs he can shout 'Knick-knock', which entitles him to use his knees to prevent the ball going through; or he can shout 'Kicks'

Plate I. KINGY

which means he may kick the ball away, provided others have not already shouted 'No knick-knocks' or 'No kicks'. Likewise, should the ball touch someone else's foot before passing through his legs the player can shout 'Rebounds', and the ball has to be picked up and dropped in the centre anew. Or again, should somebody say 'Tricks' before anyone has declared 'No tricks', the one in the middle may aim the ball through whose legs he chooses, and that person straightaway becomes 'it'.

This selection of the chaser by the fortuitous rolling of the ball is customary throughout England and Wales, and much care is taken to see that the ground is flat, so that nobody will be at a disadvantage. In some places, particularly in Wales, there is the difference that the players stand in a tight circle, sometimes having their arms round each other's shoulders, and each puts his right foot forward. It is enough then that the ball touches someone's foot for that person to be 'it'. In Aberystwyth if the ball is dropped in the ring three times and does not touch anybody's right foot, the player who has been dropping the ball becomes the chaser. In Welshpool, however, he hands the ball to the person called 'Second King', who was second last out in the previous game, and he himself 'goes for a walk', which ensures that he will not be touched by the ball and become the new chaser.

In the Walworth district of London the boys sit on the kerb with their feet apart. One boy rolls the ball towards them from across the road, and the one whose legs it goes between is He. In Wandsworth, in much the same way, the players line up facing a wall with their legs apart. And in Cleethorpes, Lincolnshire, they stand in a row but with their heels together and toes apart. The ball is rolled towards them and the person whose feet the ball touches is 'it'.

Throughout most of Scotland, although not in Edinburgh, the players form a circle and hold out their clenched fists in front of them. The player in the middle throws the ball to somebody and he catches it between his fists and throws it to someone else in the circle, who throws it to someone else, all with closed fists. When somebody drops the ball that person is 'hit' or 'het'. Sometimes the person throwing the ball is allowed to pretend to throw it to one person and in fact throw it to another. In Forfar this is known as 'jinkies', and can be prevented by the cry 'No jinkies'.

The Rules. Although the ways of choosing the chaser are numerous, the game itself is played with little variation. Reports from more than fifty places have been so similar, it is as if a mimeographed sheet of rules was carried in every grubby trouser pocket. Such a set of rules would read as follows:

1. The number of players shall be not less than six or more than twenty: the best number is about twelve.

2. The boundaries of the game shall be agreed on before the game begins. A flat area of 20×20 yards, or a length of street of about 20–30 yards, depending on the number of players, is ample.

3. One person shall be chosen chaser, and the game shall start immediately he is chosen. The chaser shall, however, bounce the ball ten times before he throws it at anyone, to give the players time to scatter.

4. The chaser may not run with the ball; but while he is the sole chaser he may bounce the ball on the ground as he runs.

5. A player shall be 'out' when the ball hits him on the body between his neck and knees (or, as may be agreed, between his waist and ankles). It shall be determined beforehand whether a hit shall count if the ball has first bounced on the ground or ricocheted off a wall; or whether only a direct hit shall count.

6. As soon as a player is 'out' he shall assist the chaser in getting the other players out.

7. When there are two or more chasers they may not run with the ball, but may manœuvre as they wish by passing it to each other.

8. Players being chased may take what action they like to avoid being hit by the ball, including 'fisting' it, i.e. punching it away with their fist. They may also pick up the ball between their fists and chuck it away.

9. Should a chaser catch the ball when it has been 'fisted', or touch a player while he is holding the ball in his fists, the player shall be 'out'.

10. Should a player kick the ball, or handle it other than with his fists, he shall be 'out'.

11. Should a player run out of bounds when trying to avoid being hit by the ball he shall be 'out'.

12. The last player left in shall be 'King', and shall officiate at the selection of the next chaser.

A few local practices may be noted. In parts of West Ham they do not make the chaser bounce the ball ten times before he begins throwing, but appoint a 'bunger' whose duty it is to seize the ball when it has passed through someone's legs, and throw it out of the chaser's reach. In West Ham, too, if a player is hit on the head or foot, the chaser, says an 11-year-old, 'has a free bung, and everyone that's not "he" shouts "Miss him, miss him, miss him" '.

At Rosneath in Dumbartonshire, if the ball is being thrown too hard, and somebody does not want to be hurt, he may shout 'No trade marking', and the ball must not then be thrown so fiercely.

In St. Andrews, Fife, if somebody thinks he is not out, and the others disagree, he must submit to 'Blind Shot'. He has to stand against a wall with his arms spread out. One of the throwing side takes aim about six yards away, has his eyes covered, and then throws. If the person is hit he is 'out'; if he is not hit he is not 'out'. A similar rule known as 'Free Shot' is observed in Orkney.

It remains to say that while 'Kingy' is the usual name in England, and 'King Ball' in Scotland, the game is sometimes known in the midlands as 'Hot Rice' (occasionally corrupted to 'Horace'), and in Lincolnshire as 'Dustbin', due to the players being allowed to fend off the ball with bats, pieces of wood, or dustbin lids. There are also the local names 'Buzz' (Enfield and Croydon), 'Fudge' (Basildon), 'Cheesy' (Exmouth), 'Peasy' (Cleethorpes), 'Punch' (South Elmsall), and 'Fisty' (the usual name in the Orkney Islands). The game is occasionally known as 'Ball Tig' or 'Dodge Ball' which are generally taken to be other games. In Scotland girls sometimes call the game 'Queen Ball', and the girl who stays in longest is 'Queen'.

§ *Suspense Starts*

In some games part of the sport is that the players do not know when the chaser is going to begin chasing. Yet they may have to remain close to him, suffer his jokes and clowning while they wait, and even have their suspense pricked with false starts. It is natural that in these games the start becomes the dominant feature.

POISON

In 'Poison' or 'Bottle of Poison', which is immensely popular with boys of 10 or 11, all players are actually touching the chaser when he begins chasing. He holds out his hands, sometimes with arms crossed, and each player takes hold of a finger and stretches as far away from him as he can, preparing to run. The chaser says 'I went to a shop and I bought a bottle of *vinegar*' (or any other substance). 'I went to a shop and I bought a bottle of *p-p-p-Pepsi*. I went to a shop and I bought a bottle of POISON!' The word 'poison' is the signal for everyone to run. In Wolstanton the players first ask the one they are holding on to 'What's in the bottle when the cork goes pop?' and he replies as he chooses '*tea*', or '*powder*', or 'POISON'. Should anybody run before he says 'poison', for instance, when he says the 'p' of 'powder', that person has to take the place of the chaser, or even be out of the game if he does it twice.

Sometimes this start so overshadows the subsequent chasing that well-known games acquire entirely new names. For example, in Bristol and Kilburn 'Release Touch' is called 'Bingo' (the chaser saying, perhaps, 'B for Bat, B for Ball' and nobody may run until he says 'B for BINGO!'). In Alton the game is called 'Jumbo' (the chaser recites the names of the players in turn, suddenly naming one of them 'Jumbo'); in Langholm it is 'Peapod Poison' ('My mother went down the street to buy some *peel*, to buy some *peat*, to buy some PEAPOD POISON!'); and in Inverness 'Sticky Glue' (Sticky *jam*, sticky *gloves*, sticky GLUE!'). In York 'Chain He' becomes 'Black Jack' ('Black *hat*, black *cat*, black JACK!'); in Fulham 'Black Magic'; and in Northampton 'Black Man's Chimney'. In Accrington 'Help Chase' becomes 'Jimmy Jack Fly'; and in Liss, where 'Help Chase' is called 'Apple Plum Pudding', a young boy describes how they take hold of any part of the chaser they can:

'We all say someone's he and we all take hold of them. We hold their trousers or pullover. Sometimes they say "No flesh or hair" so that means we are not allowed to hold flesh or hair. Sometimes they say "No shoe laces" so we are not allowed shoe laces. Then they say "Apple, Plum, *something*" and if they say banana or pie or anything like that and someone leaves go they're he. If he says "Apple Plum Pudding" we all run.'

CRUSTS AND CRUMBS

In 'Crusts and Crumbs' the players do not even know whether they are going to be runners or chasers. They divide into two sides, 'Crusts' and 'Crumbs', and face each other a few feet apart on the crown of the road. One person, who is not in the game, calls out either 'Crusts' or 'Crumbs'. If he calls 'Crusts', the Crusts chase the Crumbs to the pavement behind them; if he calls 'Crumbs', the Crumbs chase the Crusts to their pavement. Those who are caught before they reach the pavement join the other side. Sometimes the caller hangs on to the 'Crrr' of 'Crusts' or 'Crumbs' for as long as his breath will hold, keeping the players in suspense about the direction they will have to run; and sometimes he causes chaos by calling 'Crrr-umpets'.

This game is often adult-organized, and it is certainly improved by having a good caller. Several children describe it being played in the street in modified forms, for instance an 11-year-old girl in Fulham:

'About eight people play the game. It is a very catchy game. You have an outer, who stands on one side of the road. On the other side the children stand. The pavement is crusts and the road is crumbs. The outer shouts out either crusts or crumbs. If he calls out crusts you jump onto crusts. If he calls out crumbs you immediately jump onto crumbs. If you are not quick enough you are out.'

In Stoke-on-Trent the game becomes a kind of race. All the children are on one side of the road, and when the call is 'Crusts' they run across the road and back again; but when the cry is 'Crumbs' anyone who moves is out.

Other names: 'Rats and Rabbits', 'Soldiers and Sailors'.

*** 'Crusts and Crumbs' often appears in the games manuals, e.g. Barclay's *Book of Cub Games*, 1919, and Knight's *Brownie Games*, 1936. In Smith's *Games and Games Leadership*, published New York, 1932, pp. 215–18, it is described under the names 'Black and Blue', 'Crows and Cranes', 'Rats and Rabbits', 'Heads and Tails', 'Wet and Dry', and 'Black and White'. In Bancroft's *Games for the Playground*, New York, 1909, pp. 52–3, only the name 'Black and White' is given. The leader was provided with a flat disc hanging on a string which was white on one side and black on the other. He twirled the disc, and if it stopped with the white side visible the 'Whites' tried to tag the 'Blacks'; if the black side was shown the 'Blacks' tried to tag the 'Whites'. Although neither Bancroft nor Smith seems to have been aware of it, the game they describe is more than 2,000 years old. Pollux, under 'Ostrakinda' (The Game of the Shell), tells how boys in Greece took a shell, and smeared one side with pitch, calling that side 'Night', while the other side, which remained white, was 'Day'. They drew a line, picked up sides, and decided which side should be Night and which Day. The shell was twirled, and the side whose colour came uppermost chased the other party; anyone caught was denominated 'ass', which meant (as is known from other references) that the boy had to carry his catcher on his back (*Onomasticon* ix. 111). Pollux himself alludes to the curious passage in Plato where the philosopher remarks how a lover and a loved one often change roles. 'The shell being turned again', the lover or pursuer, as he was, becomes the pursued and strives to flee, while the object of his love tries to catch him (*Phaedrus* 40, written *c*. 365 B.C.). Plato again refers to the game, or so it seems, in his *Republic* (vii. 521); and it appears that the game was so well known, anyway by the second century A.D., that it was proverbial for something to change 'at the turn of a shell'. It is interesting that the game continues to be played in Italy, and to be called 'Giorno e notte' (Day and Night), although no longer played with a shell; that in France it is 'Le jour et la nuit'; that in Austria it is 'Schwarz-Weiss'; and that it is an old game in Germany, 'Tag und Nacht' being fully described by Gutsmuths, *Spiele für die Jugend*, 1796 (1802; pp. 266–8), the leader tossing a disc or coin.

How well the game was known in England in the past is uncertain. John Higins in his *Nomenclator*, 1585, described 'Ostrakinda' as a play 'not in use with us' in England. Yet it seems to have seeped through the centuries at gutter level. In *London Street Games*, 1916, p. 47, Norman

Douglas mentions a lowly game played by girls called 'Rolling Pin' in which two parties 'decide which of them has to chase the other by the red or blue colour marked on a rolling pin which is rolled between them'.

LITTLE BLACK MAN

To the younger children, naturally enough, a chasing game is immeasurably more exciting if the chaser is not thought of as a playfellow but as someone strange and fearsome; and the thrill is even greater if the strange character exchanges pleasantries with them before he gives chase. Half a century ago it was not uncommon for the start of a chasing game to be delayed while the players joined in a song or took part in a set dialogue. This still occurs in the neighbourhood of Welshpool in the game called 'Little Black Man'. When a girl has been selected chaser, the players make a circle round her, and she asks 'Where have you been?'

'Down the lane', they reply.

'What did you see?'

'A little white house.'

'Who is in it?'

'A little black man.'

'What did he say?'

'Catch me if you can!' they shout, and speed off in all directions, with the chaser after them.

*** Such a beginning to a game is a 'delayed start' rather than a 'suspense start'. The dialogue is of a set length, and the runners themselves give the signal for the chase to begin. In general today in games of this type the preliminary dialogue not only sets the scene and delays the start, it intensifies the suspense, for the player who decides when the chase shall start is the chaser. Thus, as can be seen in the games which follow, the players who have to converse with the chaser, and remain near him while they converse, become increasingly apprehensive the longer the dialogue continues.

WHAT'S THE TIME, MR. WOLF?

This game is extraordinarily popular. One child is Mr. Wolf and walks along the road rather haughtily, while the rest follow in a group as close behind as they dare. They call after him, 'What's the time, Mr. Wolf?' Mr. Wolf does not turn round. He replies in a gruff voice 'Eight o'clock', or any other time, and keeps walking. The children follow along a little further and call again, 'What's the time, Mr. Wolf?' The wolf replies, perhaps, 'Five o'clock'. The children continue to follow the wolf, and begin to pester him to know the time. Suddenly he cries it is 'Dinner time!', turns round, and chases them. The children rush back to the safety of the starting-place (usually screaming as they do so), and if Mr. Wolf catches one of them before they reach home that person is wolf next time.

Names: 'Mr. Wolf' or 'What's the time, Mr. Wolf?' (many places), 'Mr. Fox' or 'What's the time, Mr. Fox?' (Coulsdon, Cruden Bay, Thame, Stoke-on-Trent), 'Foxy' (Stenness in Orkney). Amongst Brownies, 'What's the time, Mr. Bear?'

*⃰ Correspondents recall playing 'What's the time, Mr. Wolf?' at Midgley, near Halifax, *c.* 1895; Ilminster, *c.* 1920; and Abergavenny, *c.* 1930. The game appears to stem from versions of 'Fox and Chickens' ('Chickamy, Chickamy, Chany Trow', 'Old Dame', etc., pp. 310–12) in which the players used to ask the predator the time, although in those days a set dialogue was normal, which gave the chase a delayed start rather than a suspense start. However, the connection between the two games is apparent in *The Home Book*, 1867, p. 16, where Mrs. Valentine gives a game called 'Twelve o'clock at Night'. In this one player, a hen, with other children, her chicks, clinging behind her, approaches a fox in his lair and asks the time. When the fox replies 'Twelve o'clock at night' he tries to seize one of them.

Similar games in which a wolf is asked questions are current in France, Italy, Germany, Spain, and South America; and this type of game is probably universal. In Cairo children make a circle round one player, addressing him 'O wolf, O wolf, what are you doing?' and the wolf replies that he is washing, that he is brushing his hair, and such-like innocuous occupations. Finally he replies '*Chasing you*', and does so. It is possible that the wolf's replies in 'Little Red Ridinghood' belong to the same tradition.

In Banbury children play a game called 'Weird Wolf' (Werewolf) in which one person pretends to be an old man. An 11-year-old girl writes:

'The rest have to be children and go up to the old man and ask if he will tell them a story. So he tells them a story which is, One day a long time ago I had to go in the forest and kill a weird wolf but the weird wolf escaped and scratched my arm, and the man shows the children the scratch on his arm. He then says, Every full moon I change into a weird wolf. But the children don't believe him. Then he says, There's a full moon now, and he changes into a weird wolf. The children run away and the wolf chases them. When he has caught one of them that child helps him to chase the rest.'

I'LL FOLLOW MY MOTHER TO MARKET

In this trivial drama the mother tells her children she is going to market, and gives them jobs to do about the house while she is away. As soon as she has gone, the children stop doing their jobs, band together, and cautiously follow their mother to market. In Somerset they sing as they do so, not altogether logically:

> I'll follow my mother to market
> To buy a ha'penny basket;
> When she comes home, she'll break my bones,
> For falling over the cherry stones.

The mother goes into a shop, and the children creep up as close as they dare to hear what she is buying. She asks the shopkeeper for household goods, such as a dustpan and brush, a mop, and some washing-powder. Then she asks the shopkeeper for 'a cane to beat the children with'. The children turn and run for home with the mother chasing after them. In Langholm the game is called 'The Cane'.

Several similar games are played, all of them apparently traditional. In the village school at Chawton, opposite Jane Austen's house, one of the children pretends to be 'Old Mother Hubbard', and the rest of the children shout 'What have you got in your cupboard, Mrs. Hubbard?' At first she says she has butter in her cupboard. When asked again she says, perhaps, 'Rice'. When she says that her cupboard contains 'BONES' the children flee, as if for their lives. At Roe, Shetland, one girl is selected to be 'Lucy Anna' and is sent off on her own. The rest follow her a little way behind roaring without restraint, 'Lucy Anna, Lucy Anna', and they keep up the cry (which is part of the fun) until she pretends to be provoked into chasing them. In Golspie, and elsewhere, the player they taunt is 'Black Peter'. He stands with his back to the others and calls out 'Who's afraid of Black Peter?' The others shout back 'Not I' and edge forward a little to prove it. He calls out again 'Who's afraid of Black Peter?' The children respond 'Not I' and creep forward yet further. When Black Peter thinks they are close enough he turns and chases them. (In Aberdeen the game is 'Who's Afraid of the Big Black Beetle?') Similarly, in a number of places, there is a game called 'Big A, Little a' or 'The Cat's in the Cupboard'. One child is a cat and goes to the far side of the road, turning away and closing her eyes. The other children gradually advance across the road 'making faces', says a Helston girl, and pointing at the person by the opposite wall. As they do so they chant:

> Big A, little a, bouncing B,
> The cat's in the cupboard and can't see me.

'Either during or just after the chant the catcher *suddenly turns* and tries to catch you.'

✱✱ Previous recordings: *Folk-Lore Record*, vol. v, 1882, p. 84, 'Going to Market' (Hersham); *Folk-Lore Journal*, vol. vii, 1889, p. 231, 'Basket' (Dorset); *English Folk-Rhymes*, 1892, pp. 392–3 (Warwickshire); Gomme, vol. i, 1894, p. 24; *Rhythmic Games*, 1914, p. 19, 'Follow my Mother to Market' and 'Old Daddy Wiggin'; *London Street Games*, 1916, p. 45, 'Who's Afraid of Black Peter?'; Macmillan MSS., 1922, 'Big A, Little a' (Somerset); correspondent, 'Daddy Mick has lost his stick' (Staffordshire, *c*. 1915).

In the United States: *Games of American Children*, 1883, pp. 143–5, 'Old Mother Tipsy-toe' and 'Old Mother Cripsy-crops'; Howard MSS., 1947, 'Grandmammy Tippytoe, lost her needle and couldn't sew' (Maryland).

JOHN BROWN

An intensely dramatic start to a chase occurs when the person who is 'on' feigns that he is dead, or asleep, or an inert object. In Scarborough the game is called 'John Brown'. One person has the part of John Brown and lies dead on the ground. The others walk about near him, intoning 'This is the body of John Brown'. John Brown then kicks someone, and the person who is kicked pushes whoever is nearest him, saying 'Stop kicking'. 'I didn't kick', says the other. 'You must have kicked', says the first, 'no one else was near me except the body of John Brown.' John Brown then kicks another person, and starts a second argument, and perhaps kicks a third person so that there is a further argument. When a number of the children are having mock arguments John Brown leaps up, the children scream, and John Brown tries to tig one of them.

At Ruthin in Denbighshire the Welsh-speaking children have a similar game called 'Nain Gogo' (Grandmother Gogo). In this, Grandmother sleeps on the ground and the children come to her several at a time and tickle her face with grass until she jumps up and tries to catch them. In Swansea they play 'Spider in the Corner'. One player crouches in a corner, and at first takes no notice when she is touched or even prodded, but suddenly jumps up and tries to catch as many players as possible. And at Roe, in the Shetland Islands, the one who slumbers is a giant, and when he catches someone he puts them in his den, and 'the game of "The Giant" goes on until the whole of us is caught' (Girl, 8).

** A version of this game called 'Only a Stump of a Tree' was described by Mrs. Craik (Dinah Maria Mulock) in *Our Year*, 1860, pp. 290–1:

'Somebody sits in a corner, while all the rest make believe to be taking a walk, come up to him and touch him and shake him and pull him about saying, "Oh, this is only a stump of a tree",—till suddenly the Stump comes alive—catches anybody he can, and runs after the rest, and there is such screaming and laughing! The grand object is to keep a sharp watch when the Stump is about to rise up—a good Stump will be very cunning and let himself be pulled about for a long time before he offers to stir.'

For continental equivalents see under 'Dead Man Arise'.

DEAD MAN ARISE

'Dead Man Arise' or 'Green Man Arise', played in the Manchester area, is the most sepulchral of the games in which a prostrate figure becomes the chaser. One player lies on the ground and is entirely covered with a blanket or cloth, preferably green, or with a pile of coats, or, as available, with grass or hay, or with sand if the game is played on the beach. The children process round the heap calling solemnly, 'Dead man, arise . . . Dead man, arise.' But they do not touch the heap, and they pretend not to look at it. Then, when least expected, the 'dead' man answers their call. He rushes after those who have resurrected him, trying to touch one of them, and make him the dead man in his place. In St. Helier, Jersey, the game is known as 'Green Man Rise-O'.

** This game was much played in the first quarter of the century, especially it seems, in the poorer districts of London (Bermondsey, Islington, and Bethnal Green), in Glasgow, and in the mining villages around Durham. During the Second World War a yet more ritualistic version was played in Manchester. Young girls joined hands and skipped round the recumbent figure chanting, 'Green lady, green lady, your breakfast is ready.' (No reaction.) 'Green lady, green lady, your dinner is ready.' (No reaction.) 'Green lady, green lady, your supper is ready.' (No reaction.) 'Green lady, green lady, your house is on fire!'—at which the recumbent figure leapt up and pursued the other players, trying to touch one of them.

This game seems to be an example, and not the only one, of a children's diversion being the enactment of an ancient horror story. In the repertory of English folk-tales is one (recorded in the *Journal of the Folk-Song Society*, vol. vi, 1919, p. 83 n.) in which a little girl takes service with a 'Green Lady'. The first morning, after preparing breakfast for her mistress, the girl calls up the stair (and in telling this story the words are chanted rhythmically):

'Green lady, green lady, come down to your break-fast!'

But the green lady does not come down. After preparing dinner for the green lady the maid calls up the stairs again, but the green lady does not come down; and she calls a third time after the preparation of supper, but still the green lady does not appear. At last the little servant girl goes upstairs to the chamber door and, urged by curiosity, looks through the key-hole. Inside the room she sees the green lady *dancing in a basin of blood*.[1]

Analogous to 'Green Man Arise' is the well-known game in Germany and Austria 'Nix in der Grube'. One child (the Water-sprite) crouches on the ground, and the other children form a ring, and sing as they circle round him:

> Water-sprite in the ditch,
> You are a bad lad;
> Wash your bones
> With precious stones,
> Water-sprite make a grab!

At this the water-sprite leaps up, and tries to catch one of them, who in his or her turn becomes the water-sprite. Similar, too, is the old game of 'The Tortoise' played in Greece; and the game of the snail 'Butta-butta corni' played in Italy. In Sicily an even more similar game was 'A Morsi Sanzuni'. One child lay down pretending to be dead while his companions sang a dirge, occasionally going up to the body and lifting an arm or a leg to make sure the player was dead, and nearly stifling the child with parting kisses. Suddenly he would jump up, chase his mourners, and try to mount the back of one of them (Pitrè, *Giuochi fanciulleschi siciliani*, 1883, pp. 265–6). In Czechoslovakia children played games called 'Schämpelän Dît' or 'Prinzessin erlösen'. In these the recumbent player was covered with leaves, or had her frock held over her face. The players then made a circle and counted the chimes of the clock, but each time 'Death' replied 'I must still sleep'. This continued until the clock struck twelve when, as in some other European games, the sleeping player sprung to life, and tried to catch someone. (See Böhme, *Deutsches Kinderspiel*, 1897, pp. 565 and

---

[1] It is perhaps pertinent that in parts of Glasgow today children initiate a game of hide-and-seek with the cry:

> Green lady, green lady, come doon for thy tea,
> Thy tea is a' ready an' waiting for thee—Coo-ee!

576–8; Kampmüller, *Oberösterreichische Kinderspiele*, 1965, p. 139; Lumbroso, *Conte, cantilene e filastrocche*, 1965, p. 23.)

A related game in England, now played only rarely (a single account) but popular at the beginning of the century, was the guessing game 'Dead Men, Dark Scenery' or 'Who is the Green-Eyed Man?' In this a member of one side is covered with coats or a blanket by his companions who then disappear, summoning the other side. The newcomers have to guess who is the green-eyed man, and if they guess correctly he jumps up and tries to catch the one who named him.

BOGEY

In 'Bogey', which is one of the classic examples of children enjoying being scared (provided that the situation is of their own arranging), the suspense is accentuated by the fact that the game is played in the dark; and that the children who are going to be chased do not even know the whereabouts of the chaser until he jumps out at them. Traditionally the game is played on winter evenings, while walking home from school. The child who is to be 'bogey' goes on ahead and hides in a dark place, as he thinks best, in a doorway, on top of a wall, or behind a pillarbox. The others, after counting perhaps 'five hundred in tens', follow along with an air of unconcern, but in reality successfully scaring themselves by singing out loud:

> Moonlight, starlight,
> The bogey man's not out tonight.

Or,   Tonight's the night, a very fine night,
>   I hope there'll be no ghosts tonight.

When, without realizing it, they come to the bogey's hiding-place, he jumps out at them. Everybody screams to frighten everybody else; and whoever is caught becomes next bogey. Sometimes there is a den or safe place that they can run back to, and sometimes, as in the game 'Ghost Train' in Liverpool, the bogey or ghost takes captives and makes them into 'ghost's assistants'.

Other names: 'Bogey Won't Come Out Tonight', 'Ghostie', 'Moonlight, Starlight', and 'Stealing Witch's Geese' (Accrington).

⁂ Infantile as this entertainment may be, it has been performed on dark nights in country places for the best part of a century. Numerous correspondents have recalled it, with scarcely any variation on the poetical side. Flora Thompson, who was born in north Oxfordshire in 1877, was one who played it (*Lark Rise*, 1939, p. 26), and since gipsies then seemed as alarming as any hobgoblin, the children sang:

I hope we shan't meet any gipsies tonight!
I hope we shan't meet any gipsies tonight!

'Bogey' has also long been played in Germany under such names as 'Das böse Ding', 'Der böse Mann', and 'Der böse Geist'. According to Meier, *Deutsche Kinder-Spiele aus Schwaben*, 1851, pp. 102–3, after the evil spirit had gone off and hidden around a corner, the rest followed, reassuring themselves, as in England, by singing:

> Wir wollen in den Garten gehn,
> Wenn nur der böse Geist nicht war!

And when 'der böse Geist' sprang out they had to run back to their *Bodde* or sanctuary; and whoever was caught had to be the next evil spirit.

§ *Players being Chased Assist each Other*

In the chasing games described so far each of the free runners has thought only of saving himself. In the following games, which tend to be more exasperating for the chaser, the runners are able to assist or rescue each other.

CROSS TOUCH

The game starts as ordinary 'Touch', but when the chaser is running after somebody his pursuit can be diverted simply by another player running between him and his quarry. The chaser is now bound to follow the person who crossed in front of him; and as soon as there is a wide enough gap another player is likely to dart between them so that the chaser is again obliged to set off after a fresh runner. 'Cross Touch' is thus a more lively game than ordinary 'Touch'; and it is not always as disheartening for the chaser as it might seem, for in practice the player who runs across is often over-confident, and finds to his surprise that he has been touched.

** This form of 'Touch' is described in *Juvenile Games for the Four Seasons*, c. 1820, pp. 20–1, where it is called 'Puss', but there is no other authority for the name which has here probably been taken from the French 'Le chat coupé'. At Sedgley Park, about 1803–10, the game was called 'Cross-tag' (Husenbeth, *The History of Sedgley Park School*, 1856, p. 107); in *The Playground*, 1858, p. 13, and in many other juvenile books of the period, it is 'Cross Touch'; in *Games of Argyleshire*, 1901, it is 'Tig and Relieve'. Presumably 'Cross-dadder' in Hardy's *Under the Greenwood Tree*, 1872, ch. ix, is this game. In the United States it is 'Cross Tag' or 'Turn Tag' (Brewster). In Rome it is 'Taglia-salame', the players calling out 'I cut the salame' as they run between chased and chaser (Lumbroso).

STUCK IN THE MUD

If the touch of the chaser in 'French Touch' appears to have a morbid effect on the part of the anatomy touched, the touch in 'Stuck in the Mud' paralyses the player altogether, except, that is, his vocal chords. ('It's a game that makes you sweat sometimes', comments a 9-year-old.) Once a player has been touched he must stand upright where he is, with his arms outstretched, and he has to keep like this, shouting to the others for help—'Releaso! Releaso!' or 'S.O.S.!'—until one of the free players manages to touch and release him. Meanwhile the chaser attempts to touch and transfix the rest of the players. Thus there is a second element in the game; for the chaser is not only chasing the others, but having to guard the person or persons he has already immobilized. The game ends, or rather, begins again, if the chaser manages 'to get all the players with their arms outstretched', or if one of the players has been twice caught and released, and is then caught for a third time. It may be remarked that the game is not successful if there are a large number of players and only one chaser.

It is a measure of the game's popularity that it is known by a variety of names, of which 'Stuck in the Mud' or 'Stick in the Mud' is the most widespread.[1] Other descriptive names are: 'Aeroplane Tig' (popular in Scotland), 'Wingbird' (Great Ellingham, Norfolk), 'Scarecrow Tig', 'Posts', and 'Lamp-posts' (many places), 'Standstill Tick' (Welshpool), 'Statues' (Manchester), 'Stone Tag' (Gloucester), 'Stooky Tig' (New Cumnock), 'Tick Frozen' (Plympton St. Mary), 'Freeze Tag' (Bristol), 'Frozen Tig' (Aberystwyth), 'Jack Frost' (Bristol and St. Andrews), 'Frost and Snow' (Sleaford), 'Coolee' (Wootton Bassett), 'Petrified' (Ipswich), 'Stop and Go' (Newbridge), 'Sticky Buds' (Moreton-in-Marsh), and 'Sticky Glue' (a number of places). Names which reflect the rescue element are: 'Release', 'Releaster', 'Reliev-i-o', 'Tig and Relevo', 'Tiggy Release', 'Stick and Release', 'Free Me' (Whalsay, Shetland), and 'S.O.S.' In Aberdeen the game is called 'Stone and Relieve' (often corrupted to 'Stone a Leaf'), the touching by the chaser being known as 'stoning', and in Perth it is 'Stoney Free'. Sometimes the game is played with a suspense start, and the names are the same as for 'French Touch', e.g. 'Poison', 'Poisonous Tick', and 'Sticky Touch' (common). There is sometimes a feeling of sorcery about the game. At Etton on the Yorkshire Wolds it is known as 'Stone Fairy'; in the Avon Valley it is 'Witch'; in Belfast 'Witches' Tag'; in Blaenavon 'Fairies and Witches'; in Spalding 'Cats and

---

[1] The name 'Stick in the Mud' is listed in *London Street Games*, 1916, p. 32.

Witches', and in Westmorland 'Witches and Trees'. At Knighton in Radnorshire, where much ancient lore mixes with the new, the game is known as 'Will o' the Wisp', and once a player has been touched there is no releasing him. A 14-year-old girl writes: 'For this game you need not less than six boys and girls of which two are picked to be wizards. The others are given one minute in which to get away, then the wizards chase them and must touch them on the *shoulder*. Then they immediately become stuck in the position and stay there until everyone is caught. Then the first and last caught become the wizards and so it goes on.' A further dramatized version, called 'Sun and Frost', is played at Stornoway in the Isle of Lewis, and embodies a peculiar method of choosing the chaser:

'You all stand in a row and one person picks the nicest face for the sun, and the rest thats left they all put on ugly faces and pick the ugliest one. The people thats left all go out and the frost goes after them and if theyre caught they have to stand still till the sun tips them and they will get free. Thats how you play the Sun and the Frost.'   *Girl, 11*

⁎⁎⁎ British children are not alone in finding the game uncanny. In Berlin, where it is known as 'Hexe geh — Hexe steh', three witches do the chasing (Peesch, *Das Berliner Kinderspiel*, 1957, p. 39). In Austria, where the game is variously called 'Versteinern', 'Steinerne Hex', or 'Erlösen', the chaser is also a witch, and is said to turn the children into stone (Kampmüller, *Oberösterreichische Kinderspiele*, 1965, p. 137). And in Italy, too, there are witches, wizards, and enchantments, the names for the game being 'Strega che impala', 'Mago libero', and 'Incantesimo' (Lumbroso, *Giochi*, 1967, pp. 310–12).

UNDERGROUND TIG

This is like 'Stuck in the Mud' (and is sometimes so called) but it is more difficult to release a person who has been touched. 'The rule of the game is there's one person on and if he tick you, you have to stand still with your legs open and somebody has to go under your legs and then you are free' (Boy, 9, Wilmslow). 'It is rather dangerous', adds another 9-year-old, 'because it is easy to knock the person over when one is scrambling through in a hurry.'

Confusingly, the game is often called 'Sticky Glue' or 'Releaso', but special names include: 'Bunny in the Hole' (Fulham), 'Crawly Tig' (St. Andrews), 'Ice Block' (Liverpool), 'Jam Tarts' (Chelsea), 'London Bridge' (Bristol), 'Policeman Tig' or 'Stick a Policeman' (Accrington), 'Stride Tag' (Stornoway), 'Ticky Underlegs' (Wigan), 'Underground Tig' or

'Underground Tick' (Aberdeen, Liverpool, Welshpool), and 'Underleg Release-i-o' (Fulham).

TUNNEL TOUCH

'Tunnel Touch' is similar to 'Underground Tig' but when a player is touched he has to stand with one arm against a wall, or, occasionally, with one leg lifted up against a wall, and he can only be released if a free player manages to run underneath. Release is thus more difficult to effect than in 'Stuck in the Mud', but easier than in 'Underground Tig' since the players waiting for release are in a line along the wall, and sometimes several can be released at once. The game is not infrequently played with two chasers, one of whom guards the 'tunnel'.

Names: 'Arch Tig' (Oxford), 'Bridge Tick' (Welshpool), 'French Release' (Euston), 'Frenchman's Tuck' (Camden Town), 'Ticky Under Arm' (Wigan), 'Tunnel Touch', 'Tunnel Tig', etc. (many places), 'Under Arm Tick' (Perth), and 'Wall Tig' (Castel, Guernsey).

TICKY LEAPFROG

'Another variation of the ticky game is "Ticky Leapfrog"', writes a Manchester boy. 'On being tuck a player has to crouch down in the leapfrog position to await the person to jump over to free him. If he is tuck three times he is on.' In Peckham, somewhat similarly, they play what they call 'Chinese Touch'. 'When a person has you,' says an informant, 'you have to stand still, and to be released a person has to jump on your back.' In both these games it is necessary that the chaser be fairly busy elsewhere before a player can be released; and this is even more necessary in 'Circular Touch', played by girls in Swansea, where a girl does not become free unless another girl runs round her three times.

GLUEPOTS

In this game, played mostly by girls, the chaser has to have a place where she can put her 'captives', as she calls them. In versions of the game known as 'Tick Corner' (Welshpool), 'He in the Box' (Peckham), 'Release' (Norwich), and 'Ghostie' (Greenwich), she chooses a convenient recess in a building, or a grating, or the worn ground in front of a gate. In the form of the game known as 'Gluepots' or 'Stewpot' (Accrington, West Ham, St. Leonards-on-Sea, and elsewhere), the chaser, who is thought of as a witch, marks out circles on the ground with a chalk or a stick, one circle for each person who is to be chased, and designates these places her 'pots'. The chasing takes place in the neighbourhood of the pots, and when the witch touches someone she leads her to a pot where the person has to stand with her arms outstretched hoping to be rescued. The rescuers, however, have to be careful, for should they accidentally step in one of the pots they must remain in it. In Wigan, as also in West Ham and North Acton, the young children who play 'Witch's Glue' or 'Witches in the Gluepots' feel that the witch, living in a gluepot, must be sticky, and that when she puts a person in a gluepot that person becomes sticky too. As a result, the captive cannot be rescued. On the contrary, the captives lean out of their gluepots and attempt to touch anyone who passes by, becoming an additional hazard to those who are still free. Further, as an 8-year-old states, 'When all are caught they have a punishment. Then the last one who is caught is the witch.'

⁂ When watching young children play 'Gluepots' or 'Witches in the Gluepots' it is easy to assume that they have made the game up themselves. But more than eighty years ago Newell described American children playing a game called 'Witch in the Jar'. One of the children was selected for a witch. She marked out circles on the ground with a stick, as many circles as there were players, and called these circles 'jars'. A game followed in which those who were caught were put in the jars, and could not escape unless someone else chose to free them by touching (*Games of American Children*, 1883, pp. 163–4). In a note Newell remarked that the children who played this game imagined that it was they who had invented the game; and he pointed out, in his turn, that the game was identical to 'Die Hexe' (The Witch), described by Heinrich Handelmann in *Volks- und Kinder-Spiele aus Schleswig-Holstein*, 1874, p. 65.

§ *Chaser at Disadvantage*

In some games it is customary for the chaser to be discomfited, or to suffer a form of initiation, before he begins chasing. In 'Bully Horn', as we have seen, he is escorted back, none too delicately, to the starting-line. In the game called 'Dumping', at Forfar, the new chaser is lifted up by feet and arms and dumped around a corner, or in a ditch, and while he picks himself up the others run away. And in 'Ticky Leapfrog' at Sale, he has to bend down and allow all the other players to jump over him, before he can take on his new role. In other games, already described, the chaser may be set at a perpetual, but usually minor, disadvantage while he chases, being obliged to run with his hands clasped (in 'Bull'), or with one hand covering some part of his body ('French Touch'). In the following games the chaser is at so great disadvantage that the players have little fear of him, and tend to gather round and goad him, rather than run from him. Thus the initiative passes in large measure to those being 'chased', as it sometimes does in catching games where the catcher's field of action is restricted (see Chapter 3). It is interesting that the games that follow are all old; and that none of them are played as often today or with as much verve as they used to be. It seems that games in which one player remains vulnerable and at a disadvantage for some length of time, are not now felt as amusing as formerly. But whether this niceness is due to compassion for the luckless player or to fear of having to take his place is an open question.

CAT AND MOUSE

One player is chosen 'cat' and one 'mouse', the rest form a circle 'holding hands tightly'. The game begins with the cat on the outside of the circle, the mouse within. At Windermere the cat asks: 'Is the mouse at home?'

Mouse: 'Who wants to know?'

Cat: 'The cat wants to know.'

Mouse: 'Yes, the mouse is at home.'

Cat: 'What o'clock is it?'

Mouse: 'Time the mouse was gone!'

The cat then attempts to catch the mouse, but the players forming the ring are on the mouse's side, they do not want the cat to get into the ring, and wherever the cat attempts to break through they 'push against each other' to prevent him. If the cat does get through they let the mouse out, and, says a player joyfully, 'the cat gets stuck in the ring'. The chaser is thus continually obstructed. Only if the children are small is the cat likely to get in and out of the ring without difficulty, and the chase speeds up. Indeed, small children in their excitement are liable to let the cat through and bar the way to the mouse. But if the children are older, the mouse may feel so secure on his side of the ring, he will dance about just in front of the cat; while the cat becomes more and more weary trying to break through, and sometimes never catches the mouse. 'If the cat fails to catch the mouse after a long time, the mouse has won,' explained a West Ham girl, 'but if he catches the mouse the cat has won because he pretends to eat the mouse up.' In this game the players making the circle have almost as active a part as the runners.

In Shetland the game is called 'A Mouse in the Meal Barn' or 'Moose in da Meal Barrel'.

⁎ Strutt in his *Sports and Pastimes*, 1801, p. 285, gives the name 'Cat

after Mouse', but the game he describes is a form of 'Kiss in the Ring' in which a player, striking one of the circle on the back, is chased in and out of the ring, without either of them being deliberately hindered. This is the French 'Le chat et le rat', the Spanish 'El gato y el ratón', the Italian 'Topo e gatto'. But when children in Germany play 'Katze und Maus', the cat starts outside the circle and the mouse within. The cat calls out 'Mouse, mouse, come here or I will scratch your eyes out'. The mouse comes out, and when the cat gives chase the mouse can take refuge within the circle and the other children prevent the cat from entering (Peesch, *Berliner Kinderspiel*, 1957, p. 22). In Moscow, according to a Russian correspondent, the name of the game is 'Cats and Mice', and the children help the mouse and stop the cat exactly as in England, lifting their arms to let the mouse through, and lowering them to block the way for the cat. Tolstoy refers to the amusement in *Anna Karenina* (v. 28). In Yugoslavia, according to Brewster, the sport is 'Măcka in Mïs', and in Rumania 'De-a Pisica si Scaracele', and in both these countries the players show their sympathy for the mouse by hindering the cat (1953, p. 63). In the United States children sometimes play one form of the game and sometimes another; but as long ago as 1832 Mrs. Child in *The Girl's Own Book* said that the players favoured the mouse. However, since, according to Mrs. Child, the circle was 'obliged to keep dancing round all the time' the cat soon found a weak link to break through.

A further form of 'Cat and Mouse' is a gymnasium game. The players are formed up in about six lines of six players. Each line joins hands, and the cat and mouse run up and down between the lines, neither of them being allowed to break through. There is a caller who periodically commands 'Turn'; and when he does so each child in the lines makes a right turn, and links hands with the players who were formerly in front and behind him, which usually means that extra distance is put between the mouse and the cat. This procedure gives much satisfaction to the caller, and to those children who are geometrically minded.

FOX AND CHICKENS

In this game the fox or chaser is at considerable disadvantage since, to quote a 9-year-old Widecombe boy, 'the fox has to hop on one leg and the chickens can run with two'. In consequence there is an amount of acting as well as activity: the fox has a 'hole', he comes out of his hole, and also takes refuge in his hole; and the chickens behave as foolishly as chickens do, so that sometimes the fox is able to catch one of them. When he does he takes the chicken back to his hole and that person obtains the doubtful privilege of becoming the new fox.

** In the 1960s this game has been found only among the children of Widecombe-in-the-Moor, the Devonshire village renowned in song. Gomme, in 1894, did not hear of the game in England, only in Cork, under its old name 'Fox in the Hole', the players striking at the fox with handkerchiefs, while the fox, who had to keep hopping, attempted to strike one of them back to make him his victim. Norman Douglas, however, seems to have found it in London in the twentieth century. 'Fox Come Out of Your Den', which he lists as a game with caps, is almost certainly this sport. (The caps would have been used for hitting the fox.)

In the sixteenth and seventeenth centuries the game seems to have been well known. 'Fox in the Hole' or 'Fox in thy Hole' was mentioned by Florio in 1611; by Herrick twice, in *Hesperides*, 1648; and by the author of *The Tragedye of Solyman and Perseda*, *c.* 1592. Surviving descriptions leave no doubt that the Elizabethan pastime is the ancestor of the game played at Widecombe. Indeed it can be traced back to antiquity. Classical scholars have long remarked a similarity between 'Fox in the Hole' (the game they knew), and the 'Empusae ludus' of the Romans and the 'Ascoliasmos' of the Greeks, where one boy who hopped had to catch others who had the use of both feet (Pollux, ix. 121). John Higins noted it in his *Nomenclator of Adrianus Junius*, 1585, p. 298 ('A kinde of playe wherein boyes lift vp one leg, and hop on the other: it is called fox in thy hole'); and so did Francis Gouldman in his *Dictionarium*, 1664 ('*Ascoliasmus*, *Empusae ludus*: a kind of play wherein boys lift up one leg and hop with the other, where they beat one another with bladders tied to the end of strings. Fox, to thy hole'). The Revd. J. G. Wood, who was a pupil at Ashbourne Grammar School in Derbyshire, 1838–43, states that the game of 'Fox' was extensively played in his day, the players being armed with twisted handkerchiefs, 'one end to be tied in knots of almost incredible hardness', and the fox had a den from which he must hop (*Every Boy's Book*, 1856, p. 7). J. O. Halliwell, whose account of the game (*Popular Rhymes*, 1849, pp. 131–2) most closely corresponds with present-day practice, says that only the fox has a knotted handkerchief, he must always hop when he leaves his home, and the other children are geese. 'Whoever he can touch is Fox instead, but the geese run on two legs, and if the Fox puts his other leg down, he is hunted back to his home.'

The game is also exactly described in *School Boys' Diversions*, 1820, pp. 12–13, but under the name 'Devil in the Bush'. It seems probable that the description and name are here taken from a French source (as are some other parts of the book), for the game has long been played in identical manner in Europe, being known in France under such names as 'Le diable boîteux' and 'La vieille mère Garuche', in Hungary as 'Sánta Róka', and in Germany, as in England, as 'Fuchs ins Loch'. Gutsmuths in his *Spiele für die Jugend*, 1796 (1802, pp. 268–70), described 'Fuchs zu Loche' as 'fairly common with us'; and Böhme, who gives several nineteenth-century names (*Deutsches Kinderspiel*, 1897) suggests that Fischart listed the game in his *Geschichtklitterung von Gargantua*, 1590, under the name 'Wolf, beiß mich nicht'. In support of this attribution Böhme quotes a song in Hainhofer's *Lautenbuch*, 1603, which may have been the game-rhyme in Fischart's day:

> Fuchs, beiß mich nicht, Fuchs, beiß mich nicht!
> Du hast ein g'hörig großes Maul.
> Du hätt'st ein guten Schuster g'geb'n,
> Du hast die Borst im Maul.

A friend has rendered this:

> Fox, bite me not, fox, bite me not!
> You have a proper great gob.
> You'd have made a good shoemaker,
> You have the bristle in your mouth.

The bristle referred to would be the one the shoemaker uses as a needle.

BLIND MAN'S BUFF

'Blind Man's Buff' is one game that needs little description, not because of the frequency with which it is played but because of its general renown. Even an 11-year-old, telling us that 'Blind Man's Buff' was her favourite game, added (correctly): 'People in the fourteenth century used to play it.' Indeed the thought of a chaser being at such a disadvantage that he is unable to see those he chases is, in itself, highly agreeable to the juvenile imagination. Thus great care is taken over the blindfolding, which is usually done with a scarf. It is tied tightly over the person's eyes, and he is repeatedly asked if he can see, and is tested with questions, 'What colour is my coat?' 'Who is the tallest here?' To ensure his confusion, willing hands turn him round three times, or spin, or twirl him, or in Norwich 'twistle' him, until he is 'quite dizzy'. Then the blind man stumbles off, stretching his arms out in front of him, and hoping that they will come into contact with someone. The rest of the players amuse themselves by dodging under his arms, making noises behind his back, and pushing each other towards him. If the game is being played outside, whoever he touches usually becomes the new blind man without further formalities. But indoors, at a party, or family gathering (the game is best played in the confined space of a crowded room), the blind man invariably has to guess whom he has caught. He feels the person's hair, his face, his clothes, to help him guess correctly; and if his guess is wrong he has to let the player go and try to catch someone else. Sometimes (even in the playground) the players change clothes to confuse the blind man, and deliberately let themselves be caught, trusting to their disguise to deceive him and keep them from having to take his place.

⁑ Under its various names 'Blind Man's Buff' is probably more often mentioned in English literature than any other informal game. In 1565 Thomas Cooper referred to 'a childish play called hoodman blind' in his *Thesaurus* (s.v. *Mya*). In 1573 John Baret listed 'the Hoodwinke play, or hoodmanblind, in some places called blindmābuf' in his informative *Dictionarie* (H. 566). About 1602 the anonymous author of *The Second Part of the Return from Parnassus* described his play as 'a Christmas toy indeed, as good a conceite as stanging hot cockles, or blinde-man buffe'. And Hamlet, it will be remembered, demanded of his mother: 'What Diuell thus hath cosoned you at hob-man blinde?' (printed 'hoodman-blinde' in the Folio, III. iv).

In the seventeenth century writer after writer names the game, for example Robert Armin in 1609 ('Hud-man blind'); Cotgrave in 1611 ('Hodman blind, Harrie-racket'); and Florio in the same year ('Blind-hob, or blind-man's buffe, or hood-man-blind'); as also Drayton, John Taylor, Heywood, Bramhall, and Randle Holme. Pepys, on 26 December 1664, records that he went to bed leaving his wife and household 'to their sport and blindman's buff', which they did not leave off until four in the morning. William Hawkins, in 1627, shows it was already the custom to turn the blind man round three times: 'You are tyed, now I must turne you about thrice'; and Davenant, in 1669, suggested it should be 'Twice for the maids, once for the men'—perhaps reckoning three turns in all.

It may be asked how it has come about that a game that was so popular, and for so long, is now played only occasionally, and only by small boys and girls. In the eighteenth century, it is clear, adults and children alike continued to make merry with 'Blind Man's Buff', as Gay shows in *The Shepherd's Week*, 1714; Goldsmith in *The Vicar of Wakefield*, 1766; and Blake in his *Poetical Sketches*, 1783. In the nineteenth century, too, Mr. Pickwick went 'through all the mysteries of blind-man's buff'; and during the Christmas gambols at Farringford in 1855, even Palgrave, Jowett, and the Poet Laureate diverted themselves with the game, and did so *after* the children had been put to bed (Emily Tennyson writing to Lear).

The fact that the game was popular throughout Britain is attested by the number of regional names it has had: 'Biggly' in Cumberland, 'Blind-Bucky-Davy' in the west country, 'Blind Hob' in Suffolk, 'Blind-Merry-Mopsey' in the North Riding, 'Blind Sim' in East Anglia, 'Blindy-Buff' in the West Riding, 'Willy Blindy' in Durham, 'Hoodle-cum-blind' in Northamptonshire, and 'Blufty' in the midlands. In Scotland: 'Belly Blind', 'Billy Blind',[1] 'Blind Harry' (also in northern counties of England), 'Blind Palmie', 'Glim Glam', 'Jockie Blind Man' (anciently 'Chacke-Blynd-Man'), and 'Jockie Blindie', a name still known in Angus today.

The reason for the game's obsolescence is not hard to find: 'Blind Man's Buff' has progressively softened (Blake pleaded that too much advantage should not be taken of the chaser's blindness); it has lost the buffeting referred to in its name, and consequently its *modus operandi*. In the 600-year-old 'Romance of Alexander' (MS. Bodley 264, fols. 70ᵛ, 130ʳ and ᵛ) three marginal pictures not only show that the chaser was blinded by having his hood reversed on his head (hence the names 'the Hoodwinke playe' and 'Hoodman Blind'), but that the rest of the players, men, women, and boys, swarmed about him for the sweet pleasure of giving him a buffet with their own well-knotted hoods. Thus the blind man had a chance of seizing one of them; and although the game was violent it was viable. It had in fact already been played like this for a thousand years. In classical times, according to Pollux (ix. 123), it was called 'Chalke muia' (The Brazen Fly). One boy's eyes were covered with a bandage. He shouted out 'I shall chase the brazen fly'. The others retorted, 'You may chase him, but you won't catch him', and they hit him with whips made from papyrus husks until one of them was caught.

In one form or another the sport seems to be part of the social history of the world: in Italy the game continues to be known as 'Mosca cieca' (Blind Fly), in Germany and Austria it is 'Blinde Kuh', in Sweden 'Blind bock', in Denmark 'Blinde-buk', in France 'Colin-maillard'; and it is also played in, for instance, Finland, Russia, China, Korea, Japan, India, and Ethiopia. However, two forms of the game are traditional. In one, as we have seen, the players buzz about the blind man so that they can hit him; in the other they do not hit him but remain close to him because

---

[1] 'Belly Blind' or 'Billy Blind' may originally have been a form of night time buff or 'Blind Tiggy'. Thus in Henryson's *Fabillis of Esope*, *c*. 1450, 77:

> Thou playes belly blind,
> Wee seeke all night, but nothing can wee finde.

they cannot get away. When Master Picquet played 'à la mousque' in the judicial chamber, and was himself the fly (Rabelais, III. xl), he laughed at the way the gentlemen were spoiling their caps 'in swindging of his Shoulders', so the game they were playing was clearly a version of 'The Brazen Fly'. But Pollux also described a game called 'Muinda', in which the chaser could not see and yet, apparently, was not hit. In *Il Pentamerone*, 1634, one of the games played by the royal household while waiting for dinner on the second day was 'Gatta cecata' (Blind Cat). Although 'Mosca cieca' is the usual game and name in Italy, 'Gatta cecata' survives in the Neapolitan dialect, and in this game there is no hitting of the blind cat. The players hold hands in a ring around her (this is a girls' game), and are wholly occupied in dancing out of her way when she moves first in one direction and then another. Not that it altogether matters when one of them is touched. The blind cat has also to guess the player's identity if she is to be relieved of her bandage, and take the other's place. This graceful game (as Goya depicted it) is popular today in countries such as Greece, Spain, and Uruguay. And had it come to Britain instead of the pugnacious variety of 'Blind Man's Buff', it would probably still be popular here. In fact, amongst guessing games there is one called 'How Far to London?' (pp. 301–3) that is a game of this type, apparently related to 'Muinda', and it is still in vogue in the north.

JINGLING

In 'Jingling', known as 'Jingle Chase' in Edinburgh, and 'Bell Man' in Inverness, all the players but one are chasers, and the chasers are all blinded. A semblance of fairness is achieved by making the one who is chased carry a bell or rattle which he has to keep sounding; and the game is played in a confined space so that the jingler cannot move far away, sometimes not outside a circle chalked on the ground. He must thus dodge about amongst his pursuers, continually ringing his bell, and this is a laughable game to watch: the sightless chasers bumping into each other, falling over each other, involuntarily embracing each other, and frequently grabbing each other with the cry 'I've caught you', under the impression that they have secured the jingler, only to hear the bell ring again at a distance from them. Should a blind man succeed in catching the jingler he changes roles with him; but an agile jingler can sometimes elude capture for a long time, and after a period he may be rewarded by being made a spectator, while another takes his place.

⁂ 'Jingling Matches' were a popular diversion at fairs and country wakes in the eighteenth and early nineteenth centuries, the jingler receiving a prize if he could remain free a certain length of time, 'commonly about twenty minutes'. Accounts of the entertainment appear in Strutt's *Sports and Pastimes*, 1801, p. 277; *The Sporting Magazine*, May 1810, p. 63; Pierce Egan's *Book of Sports*, 1832, p. 265; and in Thomas Hughes's *Tom Brown's School Days*, 1857, pp. 35–6, and *The Scouring of the White Horse*, 1858, pp. 110 and 149–50, where Hughes quotes a handbill of 1780 announcing that one of the attractions at a Scouring was to be:

'A jingling match by eleven blindfolded men, and one unmasked and hung with bells, for a pair of buckskin breeches.'

The informal game is referred to by William Hawkins in *Apollo Shrouing*, 1627, p. 51: 'You must have this Morice-bell tied to your point, that I may heare where you goe. Else you will haue too much oddes of me'; by Jamieson in his *Scottish Dictionary*, 1825, under 'Blind Bell'; in *The Boy's Own Book*, 1855, pp. 43–4; in Dean's *Alphabet of Sports*, 1866, under 'Rat and Bell'; and in *The Modern Playmate*, 1870, pp. 7–8. Games on the same principle, but in which there are only two players, are 'Cat and Mouse' in *The Boy's Own Book*, 1829, p. 28; 'Jacob! Where are You?' in *The Girl's Own Book*, 1832, p. 50 (called 'Jacob and Rachel' in *London Street Games*, 1916, p. 47); and 'Baiting the Badger' in *The Book of Cub Games*, 1919 (1958, p. 53).

FROG IN THE MIDDLE

Few chasers can be at greater disadvantage than the 'Frog' in this game, who is not permitted to move about or even rise from the ground, but must sit or squat where he is while the others dance round in a circle singing:

> Hey, hey, hi! Hey, hey, hi!
> Frog in the middle and there shall he lie.
> He can't get out, he can't get in,
> Hey, hey, hi! Hey, hey, hi!

'Then,' says a 13-year-old girl in Bristol, 'the frog has to try and touch the people who come and poke him or her, but the frog is not allowed to get on his feet. When he has touched somebody, then that person is the frog.' ⁂ The song 'Hey, hey, hi!' is almost identical to one given by Gomme, vol. i, 1894, p. 146, which may indicate a literary rather than oral continuity. On the other hand, the game itself had already been played for

1,700 years before Gomme found it, so it can well have survived for a few more decades. In the sport called 'Chytrinda', described by Pollux (ix. 113), one boy, termed the *chytra* or 'pot', sat on the ground while the others ran round in a circle, plucking him, or pinching him, or striking him, as they went. But if the *chytra* succeeded in catching one of them, the person caught took his place. As Strutt remarked in 1801: 'I scarcely need to add that "The Frog in the Middle", as it is played in the present day, does not admit of any material variation.' 'Frog in the Middle' is also described in *The Girl's Book of Diversions*, 1835, p. 19, and in *Games and Sports for Young Boys*, 1859, p. 75, the players baiting and hitting the frog, and calling 'Frog in the middle, you can't catch me'. The game appears to be depicted in the medieval manuscript, 'The Romance of Alexander' (fol. 97ᵛ), which was completed in 1344. And in the sixteenth century the names for the game may have been 'Selling of Peares' and 'How many plums for a penie', since these are the English names John Higins supplies for 'Chytrinda' in *The Nomenclator of Adrianus Junius*, 1585, p. 298.

Many related games were formerly common, both in Britain and on the Continent: notably 'Brunnenfrau' or 'Frau Holle' in which the lady of the well, seated on a stool, had to catch one of the players who were teasing and tugging at her (Böhme, *Deutsches Kinderspiel*, 1897, p. 579); 'Sling the Monkey', in which a boy suspended by the waist from the branch of a tree tried to touch one of the players who were hitting him with their knotted handkerchiefs ('the best of the basting games', said *The Boy's Own Paper*, 12 November 1887); and the well-known game of 'Baste the Bear', 'Badger the Bear', or 'Badger the Bull', described in most books of boys' games in the nineteenth century, from *Youthful Sports*, 1801, onwards.

In 'Baste the Bear', played in the United States under the name 'Watch-dog' (Brewster, 1953, p. 183), and still current in Britain in the 1920s, a keeper holds the bear, who is on all fours, either with a rope or by the back of his coat, and tries to catch one of the players whose sport it is to lash at the bear with their caps or knotted handkerchiefs whenever they can get near. This game is international. It is depicted in *Les Jeux et plaisirs de l'enfance*, 1657, under the name 'La Poire', which may relate it to the Elizabethan 'Selling of Peares' above; it appears to have been well known in classical times, being shown in wall-paintings of both Pompeii and Herculaneum; and, as a correspondent reported in *The Athenaeum*, 29 December 1883, it was still being played in the Greek island of Samos in the nineteenth century.

'Nothing I ever saw played can equal in roughness γλυκὺ κρασι, "sweet wine", as they euphoniously name it. A boy sits in the middle with one end of a long rope in his hand; another boy takes the rope after the fashion of a whip. The object is for the boys around to belabour the boy in the middle without getting hit with the rope. Whilst playing this game I have seen many ugly blows given and received, but, I am bound to say, with the greatest good nature.'

Since this game is ancient, and is here called 'Sweet Wine', it seems possible that it is as much a descendant of Pollux's 'Game of the Pot' as is 'Frog in the Middle'.

3

Catching Games

'If he is cot you have to lift him up or drag him to the side. He is aloud to kik or punch. It is rather a ruf game but most people like it.'

Boy, 9, Edinburgh

CATCHING games differ from chasing games in that the runners' chief object is usually to reach a designated place, or accomplish a particular mission, rather than keep out of the chaser's reach. Thus the catcher does not so much run after the other players as intercept them. Very often, too, he has considerable control over their movements. He is able to order when they shall run, is allowed to place himself in position before he gives the order, and sometimes has the power to name individual players (usually poor runners) who must attempt the run on their own. On the other hand, the catcher often has to do more than touch a runner to make him a captive. He may have to keep hold of him for a prescribed length of time, or perform some ritual action on his body, or even force him to the ground; and the runner's response, needless to say, is unlikely to be passive. In consequence some of the catching games take on the appearance of a series of dog-fights, and it is remarkable that they ever manage to remain games. Indeed, when it is remembered that the majority of catching games consist of no more than the crossing and recrossing of one small piece of ground (the basic pavement-to-pavement street game), and that they are played without equipment or preparation, or the approval of other road-users, it may be felt that the number there are of these games, and the gusto with which they are played, is no small testimonial to a people compelled to be city-dwellers.

RUNNING ACROSS

'Running Across' is one of those games, they say, that 'requires a great deal of energy'. The players split into two groups, one group going to one side of the road or playground and one to the other, with the catcher midway between them. When the catcher shouts 'Change' (or in Spennymoor, for some reason, 'Lamp oil'), the two sides rush across the ground to change places, getting in each other's way when they meet and giving the player in the middle an opportunity to catch at least one of them, sometimes more. Whoever is caught joins the catcher in the middle, and the running back and forth continues with an ever-thickening line of catchers until everyone is caught. Then the first person who was caught stays in the middle, and the game recommences: all the players running back and forth again at command 'until you are very tired and need a rest'. It is, in consequence, an inviolable rule that a player may not drop out at the end of one game when he is due to be the catcher in the next game.

Other names: 'Foxes' (Lydeard St. Lawrence), 'German Bulldog' (St. Leonards), 'Hunter and his Dogs' (Bristol), 'Lamp Oil' or 'Lamb Boys' (Spennymoor), and 'Running Across He' (Brightlingsea).

⋆ The game has also, in the past, been known as 'Katie on the Landing' (Halifax, c. 1925), and 'Getting Across' (Melbourne, Australia, c. 1916); and there is some evidence that the game has an unusual history.

In North Carolina a game called 'Molly Bright' is played in virtually the same way, except that the player in the middle is held to be a witch. The game starts with a player on one side calling to the other side: 'How many miles to Molly Bright?'

The opposite side replies: 'Three score and ten.'

The first player asks: 'Can I get there by candlelight?'

The other side replies: 'Yes, if your legs are long and light, but watch out for the old witch on the way!'

Both sides then rush across the open space, trying to avoid the witch and reach the other's base (Brewster, *American Nonsinging Games*, 1953, p. 52). This form of the game seems to be traditional in the United States, being given under the name 'How many miles to Babylon' in Eliza Leslie's *Girl's Book of Diversions*, 1835, pp. 6–7, where the catcher is also a witch. The verse 'How many miles to Babylon' (alternatively 'How many miles to Bethlehem', 'Burslem', 'Banbury', 'Barney Bridge', 'Barley-Bridge', and 'Marley Bright') has long been known in Britain where it has added wonder to various children's amusements, including this very game (Somerset, 1922). Further, Mactaggart in his *Gallovidian Encyclopedia*, 1824, p. 300, described a game called 'King and Queen o' Cantelon' (a 'chief school game' in Galloway), which also turns out to be a version of 'Running Across'. Two boys, he says, stood between two 'doons' or places of safety, and had to try and catch the rest of the players when they ran from one doon to the other, after being addressed with the rhyme:

> King and Queen o' Cantelon,
> How mony mile to Babylon;
> Six or seven, or a lang eight,
> Try to win there wi' candle-light.

Likewise a game known as 'King Caesar' or 'Cock and Chickens' was played in Cheltenham in the nineteenth century with the chant:

> Warning once, warning twice,
> A bushel of wheat and a bushel of rye,
> When the cock crows, out jump I.

Here, too, the 'cock' or chaser who repeated these words stood between two sets of players ranged at opposite bases; but there was this difference that the cock had to hold his hands clasped together while he chased, and not loosen his clasp until he caught someone, when he joined hands with that person, and remained linked to him while chasing the rest (Bodley MS. Eng. misc. e 39–40). This *modus operandi* seems not unlike that in the celebrated Tudor game 'Barley-break', where the players also came from opposite ends of the ground, and two chasers in the middle had, in similar fashion, to keep their hands linked while chasing (see under 'Stag' pp. 129–30). Thus there may be linguistic connection between 'How many miles to Babylon' or 'How many miles to Barley-Bridge', and the name 'Barley-break' of 400 years ago.

CHINESE WALL

Two parallel lines are drawn across the middle of the playground about a yard apart. This is said to be the 'Chinese wall', and one player or some-times two stand between the lines, and may not go beyond them. The others have to run across the wall without being touched. If they are touched while crossing they join the catchers on the wall. This is often an organized game, the instructor on the side-line taking upon himself the not over-fatiguing task of commanding when the children shall rush one way across the wall, and then the other way, and so on until all have been caught.

Additional names: 'Giant on the Wall' (Thirsk); 'Over the Wall' (Market Rasen).

WALL TO WALL

'Wall to Wall' is often played in the school yards of junior schools, or at home in the evening across a road. One player stands mid-way between the two walls and the rest ('there need to be more than twenty children for a decent game') line up at one wall and have to run across to the other, keeping within agreed boundaries. Anyone touched while running from one wall to the other joins the catcher in the middle, as in the previous games, so that the balance gradually changes in the favour of the catchers, and the last two or three players are hard put to it to get from one wall to the other without being caught. Nevertheless the runners are not allowed to remain at either wall for more than the count of ten, nor may they turn back once they have left a wall, but must keep running to the opposite side even if they are bound to be caught.

Other names for this much-played game include 'Den to Den' (Bristol), 'Wall to Wall Tig' (Bacup), 'Brixers Last Up' (Cruden Bay, Aberdeenshire), 'Charlie' (Knighton), 'Cross Channel' or 'Dicky Birds and Breadcrumbs' (Croydon), 'Fox and Hounds' (Widecombe-in-the-Moor), 'King Alely' (Ponders End), 'Lollipop' (Glastonbury), 'Old Grannie Witchie' (Edinburgh), 'Poison' (Knighton), 'Running Across' (Retford), 'Touch Road Must Go Over' (Enfield), 'Onefootoveryoumustgo' (Norwich).

At Market Rasen the game is sometimes played between the two semicircles of a netball court and then called 'Top and Bottom'. At Meir, Stoke-on-Trent, where the name is 'Press Button', the catcher stands by a 'clod of grass' placed in the middle of the road, and everyone has to run when he puts his foot on it. At Wilmslow the players run across when the one in the middle shouts 'Tally-ho', and the game is thus called 'Tally-ho'. In Ipswich and Troutbeck, near Windermere, it is called 'Boiler's Bust'. Everyone must run across when the catcher or caller says 'Boiler's bust', which may come as part of a story (perhaps a Brownie version). In New Cumnock the name and cry is 'Peas and Beans', and an old-style ending to the game is maintained: 'When the last person is caught he tries to tig somebody before they get to the starting-place. If he tigs one they start off the next game [i.e. become the catcher], but if he does not he starts the next game himself.'

Sometimes the game is played with everybody hopping, and it is then called: 'Half Loaf' (Eassie), 'Hauf the Loaf' (Cumnock), 'Hop and a Stag' (Leicester), 'Hopping Red Rover' (Enfield), and 'Yorkshire Pudding' (Knighton). Compare 'Cockarusha' pp. 136–8.

_** Children seem to have been running back and forth across the road playing this game ever since Georgian days. Jamieson, in 1825, described a game called 'Rin-'im-O'er', played by children in Roxburghshire, 'in which one stands in the middle of a street, road, or lane, while others run across it, within a certain given distance from the person so placed; and whose business it is to catch one in passing'. The only difference from

the present-day game seems to have been that when someone was caught 'the captive takes his place', which is now rare. Jamieson says that the game was also known as 'King's Covenanter'. Other names have been: 'Bristol' (*Yorkshire Folk Lore*, vol. i, 1886, p. 46); 'Cock-a-Reedle' in Nottinghamshire (*EDD*, vol. i, 1898); 'Dyke King' in Tyrie, 'Rax' or 'Raxie-boxie, King of Scotland' in Ballindalloch, Banffshire; and 'Red Rover' in Liverpool (*Traditional Games*, vol. ii, 1898, pp. 106–7); 'Rex' in Perthshire, and 'Kinga be Low' (*Games of Argyleshire*, 1901, pp. 209–10); 'Pirates' (*Folk-Lore*, vol. xvii, 1906, p. 95); 'Lockit' (Bristol, *c.* 1920); 'Middlers' (Crewkerne, 1922).

In both the United States and Canada, where the game is often played on the ice, the common name is 'Pom Pom Pull Away', the one in the middle crying:

> Pom pom pull away,
> If you don't come I'll pull you away.

In Missouri it is 'Wolf Over the Ridge' (Brewster, 1953, pp. 53 and 76); and a similar game is, or was, 'Black Tom' played in Brooklyn, and known in the South as 'Ham, Ham, Chicken, Ham, Bacon' (Bancroft, *Games for the Playground*, 1909, pp. 54–5).

STAG

This game is played like 'Wall to Wall' with the difference that when the player in the middle has shouted 'Cross', and succeeded in catching someone who was crossing, he links hands with him. The pair then shout 'Cross' together, and have to remain attached to each other while they catch someone else crossing. The next person caught links up between the first two catchers, so that the first two continue on the outside (this is felt to be important), and anyone else caught joins, similarly, in the middle of the line. The two end players remain the only two with the power to catch, no matter how long the line becomes. If there are a large number of players the last few remaining free will find themselves faced by a great chain of catchers spreading out in both directions; but since only the player at each end can catch them, and even these two only when the chain is unbroken, the last few players usually charge at the middle of the line, hoping to tear their way through before the ends can curl round and reach them.

'Stag' is particularly popular in Montgomeryshire and Radnorshire, where it is also known as 'Stag Tick'; but elsewhere it is not played as frequently as it was before the war. In West Ham, where it is called 'Sheep

Dogs', the game starts with two catchers in the middle holding hands. The first person they catch joins hands with them to make a threesome; but when a further person is caught and there are four catchers, they split into pairs, and the crossing thus becomes considerably more difficult for the free players. Compare 'Chain He' which is similar, but the free players are at liberty to run when and where they like within the boundary of the game.

** It appears that an ancestor of 'Stag' is the renowned game of 'Barley-break' or 'Last Couple in Hell'. Unfortunately, despite Barley-break's great popularity in Shakespeare's day, no precise account exists of how it was played; and subsequent expositions by scholarly but unathletic commentators (notably William Gifford) have tended to obscure rather than clarify the rules of the game. Yet, by ignoring the rules of more recent games that happen to have similar-sounding names, and compounding the poetical descriptions in Sidney's *Arcadia* (written in the early 1580s), Nicholas Breton's *Barley-Breake, or a Warning to Wantons*, 1607, and Suckling's allegorical piece 'Love, Reason, Hate' in *Fragmenta Aurea*, 1646, it is clear that Barley-break was a game for six players, three of whom were boys and three girls, who divided into pairs of boy and girl; one pair going to one end of the ground, one pair to the other, and the third pair taking the middle position known as 'Hell'. It appears that the two pairs on the outside had to 'break', and attempt to change partners with each other, while the pair in hell, coupled together, tried to intercept them. If the two players in the middle succeeded in catching someone, that person, together with the player he or she should have linked with (i.e. the member of the opposite sex who started at the far end of the ground), took their place in hell for the next round. But the catchers could not catch if they were not joined together; and the outside players were safe once they had linked with their new partners. Each game seems to have consisted of a series of two to three 'breaks' or 'barley-breaks', and whichever two players ended up in the middle were said to be 'Last couple in hell'. The basic operation of the game was thus not unlike 'Sheep Dogs', and also some versions of the game 'Befana' played in Italy. This concept of the game is reinforced by hints from Cotgrave (1611)—although he compared the game with 'Tiers'; by Florio (1611), who compared it with 'Pome'; and by Randle Holme's bald description in *The Academie of Armoury*, 1688:

'Barla Brakes, is a play of 6: runing, of which two that stands in the midle are to take and hold any of the other lott, and so to put them in their place of catching.'

Further allusions to the game occur in Henry Machyn's *Diary*, 1557 (1848, p. 132); Fletcher and Shakespeare's *The Two Noble Kinsmen*, IV. iii; and Ben Jonson's *The Sad Shepherd*, I. iv. It is also mentioned in the writings of Armin, William Browne, Burton, Chettle, Dekker, Holiday, Massinger, Rowlands, Shadwell, and Shirley. It is apparent from Beaumont and Fletcher's *The Scornful Lady*, v. iv and *The Captain*, v. iv; and from Brome's *The Queen and the Concubine*, IV. iv, that the term 'last couple in hell' was so well known it was proverbial.

See also under 'Running Across'.

BLACK PETER

In this game, too, the catcher, here named 'Black Peter', stands out in front, and the rest have to dodge past him to the safety of the other side. Black Peter calls out: 'Who's afraid of big Black Peter?'

The rest shout back, 'Not I.'

The player out in front calls again, 'Who's afraid of big Black Peter?' The rest reply, 'Not I.'

Black Peter calls out a third time, 'Who's afraid of big Black Peter?' and the rest chorus 'I', and rush across to the other side of the street or playground. The first person caught then becomes Black Peter, or sometimes helps Black Peter to catch the others.

Descriptions from Aberystwyth, Knighton, Langholm, and Lerwick.

** The game's present-day distribution suggests that it has long been played in Britain, but no account has been found earlier than 1922, in Somerset, when it was called 'Black Tom' (Macmillan MSS.). It has, however, a well-documented history in Austria, Germany, and Switzerland, where it continues to be popular, the catcher generally being known as the Black Man. Kampmüller (1965, p. 138) reports that in Upper Austria the catcher starts the game by demanding: 'Fürchtet ihr den schwarzen Mann?'

Children: 'Nein!'

Black Man: 'Wenn er aber kommt?'

Children: 'Dann laufen wir davon!'

Peesch (1957, p. 36) reports that in Berlin 11- and 12-year-olds have three names for it, 'Wer fürchtet sich vorm schwarzen Mann', 'Wer hat Angst vorm schwarzen Mann', and 'Schwarzer Mann', the dialogue often being:

'Who's afraid of the Black Man?'

'Nobody!'

'And if he comes?'

'Then we'll go to America!'

Black Man: 'America is all burnt up!'

Children: 'Then we'll come over', and they rush across the street, hoping to get past the Black Man.

'Der schwarze Mann' certainly goes back to the eighteenth century, when it was fully described by J. C. F. Gutsmuths in *Spiele für die Jugend*, 1796 [1802, pp. 261–3]; and it may even be the game 'Der schwarze Knab' listed by Fischart in 1590. The possible antiquity of the game, the foreboding colour of the chaser, and the fact that in Switzerland players sometimes made a ring round the Black Man while they defied him (Rochholz, *Kinderspiel aus der Schweiz*, 1857, p. 376) has given some nineteenth- and twentieth-century scholars sufficient grounds for suggesting that the game is a relic of the death dances notorious in the Middle Ages. However this may be, in Italy the game is not so morbid, being known as 'Avete paura della Befana?' (M. M. Lumbroso, *Giochi*, 1967, p. 328).

SHEEP, SHEEP, COME HOME

The situation in 'Sheep, Sheep, Come Home' is even more dramatic than in 'Black Peter', and the game is a favourite with younger children. All the players are sheep except two, one of whom is a shepherd and the other a wolf. The shepherd leaves the sheep and goes to the far end of the field or playground while the wolf hides (or pretends to hide) somewhere between sheep and shepherd. The shepherd calls: 'Sheep, sheep, come home.'

The sheep reply: 'We are afraid.'

Shepherd: 'What of?'

Sheep: 'The wolf.'

Shepherd: 'The wolf has gone to Devonshire,
    Won't be back for seven year;
    Sheep, sheep, come home.'

The sheep then run towards the shepherd. The wolf waits until they are near him and springs out, and tries to catch one of them. Any sheep who is caught either helps the wolf or takes the wolf's place, or sometimes has to wait in a den. The shepherd goes to the opposite end of the playground, and the drama recommences.

In Stoke-on-Trent they play the game without a shepherd. One player volunteers to be wolf ('to save dipping') and stands in the middle of the road, the rest are the sheep, and stand on the kerb.

Wolf: 'Sheep, sheep, come over.'

Sheep: 'We are afraid.'

Wolf: 'What of?'

Sheep: 'The wolf.'

Wolf: 'The wolf has gone to Lancashire to buy a penny hankershire.'

Sheep: 'How deep is the sea?'

Wolf: 'Try it and see.'

Sheep: 'It's too deep. How many days will it take us to cross by boat?'

The wolf then says a certain number of days. 'If the wolf says "Two days",' explains a 13-year-old, 'the sheep will cross the road to the other kerb and back, that is one day, so they will do the same to make two days. After they have walked the two days they have to run across the road before the wolf can touch them. If he touches them they have to help the wolf to catch the others and so on. The first one caught is the wolf in the next game. This game is played in our street.'

Other names: 'Who's Afraid of the Wolf' (Langholm); 'Mr. Wolf and the Sheep' (Norwich).

∗ This is another game that is exactly paralleled in Austria, the shepherd saying: Alle meine Schäflein, kommt nach Haus'!

Sheep: Wir können nicht.

Shepherd: Warum denn nicht?

Sheep: Der Wolf ist da.

Shepherd: Was tut er denn?

Sheep: Uns fangen.

Oberösterreichische Kinderspiele, 1965, p. 139

It is also traditional in Germany, 'Schäflein, Schäflein kommt nach Haus!' being described in, for instance, Meier's *Deutsche Kinder-Reime aus Schwaben*, 1851, p. 370, and Böhme's *Kinderspiel*, 1897, pp. 572–3. In Italy, at Maniago near the Yugoslav border, children play a biblical version, 'Sette Secella', in which the Lord calls his angels to run to him, but they are reluctant to come because the Devil is waiting to chase them (Lumbroso, *Giochi*, 1967, p. 220).

The game was much played in Britain in the nineteenth century, the usual name being 'Sheep, Sheep, come home' (another name was 'Wolf'), while sometimes the players were 'Fox and Geese', as at Eckington in Derbyshire where the dialogue was as follows:

Fox: 'Geese, Geese, gannio.'

Geese: 'Fox, Fox, fannio.'

Fox: 'How many geese have you today?'

Geese: 'More than you can catch and carry away.'

Traditional Games, 1894, pp. 140–1

In Arkansas the game is called 'Fox in the Wall', and the fox has to tap a goose on the shoulder three times to make a capture (Brewster, 1953, pp. 78–9). In *North Carolina Folklore*, 1952, p. 78, it is called 'Fox in the wall'. O. Henry, born in North Carolina in 1862, describes 'Fox-in-the-Morning' in *Cabbages and Kings*, 1904, ch. I.

Other forms of the game in England have been 'Old King Dick', played in Berkshire, *c.* 1920; and 'Blackthorn' which was still being played in the Yorkshire dales in 1930:

Geese: 'Blackthorn, Blackthorn,
      Buttermilk and barley corn.'
Blackthorn: 'How many geese have you today?'
Geese: 'More than you can catch and carry away.'

'Blackthorn' is named as early as 1837 in William Thornber's *Account of Blackpool*, p. 90 (*EDD*); and further descriptions occur in *Notes and Queries*, 3rd ser., vol. vii, 1865, p. 285 (recalling a Lancashire childhood), and in the *Almondbury and Huddersfield Glossary*, 1883.

FARMER, FARMER, MAY WE CROSS YOUR GOLDEN RIVER?

This is probably the most popular game in the streets of Britain today (descriptions from 127 places), being fascinating to little girls, partly, it seems, because of the way it draws attention to item after item of their clothing. One child is named the farmer and stands in the middle of the road while the rest line up on the edge of the pavement. The children on the pavement call out 'Farmer, Farmer, may we cross your golden river?' and the farmer replies, choosing a colour, 'You mayn't cross my river unless you have *blue*'. The children who have this colour on them, even if only on part of a garment, or on a handkerchief or a brooch, are allowed free passage across the river (that is to say the road), and take pleasure in walking sedately across unmolested (this is why, according to one informant, mothers get asked to knit multi-coloured jerseys); but those who are not wearing the colour have to dash across the road and risk being caught by the farmer. When the children ask if they may cross the river from the other side the farmer chooses another colour ('He may choose purple because he knows none of us wears that colour'), and the rush across the road is repeated. In some places when a person is caught he takes the place of the farmer; but more often he is out of the game until everyone has been caught, or he has to help the farmer catch the rest, so 'it is harder to cross as the game goes on'.

If ever there was a game which showed children's love of blending fancy with strenuous activity this is it, for the game, in its multiplicity of forms, has the weirdness of a fairy-tale.

In Liss the players on the pavement ask: 'Farmer, Farmer, may we cross your golden river in our silver boat?'

At Wilmslow: 'Farmer, Farmer, may we cross your golden bridge on our golden horse?'

At Cleethorpes:

> Farmer, Farmer, may we pass
> Over the hills and over the grass?

On Tyneside: 'Farmer, Farmer, may we cross your stinking dirty clarty water?'

The player in the middle may variously be known as 'Jack' (particularly in the south-west and in Wales), 'Boatman', 'Policeman', 'Mr. Duck', 'Mr. Fish', 'Mr. Jellyfish', 'Mr. Fisherman', 'Mr. Frog', 'Mr. Piggie', 'Mr. Crocodile', 'Charlie' (north-east England), 'Charlie Chaplin' (parts of Scotland), and 'Charlie Chapman' (Cumnock).

In Headington the players plead: 'Boatman, Boatman, ferry me across the water.'

In Rossendale: 'Old Mother Witch may we cross your ditch?'

In Ipswich: 'Please Mr. Frog may we cross your Chinese Channel?'

In Offham, Kent: 'Please Mr. Crocodile may we cross the water in a cup and saucer?'

In some places it is customary to give an excuse for crossing. 'Farmer, Farmer, may we cross your golden river to fetch our father's dinner?' (Featherstone). 'Please Mr. Crocodile, may we cross the river to take the Queen's dinner?' (York). 'Charlie, can I be over the water to take my father's bait?' (Spennymoor). In Sheffield they chant:

> Farmer, farmer, may we cross your waters today?
> Because we go to school this way
> To learn our A.B.C.

In Plympton St. Mary:

> Please Jack, may I cross the water
> To see the Queen's daughter?
> My mother's gone, my father's gone,
> And I want to go too.

In Walworth, where the catcher is Mr. Porter, the girls ask politely:

> Please Mr. Porter
> May we cross your golden water
> To see your fairy daughter
> And have a cup of water?

But the boys say:

Please Mr. Porter
May we cross your water
To see your ugly daughter
Swimming in the water?

And in Swansea they ask:

Please Mr. Froggie may we cross the water
To see the King's daughter
To chuck her in the water
To see if she can swim?

'It is an exciting game to play when you are bored with other games', comments a 10-year-old.

Names and variations: The names of the game are as various as the formulas, e.g. 'Boatman, Boatman', 'Mr. Fisherman', 'Please Mr. Crocodile', and 'May I Cross the River?' Sometimes two or three names may be current in the same locality, one of which may be 'Colours' or 'The Golden River', and in Scotland, from Orkney to the English border, the game is often known as 'You Can't Cross the River' or 'Ye Canna Cross the Golden Stream', because the person in the middle of the road is the 'king' or 'keeper' and starts the game by issuing a challenge. In some places the guardian of the river or field tries to dissuade the players from crossing, asserting: 'the river is too deep', 'there is a bull in the field', 'the corn is being reaped'. In Somerset 'Jack' first answers with a blunt 'No, you can't cross my river', and traditionally makes this refusal three times before he names a colour. And in Monmouth and south Wales Jack at first procrastinates, saying 'No today and yes tomorrow', and in so doing he repeats words that have been customary in Glamorgan since the beginning of the century.

⁂ Little is known of the history of this game other than that our correspondents played forms of it when they were young, e.g. 'Jack Across the Water' (Glamorgan, c. 1900); 'Charlie, Charlie, Let Me Over the Water' (Lanarkshire, c. 1902); 'Farmer, Farmer, Can We Cross Your River?' (Forfar, c. 1910). Peesch reports that children in West Berlin play it, asking 'Fischer, welche Fahne weht?' (*Berliner Kinderspiel*, 1957, p. 37); and Kampmüller describes the game, somewhat defectively, in Austria, 'Wassermann, mit welcher Farbe dürfen wir hinüber?' (*Oberösterreichische Kinderspiele*, 1965, p. 138). The game's currency in Germany and Austria may, however, be due to British occupation following the war.

BAR THE DOOR

The children say this is an 'interesting game' probably because it is in part a spectator game. The catcher who is in the middle of the road or open space starts by choosing one player to run across on his own. The rest of the players on the pavement or touch-line watch while he attempts to dodge the catcher, for if he succeeds in getting across he cries 'Bar the door!' (or in Forfar 'Schoolie!', in New Cumnock 'Squatter!') and they rush after him in a body, hoping not to be caught themselves. But if he is caught he joins the catcher in the middle and challenges a further player to make the crossing, who has now to run the gauntlet of two catchers. Thus an increasing number of catchers face the runners as the game proceeds ('Always the person who is last tug shouts out the next name'), and the winner, naturally, is the player who remains free longest, while he who was caught first has the unenvied duty of staying in the middle to be catcher in the next game.

Names: 'All Over' (Langholm), 'Bar the Door' or 'Barley Door' (Dunoon, Forfar, Kingarth, Liverpool), 'Bloaters' (Pontypool), 'Bolter' (Newbridge, Monmouthshire), 'Burning Bar' (Cumnock), 'Cross Tig' (Flotta), 'Cross and Across Tig' (one boy, Forfar), 'Levi-hi-hoe' (Pontefract), 'Run Across' (Acocks Green), 'Running Across' (Broadbridge Heath), 'Semi' (Hayes, Middlesex), 'Tally-ho' (Brightlingsea).

.•. 'Bar the Door' was being played in Argyllshire at the end of the nineteenth century (Maclagan, 1901, p. 210), also in Forfar, *c.* 1910, and in Dunedin, New Zealand, in 1870, where the catcher had to tap a runner three times on the back to make him captive. In Aberdeenshire the game was known as 'Burrie' (*EDD*, 1897). Another name in Argyllshire was 'Cock-a-Rosy' (*Folk-Lore*, vol. xvii, 1906, p. 96). And on the Scottish border, *c.* 1925, it was called 'Joukie', the players having to jouk or dodge past the one in the middle (*Southern Annual*, 1957, p. 28).

COCKARUSHA

'Cockarusha' is basically the same game as 'Bar the Door', but everybody hops, and this limitation considerably affects the character of the game. The player who is 'cocker' or 'he' goes into the middle of the road (the usual site for the game) and stands on one foot with arms folded. He challenges any player he likes (or dislikes) to cross the road, and this person, with arms similarly folded, hops forward and tries to get past him. If he does not manage to dodge him it does not matter. What counts in this game is a player's ability to stay on one leg. Only 'if the cocker barges you so that you fall over or put your other foot down have you to stay and help the cocker'. And should the cocker put his own second foot on the ground the player can continue across without further hindrance, whereon he shouts 'Cockarusha', and the rest of the one-legged players attempt the crossing. However, the cocker can now start hopping again, and he weighs into them, unbalancing whom he can; so that the next time a player is challenged to cross over, the cocker may have more than one ally beside him. Thus the game is one of charging, barging, and 'dunting', until only one player remains to batter his way through the rest, who if he succeeds is highly regarded; and if he succeeds a second time is acclaimed a 'double winner'. In fact in some places the game is more a tourney than a catching game. For instance, in Helensburgh the boy who has been challenged is not permitted to dodge, but must withstand being barged three times by his challenger. 'If after three bumps the victim has escaped unhurt he may hop across to the other side and the others may follow him.' In Forfar whoever wins the duel automatically joins the others on the pavement, so that there is only ever one person in the middle, and 'the game carries on like that until you are fed up with it'. In all variations of the game that have been noted, players are not allowed to change legs while hopping, not allowed to turn back once they have started crossing, and not allowed to hold their opponent, or push him, or trip him up. In 'Cockarusha' the power of the shoulder is paramount.

Names: 'Cockarusha' (Southwark, Walworth, Offham in Kent), 'Cockeroosher' (Camberwell), 'Cock-a-Rooster' (Swansea), 'Cock of Roosters' (Spennymoor), 'Cockeroustie' (St. Andrews, Fife), 'Cockay Duntie' (Ballingry), 'Cock Heaving' (Perth), 'Cripple Dick' (Kirkcaldy), 'Hop-a-Kicky', 'Hop Charge', and 'Hopping Bulldog' (St. Peter Port), 'Hop All Over' (Helensburgh), 'Hopping Barge' (Camberwell), 'Hop and Dodge', 'Hoppie Diggie', 'Hoppin' and Diggin'.', and 'Hoppie Dick' (Forfar), 'Hoppie Bowfie' (Aberdeen), 'Hopping Caesar' (Enfield), 'Hopping Charlie' (Cumnock), 'Hopping Jinny' (Birmingham), 'Hopping Johnny' (Manchester), 'Hopping Tommy' (Welwyn), 'Knock 'em Down' (Barrow-in-Furness), and 'Tally-ho' (Chelmsford).

₊ Previous recordings: 'Cock Dunt' (Clackmannanshire, c. 1920), 'Dunty' (Belfast, c. 1910), 'Hippy Joukie' (Scottish Border, c. 1925), 'Hop-o-Cock-Rusty' (Nottingham, c. 1920), 'Hopping Johnny' (Newton-le-Willows, Lancashire, 1930s). In Melbourne, Australia, 'Hoppo Bumpo' for two generations; in Denniston, New Zealand, 'Humpty Dumpty'

formerly 'Dunk and Davey' (Sutton-Smith, *Games of New Zealand Children*, 1959, p. 136).

CIGARETTES

It will be appreciated that in the games 'Bar the Door' and 'Cockarusha' the catcher, having his own interests well in mind, is not inclined to choose out the swiftest runner or craftiest fighter to be his opponent: he prefers someone he is confident he can overcome. In 'Cigarettes', here described by an 11-year-old girl in Edinburgh, this luxury is denied him:

'Cigarettes is a game where you all stand on the pavement and one person stand in the middle of the road. The people pick the name of a cigarette and tell each other what they have picked but they do not tell the person in the middle of the road. Here are some of the cigarettes you could pick: Black Cat, Camel, Compass, Three Threes, Prize Crop, Churchman 1 and 2, Bar One, Airman, Player's Weights, Codgent, Dunhill, Four Square, Piccadilly, and Kensitas. Then the person in the middle thinks of all the different cigarettes and if she says your one you try to run across to the other side of the road without her or him tigging you.'

In some places the names of film stars, football players, tennis players, animals, makes of cars, or numbers, are adopted. 'The person who is by himself does not know who has which number but he knows what the numbers are.' Sometimes the game is played with everybody hopping; and sometimes the person who has been caught instead of helping the catcher either takes his place or is out of the game, so that there is ever only one challenger in the middle.

Names: 'Animals' (Norwich), 'Barging' (Peterborough: players choose numbers and hop), 'Bulldog Says' (Spennymoor: players run or hop according to bulldog's instructions), 'Cigarettes' (Aberdeen, Edinburgh, Forfar, Spennymoor), 'Jungle's on Fire' (Glasgow: players choose names of animals, the one who gets across on his own shouts 'Jungle's on fire' and the rest stampede across), 'Long Lamp' (Pontypool: players choose numbers), 'Tally-ho' (Yarmouth: players choose animals).

BRITISH BULLDOG (I)

'British Bulldog' is the toughest, and the most popular, of the games in which players are waylaid while crossing a street or open space. The players line up on a pavement, within agreed bounds, and usually somebody strong, sometimes two people, face them in the middle of the road. At a signal the players rush across the road to the sanctuary of the other pavement, and the 'bulldog' tries to stop one of them, but it is not enough for him just to seize the runner. As a 10-year-old put it:

'If he tigs you but you get away you are all right. The bulldog has to catch hold of you and lift you up, and say "One, two, three, British Bulldog". You can, however, struggle and if you get free before he has shouted all the words you are all right.'

Only if the bulldog holds the person so that both his feet are off the ground while he says 'British Bulldog, one, two, three', or counts to five, or to ten, or whatever is the prescribed number, must the player submit and join the catcher. ('It is bad luck if you catch a fat person.') Alternatively, in some places, 'When you catch somebody you have to make him fall down, and hold him down for the count of ten.' In Street, Somerset, he has to be held until he gives in, or until 'everybody who is "on it" touches him'. At Whalsay, in Shetland, the capture is made by tapping the player three times on the back, and a fight usually develops as the boy strives to avoid being tapped. In Edinburgh the runner has to be lifted up, or be 'head 'n tailed', or be dragged to one of the boundaries, and the player is allowed to kick and struggle to prevent this. At Netley, 'You have to hold him up in the air and bump him three times.' In Liverpool, where 'nearly all the street plays the game except the babies and those over fourteen', the catcher or catchers lift the captive off the ground, but do not say 'British Bulldog' themselves, 'they squeeze him until their victim cries out "British Bulldog" '. Thus the game proceeds, with the number of catchers steadily increasing, so that a game which started with one boy against twelve, will end with twelve boys lifting up or piling on to one. 'The bigger boys are usually left to last,' observes a Twickenham lad. 'It is when trying to catch these that the roughness begins. You have only half finished when you have got them down because they kick and punch at everybody in sight.' 'Sometimes,' says a 13-year-old Liverpool boy, 'when about half the boys are caught the game becomes a free-for-all, with no side gaining. In the end everyone stops fighting to lick their wounds.' But then, he says, the game starts again where it left off. 'In one game an ambulance had to be called to take a boy to hospital with a broken leg. Others go home with black eyes and torn clothes, but we really enjoy the game.' As a 9-year-old commented, 'When you've finished playing and go home your mother says you're in a "terrible state".'

This game, which more than one Londoner has declared 'the most commonest game that my friends and I play', is known as 'British Bulldog' almost everywhere. The paucity of regional names is probably due to the

Boy Scout and Wolf Cub movement, although a senior Scoutmaster told us they tried to dissuade cubs from playing the game, since the younger boys were likely to get hurt. Alternative names are: 'Across the Middle' and 'One, Two, Three' (Croydon), 'Cannonball' (Fulham), 'Fox and Hounds' (Llandrinio and Welshpool), 'King Come-a-lay' (Whalsay, Shetland), 'Lolly' (Street), 'Pigwash' (Stoke-on-Trent), and 'Stampede' (Bristol). In Liverpool, if too many players are taking part to run across at once, they play a variation called 'Vicious Bulldog', in which they divide into two groups, and the parties rush alternately from opposite sides of the ground.

⁎⁎ 'British Bulldog' seems to be little different, except in name, from the Victorian schoolboys' excuse for a rough-house called 'King Caesar' or 'Rushing Bases'. In this the player who stood between the two bases was termed 'King', and (according to *The Boy's Own Book*, 1855) when he succeeded in intercepting a player 'he claps him on the head with his hand three times, and each time repeats the words "I crown thee, King Caesar" '. The apprehended player was, however, under no obligation to stay in the middle unless he was 'properly crowned'; and it was during the performance, or attempted performance, of the coronation rites that bruises were liable to be acquired. The actual formula, however, varied at different schools. At King's School, Sherborne, about 1840, where the game was known as 'King Sealing', a boy did not have to submit unless the king succeeded in holding him long enough to utter the words:

> One, two, three, four, five, six, seven, eight, nine, ten,
> You are one of the king-sealer's men.

At Foyle College, Londonderry, about 1905, where the game was known as 'Rush', the formula to complete the capture was:

> One, two, three, a man for me,
> Lock him tight, Amen.

At 'King's School, Tercanbury', in *Of Human Bondage*, where the game was called 'Pig in the Middle', the words that mystically turned a boy into a prisoner and caused him to change sides were:

> One, two, three,
> And a pig for me.

And in the town of Marlborough, where the game was known as 'Click' (*Traditional Games*, 1894, pp. 69–70), the catcher had to retain his hold long enough to say:

> One, two, three, I catch thee,
> Help me catch another.

Here, if the last player succeeded in getting across three times after all the others had been caught he was allowed to choose who should be catcher, or 'go click', in the next game; and it may be remarked that several children have reported this rule today when playing 'British Bulldog'.

Other names: 'Cock' at Nairn, where the captor had to 'croon' his captive (i.e. put his hand on his head), and 'Rexa-boxa-King' at Duthil in Inverness-shire (*Traditional Games*, 1894, pp. 72–3); 'Cosolary' at Cross Fell, *c*. 1885, where the captor had similarly to clap his captive on the head (*Journal Lakeland Dialect Society*, no. 7, 1945, p. 6); 'Fox a' Dowdy' in Warwickshire, where the captor had to cry 'Fox a' dowdy—catch a candle' while holding his captive (*EDD*); 'Lamplighter' at Chard, *c*. 1922, where the captive's head had to be patted three times; 'Pirates' in Hull, *c*. 1895, where the captive had to be 'tailed' (i.e. have both hams pinched); and 'Run-Across' at Ackworth School, Yorkshire, *c*. 1805, where the captor had to detain his prisoner for the count of ten (William Howitt, *Boy's Country Book*, 1839, pp. 219–20).

BRITISH BULLDOG (2)

'British Bulldog' is also, but rather less often, played in the manner of 'Bar the Door', with the catcher in the middle of the road first challenging some player by name to cross on his own. This makes the game initially somewhat easier for the one in the middle; but rather less pleasant for the weaker players, since however strong the boy in the middle he is unlikely to call out someone of his own size to oppose him. It is not to his advantage that the person he challenges should succeed in getting past him, for that player then shouts 'Bulldog', and the rest can swamp him in an overwhelming wave. To avoid discrimination, they sometimes give the players numbers or the names of colours, so that the one in the middle does not know whom he is calling out (cf. 'Cigarettes'); and sometimes, says an 11-year-old, the game is played at night when it is 'difficult to see the running person, and therefore makes the game more exciting'. Yet the game, however played, is a tough one. As it progresses the opposition in the middle grows more formidable; and a player who is called out in the latter part of the game knows he must charge full tilt at the barrier of boys on the road if he is to break through and avoid being piled upon. Indeed, as a Barrow-in-Furness boy commented, 'It's hard luck for the last boy to be caught because he gets quite a hammering.' It was a young lady (age 11) who commented: 'Very often the game ends in a fight *and it is a very interesting game to watch*.'

Names: 'British Bulldog' is the usual name, but this version of the game is also known as 'Cruso' (Kilburn), 'Cock a Rusha' (Southwark), 'Cocky Rusty' or 'All Across' (Wigan), 'Cock a Rooster' (Swansea), 'Cocker' or 'Cockeroustie' (St. Andrews, Fife), 'Long Range' (Coventry), and 'Ten a Foxy' (Forfar). In each of these places the player caught usually has to be held for the count of ten; or in Wigan for the count of 'Cocky Rusty, two, four, six, eight, ten, twelve'. Sometimes for a joke the game is called not 'British Bulldog' but 'French Poodles'; while in the land of the Outback, perhaps predictably, it is 'Australian Dingo'.

⁂ For an antecedent see under 'Red Rover'.

WALK THE PLANK OR JOIN THE CREW

This game, which can be as rough as 'British Bulldog' if the contestants are so minded, is played mostly in the north-east of Scotland, very often by girls. One player goes into the middle of the road, while the rest stand on the edge of the pavement. The one in the road calls someone's name, and asks 'Walk the plank or join the crew?' If the person addressed agrees to 'join the crew', he or she peaceably joins the one in the middle, and another player is asked his choice. But if the player is bold, or thinks himself a fast enough runner, he replies 'Walk the plank', and has to try and reach the other side of the road without being caught. If he succeeds he shouts 'Schoolie' or 'Overboard', and all the others rush over in a body. But if he is caught the one in the middle 'tortures him' until he agrees to join the crew. There are then two in the middle, and the game continues until everybody has, in one manner or another, been persuaded to join the crew. 'If they catch you before you reach the other side of the road, they pull off your socks and shoes and tickle your feet, and twist your ears, and pull your hair, until you join the crew' (Girl, 12, Aberdeen). 'You are allowed to do anything but bite, kick, or scratch' (Boy, 14, Forfar). 'I like best when the others walk the plank because I like making them take off something' (Girl, 13, Aberdeen). 'You can play this for hours because there is plenty of fighting and you are never cold' (Boy, 14, Forfar).

⁂ Compare 'Pressgang' in *School Boys' Diversions*, 1820, pp. 32-3:

'One of the boys represents an officer, and four or six others the gang. They catch their companions, one at a time, and, on catching one, say to him,

"High ship or low ship;
King's ship, or no ship?"

If he chooses either of the ships, they send him as a prisoner, in the custody of two of their gang, to any place they may agree upon, where he must stop a prisoner; but if he say "No ship", they must all take him by force, by his hands, legs, and arms, to their rendezvous for pressed men. When they are all pressed, the pressed-men and volunteers, by turns become Press-gang and officer.'

KINGS, QUEENS, AND JACKS

'Kings, Queens, and Jacks' has only been reported from Edinburgh. The catcher stands in the middle of the playground and calls out 'Kings', 'Queens', or 'Jacks'. If he calls 'Kings' the players have to run across to the other side of the playground without being seized by the catcher and dragged to the place where they started. If he shouts 'Queens', the players have to hop across without being knocked over by the catcher, who is also hopping. And if the cry is 'Jack', anybody who puts a leg forward or even moves, has to join the catcher in the middle, just as if the catcher had seized him or had knocked him over. 'The last man left is the winner.'

PRISONERS' BASE

'Prisoners' Base', which for centuries was the most renowned of catching or capturing games, needs some organization, and is not now much played by children when on their own: its place, particularly in the south, being taken by Relievo (p. 172). Nevertheless the principle of Prisoners' Base is ingenious, and it is certainly one of the most exciting of organized games. Two bases or camps are chalked out on the same side of the playground, or marked in a field with sticks or cricket stumps: six stumps are enough since the bases can adjoin each other. At the other end of the playground, or about twenty yards away, two prisons are marked out. Two captains pick up sides (it is best if there are some twenty players), and each side takes possession of a base, but the prison in which they hope to place their captives is the one diagonally opposite, not the one nearest them. The captain of one side sends one of his players into the middle to taunt the others and start the game. The captain of the other side sends one of his players out to catch him, and the first player has to try and get back to his own base. He is helped by the fact that as soon as someone has been sent to catch him, his own captain will send someone in pursuit of his pursuer, whereon the other captain will send someone to pursue that pursuer, and the first captain will send someone after him. Thus each player, other than the first, will be both chasing and being chased, and as soon as a player gets back to his base, he can be sent in pursuit of someone else. But a player may only chase the one person he has been sent after. If he succeeds in catching him he cannot be caught himself, but takes his captive to the prison and returns to base ready to be sent out again. Once a prisoner has been taken and put in the far corner, the captain of his side will send someone running to attempt his release, and the captain of the side who has the prisoner will send someone chasing after to prevent him, whereon the captain of the first side may send someone after him, and it

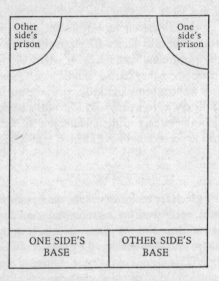

will be noticed that the player attempting a rescue, although he starts first, will have to run further than the player who is sent after him to frustrate the rescue. However, should a rescue be effected both rescuer and rescued can return to their own side unmolested. This active game, which also needs some skill and concentration on the part of the leaders, continues until all the players on one side have been made prisoner, or until an agreed time has elapsed (the side with the most prisoners being counted the winner), or, not infrequently, until the players are in such confusion about who is chasing whom, that the game has to stop. For orderliness it is helpful if the sides wear distinguishing marks, but it is not essential.

✳✳ Up to the twentieth century 'Prisoners' Base', also known as 'Chevy Chase' or 'Chivy', was one of the most-played of schoolboy games: a favourite sport at Sedgley Park about 1805; and played alike by Southey at his Bristol school and George Sturt a lifetime later at Farnham in Surrey. It was one of the games Tom Brown played with the village boys in the Vale of the White Horse even before his real schooldays began; which Tom Newcome played at Grey Friars, along with cricket, hockey, and football, 'according to the season' (*The Newcomes*, ch. ii); and which was played daily and was 'in a great measure compulsory' at Dr. Grimstone's establishment, as the unhappy Mr. Bultitude was to discover (*Vice Versâ*, ch. v).

For generations, too, the game was more than a juvenile diversion. Strutt recalls going, *c*.1770, to the fields behind Montague House (now the site of the British Museum) to see a grand match of 'Base' played by twelve gentlemen of Cheshire against twelve of Derbyshire 'for a considerable sum of money' (*Sports and Pastimes*, 1801, p. 62). Charlotte Burne records that men-servants in the eighteenth century 'were wont to ask a day's holiday to join or witness a game of prison-bars, arranged beforehand as a cricket-match might be' (*Shropshire Folk-Lore*, 1883, p. 524). Gomme reprints a ballad describing a match played at Ellesmere in Shropshire, 8 August 1764, between a team of bachelors and a team of married men, eleven a side, in which the bachelors (huzza'd by the fair maids amongst the spectators) seem to have had the best of the contest. And it will be recalled that Samuel Povey in *The Old Wives' Tale*, who had never played cricket, could yet boast of 'the Titanic sport of prison-bars' played in the Five Towns, where the teams went forth preceded by a drum-and-fife band, and the game was such that 'in the heat of the chase, a man might jump into the canal to escape his pursuer'.

Even in the eighteenth century the game was an old one. In 1598 Drayton could describe a place as one 'where light-foot Fayries sport at Prison-Base' (*Heroicall Epistles*, xxi. 200). In 1611 Cotgrave in his French–English *Dictionarie* defined *Barres* as 'the play at Bace; or, Prison Bars'. And Bace, Base, or Bars was frequently alluded to at this time, notably in *Cymbeline* (v. iii) where the youths Guiderius and Arviragus are described as—

> two striplings (Lads more like to run
> The Country base, than to commit such slaughter);

and in *The Faerie Queene* (v. viii. 5) where two knights pursuing a damsel fleeing on horseback are in turn pursued by another knight and are said to run—

> as they had bene at bace,
> They being chased that did others chase.

Indeed the game was so well known to the Elizabethans that when a person provoked someone to come after him, it was customary to say he 'bid the base'.[1]

This notoriety in the sixteenth century is not surprising, for the game seems already to have been popular for 200 or 300 years. In the preamble to the Parliamentary Statutes of 16 March 1332 the playing 'à barres' was explicitly prohibited in the precincts of the king's palace while Parliament was sitting. The game was named in Jean de Garlande's glossary of the early fourteenth century. Froissart played 'aux bares' in his boyhood at Valenciennes, about 1345 ('L'Espinette amoureuse', l. 221). And D'Allemagne in his *Sports et Jeux*, 1904, p. 56, suggests that on the Continent, at least, barres was the chief communal competitive game of the Middle Ages.

FRENCH AND ENGLISH

The great object of the players in 'French and English' is to run off with the property of the opposing party, and it is this that gives the game its

spice. Two leaders pick up sides, agree upon a line that shall divide their territories (preferably a natural feature, such as a ridge or stream), and each player deposits some possession, as cap, coat, or handkerchief made into a flag, a certain distance back within his side's territory. The members of each side then attempt to make away with the other side's possessions, the game being one of forays into enemy territory. Sometimes a single player darts off on his own when he sees an opportunity; at other times three or four players make a concerted sortie, hoping by so doing to divert attention from one another. But even if a player finds himself to be momentarily unopposed he may take only one object at a time. Once a player has crossed the dividing line he can be caught, and is kept prisoner in the enemy's camp along with their possessions. A member of his own side must then rescue him before any more booty can be taken. In this way the fortunes of the two sides can alter dramatically in less than a minute. Four people may sally out from one side and perhaps all be captured, or perhaps all four return with treasure. It is a 'busy game', as one child put

---

[1] Thus Lucetta, in *The Two Gentlemen of Verona* (I. ii), counselling Julia to fall in love, admits 'Indeede I bid the base for Protheus'; and in *Venus and Adonis* (stanza 51) the wonder horse is thought so fleet of foot he would 'bid the wind a base'.

it, a game in which every player is important; and the sport continues until one side or the other has acquired every article of property that the other side has laid out, or until (what amounts to the same thing) one side has made every member of the opposing side a prisoner.

For some reason the game is not as popular as it used to be, even in the north; and it seems possible that the briefer and more impersonal versions of the game which are currently organized for children in the shelter of gymnasiums have taken the edge off the traditional game. In organized versions of the game each side usually has to capture only a single flag or trophy.

Names: 'French and English', 'Scotch and English', 'Germans and English'. Organized versions: 'Flag Raiding', 'Capturing the Flag'.

⁂ In the eighteenth and nineteenth centuries this game, made romantic by reference to the marauding raids of the Borderers, was much played in the northern half of Britain, the sport being enriched by taunts and feigned enmity: 'Here's a leap into thy land, dry-bellied Scot'; 'Here's a leg in thy land, thieving Sassenach'. In those days, it is said, a well-contested match might last 'nearly a whole day', the young players on the losing side replacing lost property with further of their garments until each of them was approaching the state of nature. The game was indeed a 'heroic contention, imbued with all the nationality of still older days'; and those who describe it often seem to feel that they are recalling some of the happiest hours of boyhood.

As might be expected the game has been played under a great number of names, amongst the earliest being: 'Scotch and English' (W. Hutton, *History of the Roman Wall*, 1802, p. 105); 'Wadds', 'Steal-Wads', 'Rigs', and 'Tak-Bannets' (Jamieson, *Scottish Dictionary*, 1808–25); 'England and Scotland' (Cromek, *Remains of Nithsdale and Galloway Song*, 1810, p. 251); 'Scotch-and-English', 'Stealy-Clothes', and 'Watch-Webs' (Brockett, *North Country Words*, 1829); 'Set-a-Foot' on Tweedside, *c*. 1820, and 'Stone Heaps' in London (*Notes and Queries*, 4th ser., vol. ii, 1868, pp. 97, 165). In more recent times: 'French and English' at Bitterne in Hampshire, 'Range the Bus' in Aberdeen, and 'Bonnet Ridgie' at Dyke in Morayshire (Gomme, 1894–8); 'Beggarly Scots' and 'Watch Webb' in Wigton, late nineteenth century (*Journal Lakeland Dialect Society*, 1951, pp. 38–9); 'Lands' in Argyllshire (Maclagan, 1901, pp. 218–19); 'Herdie Pans' in Orkney, and 'Regibus' in Banffshire (*Folk-Lore*, vol. xvii, 1906, pp. 104–5); 'Seizing Sticks' (*London Street Games*, 1916, p. 17); 'Japs and Russians' (Chard, 1922); 'Prisoners' (Taunton, 1922, also New Norfolk, Tasmania, *c*. 1910).

In the United States, somewhat restricted versions: 'Stealing Sticks' (*Games of American Children*, 1883, p. 168), 'Stealing Sticks' or 'War' (*North Carolina Folklore*, 1952, p. 80), 'Capture the Flag' (*Saturday Evening Post*, 19 December 1964, p. 18):

In the Philippines 'Kawat-Kawat' (Brewster, p. 70). In Italy 'Guerra francese'.

4

Seeking Games

'At night is the best time to play. My friends and I went to hide. We hid in a man's cabidges. John walked by us about six times and he never saw us.'

Boy, 12, Luncarty

FEW people can feel more tense than the young player as he sets out alone to search for his companions, seeing no one, where a minute before was a mob, yet knowing that every bush and tree may be a mask for a pair of eyes. Seeking games have this peculiarity, that for much of the game the players are out of sight of each other, uncertain of what is happening, yet are all the while within hailing distance. In consequence there is much calling, and the calls, being traditional, are often curious and even poetical. At the outset of a game, when the seeker has finished counting to a hundred, or whatever number has been agreed, he announces his search by calling into the emptiness: 'Coming, ready or not', or 'Here I come, ready or not, if ye're spied it's no my fau't' (Perth and Falkirk), or 'Look out, look out, the fox is about, and he is coming to find you' (Swansea). In Wickenby, Lincolnshire, he shouts 'Tins' because, says a 12-year-old, 'this means coming'. In Norwich, before he starts, he shouts,

> Whether you run or not
> I will catch you hot,

and, as if to force meaning into the words, when he catches somebody he shouts 'Hot'. In Leicestershire he, or more likely she, calls:

> I hold my little finger,
> I thought it was my thumb,
> I give you all a warning,
> And here I come.

And amongst children in Somerset, where poetry seems to come as second nature, the traditional call is this:

> The cock doth crow, the wind doth blow,
> I don't care whether you are hidden or no,
> I'm coming![1]

There are rhymes, too, which the hiders repeat to alert each other. If a seeker has stolen up quietly on one of the hiders, and put him out of the game, the one who has been discovered instantly sets up a roar for the benefit of his companions:

> Keep in, keep in, wherever you are,
> The cat's a-coming to find you.
> > *Birmingham*

> Keep in, keep in, wherever you oor,
> The rats and mice are at your door.
> > *Helensburgh, Morpeth, and*
> > *Newcastle upon Tyne*

> Jeep in, jeep in, whatever do in,
> Da clockin hen is seekin de.
> > *Whalsay, Shetland*[2]

If the seeker is becoming discontented with his task, and beginning to imagine that the hiders have vanished in reality as well as from sight, he may make the plaintive appeal or threat:

> A whistle or a cry,   A whistle or a cry,
> Or let the game die.   Or the game gans by.
> > *Luncarty*   *Langholm*

And if the game is to be brought to an end prematurely, because the players have become tired, or there has been an argument, or, says a 13-year-old, 'because the hunter is going away for his tea', the general cry is

[1] In the United States:

> Bushel of wheat,
> Bushel of rye,
> All not hid
> Holler I.

> Bushel of wheat,
> Bushel of clover,
> All not hid
> Can't hide over.
> > *Widespread, e.g. Carolina,*
> > *Maryland, Missouri, Nebraska,*
> > *and Texas*

In France:

> C'est-i-fait,
> Minon, minette.
> > '*Les Amours de Bastien et Bastienne*',
> > *1753. Cited Rolland, 1883, p. 152*

In Germany:

> 1, 2, 3, 4, Eckstein,
> Alles soll versteckt sein,
> Hinter mir und vorder mir
> Das gibt es nicht,
> 1, 2, 3, Nun komme ich.
> > *Current Hamburg, 1956*

[2] In the eighteenth century as well as the nineteenth, the following was apparently often to be heard in the playgrounds of Edinburgh 'addressed to the secreted personage at Hidee':

> Keep in, keep in, wherever you be,
> The greedy gled's seeking ye.
> > *Blackwood's Magazine, August 1821, p. 37*

'Alley, alley in', or 'Allee-ins, not playing', or 'All the ends stop play', or 'Olly, olly in', or the wonderful liquid warble of 'All-ee, all-ee, eeeze'. In Scalloway, in Shetland, the call is appropriately nautical, 'All hands ahoy!' In Bishop Auckland: 'All in, all in, spuggy in the tin.' In Plymouth:

> All in, all in, wherever you are,
> The monkey's in the motor car.

In Bradford and Birmingham:

> All up, all up, wherever you are,
> If you don't want to play stay where you are.

In South Elmsall:

> All up, the game's up,
> Ready for Sunday morning.

In Manchester and Newcastle:

> Billy, Billy Buck,
> The game's broke up,
> And all through *Tommy Skelly*.

While in Scotland if something has gone wrong with the game, 'the game's a bogie':

> Come oot, come oot, wherever you are,
> The game's a bogie.
> > *Edinburgh, Glasgow, Falkirk, Ballingry, Langholm*

> Come oot, come oot, wherever you be,
> The monkey's up the apple tree.
> > *New Cumnock*

> Lees, lees, whit dae ye please,
> Little boys living on candle-grease.
> Come oot, come oot, where ever ye be,
> Or the gem's a bogie.
> > *Cumnock Academy, 'Those Dusty Bluebells', 1965, p. 26*

Such cries ring out frequently in Scotland, remarks a correspondent, 'because of the Scottish love of litigation and disputation over trivial points arising from rules'.

THE NAMES OF SEEKING GAMES

The study of seeking games is complicated by the fact that they have long been played in a diversity of ways, but not under a diversity of names. Even today when a child speaks of 'Hide and Seek' he may be referring to one of four different games; and in the past when a writer mentioned a seeking game he rarely thought it worth describing. It seems best therefore to bring the early references to seeking games together under a general heading, rather than attempt to distinguish which reference is to the forerunner of which particular game of the present day. When Biron felt himself to be playing a part in 'All hid, all hid, an old infant play' (*Love's Labour's Lost*, IV. iii) we know little more about how the game was played in Shakespeare's day than that 'All hid' was the cry to start the game. We receive no help from Dekker when Sir Rees ap Vaughan declares:

'Our vnhansome-fac'd Poet does play at bo-peepes with your Grace, and cryes all-hidde as boyes doe.'   *Satiro-mastix, 1602, v. ii*

Nor do we learn anything from William Hawkins when Ludio argues that Phoebus plays

'At Bo-peepe, and Hide and seeke. All night is our all hid. But in the day We seeke about.'   *Apollo Shrouing, 1627*

Cotgrave merely confused the issue in 1611 when he defined *Clignemusset* as 'the childish play called Hodman blind, Harrie-racket, or, are you all hid'; and Robert Sherwood was similarly unhelpful in 1632 when he described 'All hidde' as a game 'où vn se cache pour estre trouvé des autres'. (If the cry was 'All hid' would not more than one player have been hiding?)

Possibly Sherwood had in mind a game such as Hamlet's 'Hide Fox, and all after' (IV. ii). Pegge in his *Alphabet of Kenticisms*, 1735, defined 'Hide-and-Fox' as 'Hide-and-Seek'. In 1688 Randle Holme described 'Hide and seech' (his spelling was wild even for the period) as a game in which 'one or more to goe hide themselues, and the rest to seek them out'. And the boys and girls of Lilliput, it will be recollected, played at 'Hide and Seek' in Gulliver's hair (I. iii). Another early name was 'Winck-All-Hid', presumably referring to a player being hoodwinked while the others hid (John Davies, *Humours Heav'n*, 1609, II. iv). But most early names simply echo the dominant call, as 'Whoop' (1798), 'Whoop Oh!' (1828), 'Hoop and Hide' (1711), and 'Hoopers-Hide' (1719)—as it would be 'Cooee' in the present day.[1]

---

[1] Although now so ordinary, the cry 'Cooee' is of recent date in England, coming from Australia where it was a signal used by the aborigines. Peter Cunningham in *Two Years in New South Wales*, vol. ii, 1827, p. 23, recorded: 'In calling to each other at a distance, the natives make use of the word *Coo-ee*, as we do the word *Hollo* . . . [It has] become of general use throughout the colony; and a newcomer, in desiring an individual to call another back, soon learns to say "*Coo-ee* to him" instead of Hollo to him' (*O.E.D.*). It will be recalled that Sherlock Holmes in 'The Boscombe Valley Mystery' (1891) could presume a man to have come from Australia because he called 'Cooee'.

In Scotland one game and cry seems to have been 'Keek-Keek' (' "Te he", quod Jynny, "keik, keik, I se 30w".'—*Jok & Jynny*, *c.* 1568), hence 'Keek-Bogle' or 'Bogle Keik' (1791). In Edinburgh, at the beginning of the nineteenth century, the game and call was sometimes 'Ho spy!' (*Blackwood's Magazine*, August 1821, p. 35), and Gregor told Gomme (vol. i, 1894, p. 212) that in Keith this was abbreviated to 'Hospy'.

Further dialect names that have been recorded include: 'Beans and Butter' (Oxfordshire, 1849), from the cry to commence the search:

> Hot boil'd beans and very good butter,
> If you please to come to supper!

'Bicky' (West Somerset, 1888), 'Boggle-Bush' (Whitby, 1876), 'Cuckoo' (Northamptonshire, 1854), 'Felt' (Scarborough, *c.* 1895), 'Felt and Late' (Sheffield, 1888), 'Halloo' (John Clare, *Village Minstrel*, 1821, i. 5), 'Heddie-ma-Blindie' (Weardale, 1939), 'Heddo' (East Yorkshire, 1889), 'Hiddy' (Leeds, *c.* 1890), 'Hide-a-Bo-Seek' (Berwickshire, 1825), 'Hide an Find' (Suffolk, 1823), 'Hide and Wink' (Leicestershire, 1844), 'Hide-Hoop' (Pembrokeshire, 1888), 'Hiders-Catch-Winkers' (Hampshire, 1871), 'Hie, Spy, Hie' (Newcastle upon Tyne, 1813), 'Hy Spy' (Scott, *Guy Mannering*, 1815, xxxvi), 'Huddin-Peep' (Lancashire, 1895), 'Pee-Koo' and 'Pi-Cow' (Angus, 1887 and 1808), 'Salt Eel' (Suffolk, 1823), 'Shammy Round the Block' (Liverpool, *c.* 1925), 'Spinny Wye' (Newcastle upon Tyne, 1813), 'Spy All' (Bath, *c.* 1890), 'Spy Hole' (York, *c.* 1910), 'Spyo' (Barrie, *Sentimental Tommy*, 1896, xiv), and 'Steik-and-Hide' (Aberdeenshire, 1825).

In France, Froissart played 'à la clingnette' and 'aux reponniaus' in his childhood at Valenciennes, *c.* 1345 ('L'Espinette amoureuse', ll. 233 and 226). Gargantua played 'à clinemuzete', 'au responsailles', 'au bourry, bourryzou', and 'à la cutte cache' (Rabelais, I, 1534, xxii). And in Italy, Taddeus played 'a covalèra' and 'a vienela, vienela' (*Il Pentaphperóne*, 1634, Day II). For hide-and-seek in ancient Greece see p. 155.

§ *Dissimilar Number of Players Hiding and Seeking*

HIDE-AND-SEEK

The simplest form of 'Hide-and-Seek', the stay-where-you-are-until-found variety, is now played mainly by small children, or when only two are playing, or when the game is played indoors. The first person to be found is the seeker in the next game; the last to be found is the winner. A necessary preliminary is arranging how long the seeker shall contain himself (generally with eyes closed) before he starts his search. He is usually told to count a hundred, or 'ten, ten times', or 'five hundred in fives', or five hundred 'the shortie way'—'Five, ten, double-ten, five, ten, a hundred'—an abridgement favoured in Scotland. Sometimes the number to be counted is set according to the number of children playing, ten or twenty for each person, and twenty more for the 'den', and twenty more 'for luck'. In Edinburgh they play 'Vehicles' or 'Buses', the child who is 'het' has to wait where he is until a car or van passes by, or—if they are by a main road—until a bus is seen. And in Grimsby, 'very commonly' says a 13-year-old, they make it physically impossible for the seeker to come after them too soon. They tie the boy or girl to a lamp-post, and he has 'to release himself from his bonds' before he can start seeking. (See also under 'I Draw a Snake upon your Back'.)

'Hide-and-Seek' becomes more fun, and is considerably speeded up, when the hiders do not remain in their hiding-places, but try and get back to the starting-place unobserved while the seeker is out looking for them. Even so the game is unsatisfactory. Those who have been found, or who have made their way back safely, often weary of the game before the last person has been discovered (as H. E. Bates has remarked, there is bound to be some clever-dick who has hidden in a coal-hole and refuses to show himself), so that after a while they will be calling in those still in hiding and proposing a different game.

When older children speak of 'Hide-and-Seek' they usually mean a racing-home variety, such as 'Block' (q.v.), which is a faster and more compact game.

Names: 'Hide-and-Seek' is sometimes referred to as 'Hiding Seek' or 'Hidy'; and in Scotland it is often 'Hide and Go Seek'—as also in the United States. When played after dark, as is not unusual, it may have a special name, such as 'Cat's Eyes' (Forest Hill), 'Ghosts' (Inverarity), 'Run by Dark' (Peterborough), 'Toad in the Hole' (Forfar), 'Bug in a Rug' (Accrington), 'Spotlight' (with a torch, Knottingley), and 'Torchlight' (Spennymoor). In Ipswich, when played up trees, it is 'Chip and Chap'. When the seeker has not only to find the others, but has to try and touch them as they run back to the starting-place, the game may be differentiated by a name such as 'Hide and Tick' (Welshpool).

⁎⁎ The running-home form of hide-and-seek is described in *Every Boy's Book* by J. L. Williams, 1841, under the name 'Whoop!'.

'One player takes his station at a spot called the "home", while the others go to seek out various hiding-places in which to ensconce themselves; when all are ready, one of them calls out Whoop! on which the player at the "home", instantly goes in search of the hiders, and endeavours to touch one of them, as they run back to "home"; if he can do so, the one caught takes his post at the home, and he joins the out-players.'

In the sixteenth century players seem actually to have sought the office of seeker: the first to reach the base unimpeded acquired this honour by right, and was known as King. It was thus in the 'old schoole-boyes game' of 'King by your leave' or 'Old shewe', referred to several times by Elizabethans, and described in Huloet's *Dictionarie*, 1572:

> '*Kinge by your leaue*, a playe that children haue, where one sytting blynde-folde in the midle, bydeth so tyll the rest haue hydden them selues, and then he going to seeke them, if any get his place in the meane space, that same is kynge in his roume.'

This procedure was already more than a thousand years old, being the rule in the ancient Greek game 'Apodidraskinda'. Pollux stated (ix. 117) that one player shut his eyes, or had somebody covering them to ensure that he did, while the others ran off. This player then proceeded to look for them, while the object of each of the hiders was to reach the seeker's place and become seeker in his stead.

ONE MAN PLUS

In this game there is initially one seeker, but those whose hiding-places have been discovered, or who have been seen while attempting to reach home, join the seeker in searching for the rest, so that eventually all the players (it is best if there are not more than ten) are looking for the last person. The game is not as common as might be expected, and has no standard name. A Grimsby boy, giving the above name, said, 'We held a special meeting in our gang hut to decide what to call the game'. In all accounts received the game is played around the streets at night. In some parts of Glasgow it is called 'Pea Hot', since this is the hiders' call when they are ready. In other parts of Glasgow it is known as 'Over the Fences', the game being played entirely in other people's gardens, with this rule, that both hiders and seekers must enter each garden by jumping the gate or fence. 'It has to be played in the dark', remarks a keelie, 'so that the neighbours won't know.' In Ballingry the game is called 'Bully Horn': those found may escape, if they can, from their hiding-places, whereon the chasers are rallied with the cry 'Bully horn', for a person is not considered caught until he has been clapped on the back three times. In Glastonbury the game is called 'Multiplication Touch'.

⁎ According to informants the game was played in the Isle of Dogs, East London, *c.* 1905, under the name 'Point', and in County Kerry, *c.* 1935, as 'Hunts'. Cf. 'Barla-bracks about the Stacks' stated by Jamieson, *Scottish Dictionary*, 1808, to have been played in northern Scotland.

MAN HUNTING

In this version all seek one. Everyone hides their eyes while one person goes off and secretes himself in a place of special difficulty, as 'up a tree or somewhere'. The seekers count to as much as 500 'in ones' to give him plenty of time, and he is also allowed, if he wishes, to move from place to place while they are searching, provided that he is not seen, for to be seen is to be caught. 'If someone sees the hider and he does not come he is out of the game.' At Enfield: 'When we catch him we hit him and then let him go. Then the one who caught him goes and hides.' If he manages to get back to the home without being seen he is safe.

Names: 'Cuckoo' (Alton), 'Exploring' (Bacup), 'Find Her if You Can' (West Ham), 'Hide and Seek' (Market Rasen), 'Man Hunting' (Enfield).

⁎ This is 'Hide and Seek' as described in *School Boys' Diversions*, 1820, pp. 40–1. 'One boy is appointed to hide wherever he pleases . . . when he has secreted himself, he is to cry, "Spy all", at which signal, the rest are to search him out; and if discovered, he is to be buffeted with knotted handkerchiefs, until he can reach the goal, or starting post.' Much the same game, it appears, was 'Cock's-Odin', played at about the same period in the Scottish Lowlands (*Notes and Queries*, 4th ser., vol. ii, 1868, pp. 97 and 165); also the game of 'Cuckoo', described in Burne's *Shropshire Folk-Lore*, 1883, p. 222, although here the hider, when discovered, 'rushed out and did his best to reach "home" without being captured'.

SARDINES

'Sardines', played indoors or out, is the most popular of the games consisting purely of hiding and finding. One person goes off to hide while the others shut their eyes and count to the agreed number. The seekers split up, and search independently of each other. Indeed, if one of the seekers finds the hider he is careful not to let the others know, but slips into the hiding-place when they are not looking. Ideally the hiding-place should be somewhere that will accommodate all the players; but it seldom is, and as further players find it, and crowd in, the silent squeeze becomes tighter and more suffocating, players sometimes having to lie on top of each other. Those who are still searching gradually become aware that their fellow searchers are disappearing, and rush to the places where they were last seen, thinking that they will be near the hidy-hole. When the last person arrives he is sometimes chased back to the starting-place, but more often than not there are just sighs of relief as the sardines extricate themselves from their cramped positions, and complain of their stiffness and the length of time they have been waiting.

The game is usually known as 'Sardines', but also 'Sardines in a Tin', 'Sardines and Tomatoes', 'Squashed Sardines', and 'Squashed Tomatoes'. In Wigan it is 'Mexican Hideout'. The game is sometimes played in couples or with two teams.

I DRAW A SNAKE UPON YOUR BACK

If 'Hide-and-Seek' is named less frequently than it used to be amongst favourite games, it is because an ingenious method of starting the play has become popular, and given hide-and-seek a new appearance, and new nomenclature. There now need to be at least four players. One of them is chosen to turn his back to the others, and usually leans against a wall or lamp-post, with his face buried in his arm. The others gather round, and the leader chants:

> I draw a snake upon your back.
> Who will put in the eye?

When he has drawn the snake, another player (taking a hint from the leader) stretches forward and pokes the person's back. The person whose back has been poked then turns round and guesses who poked him, but— here lies the sport—he is not told whether his guess is correct. He has first to set a task for the person he has named; and it is only after he has done so that he learns whether or not his guess was correct. If it was, the person who 'put in the eye' has to perform the task he has been set; but if the guess was wrong, the guesser himself has to carry out his own instructions; and whichever of them it is also becomes the seeker, for while the task is being performed the others run off, and the game which follows is, as the children say, 'just like hide-and-seek really'.

This manner of starting the game is not only fun ('It's really really fun'), it is also highly ingenious. It determines, with indisputable fairness, the length of time the hiders shall have to hide in, because he who names the task will not make it too onerous for fear he himself has to undertake it, nor make it too simple since he hopes to be amongst those who hide. Thus he says:

'Run round a car five times and count to a hundred in singles.'

Liss

'Walk five hundred yards shouting "Hot peas and pies".'

Hoyland, near Barnsley

'Go to your house and ask for a piece and jam.'

Forfar

'Count twenty, then touch a pole, then touch a flower, then come back and count a hundred.'   *Spennymoor*

In some places (e.g. Bishop Auckland and Pontefract) the player suspected of putting in the eye retorts 'Where are you going to send me?' and urges the guesser to make the task more difficult, challenging: 'How many times shall I do it?'—'Hop, skip, or jump?'—'Drunk or sober?'[1] Indeed versions of the game vary only in their degree of fantasy. At Annesley the leader draws 'numerous squiggle lines' saying, 'I draw ten thousand snakes down your back, who tipped your finger?'; at St. Peter Port he says 'I draw a snake on the old man's back, two eyes, a nose, and who puts his tongue in?' In Ipswich:

> Draw a snake on a black man's back,
> Chop off his head and who did *that*?

In Gloucester the snake is drawn on a 'dead man's back', in Sixhills on an 'elephant's back', in Swanpool on a 'unicorn's back'. The game also goes under such names as 'Stroke-a-Bunny' (Liverpool, for past thirty-five years), 'Stroke the Baby' (Welshpool), and 'Smooth the Cat' (Penzance):

> Smooth the cat, smooth the cat.
> Who touches you last?

In Accrington several players engage in the draughtsmanship:

> Stroke a bunny, stroke a bunny,
> Someone's going to wake you in the morning.
> I'll draw the snake (*someone does so*),
> I'll draw the ladder (*someone else draws the ladder*),
> I'll draw the question mark (*a third draws the top part of a question-mark*),
> And someone draws the dot (*a fourth player supplies the dot*).

In St. Helier, Jersey, the game is known as 'Crow's Nest'; in Bristol, sometimes, 'Who Put the Egg in the Birdie's Nest?'; and in a number of places it is 'North, South, East, West':

---

[1] Similarly in Oldham, about 1930, as recalled by a correspondent, the player asked: 'Where do I go?'—'How many times?'—'Eyes closed or open?'—'Stockings up or down?'—'Running or walking?' whereafter he might say, 'Well go yourself because it wasn't me.'

North, South, East, West,	North, South, East, West,
Who's the king of the crow's nest;	Who's the king of the crow's nest;
Draw a snake right down her back,	Draw a ladder, draw a snake,
Who's the one to finish that?	Would you kindly finish the cake.
West Ham	*Enfield*

In Forfar, where they draw a cross on the person's back, the game is called 'I'll Cut the Butter and I'll Cut the Cheese'; in Orkney it is 'Spread the Butter on the Cheese'; in Glasgow, where words are not wasted, it is 'Cheesey'. In Edinburgh the leader chants:

> I spread the butter, I spread the cheese,
> I spread the jam on your dirty knees.
> Guess who tipped.

Here, when someone has been picked, and the task set, he who was named asks, 'How many fish in the barrel?' The guesser, who is still in doubt about the rightness of his guess, gives a number, and the task has to be performed that number of times. The most common task set in Edinburgh is to climb the stairs to the top of a tenement. 'I like this game', remarked an 11-year-old in the Canongate. 'I like the rhyme with its clowny words, and most of all I like the excitement of the child trying to guess who tipped.'

⁎⁎⁎ This game, or manner of starting a game, is also widespread in the United States under the names 'Tappy', 'Tap-on-the-Back', 'Tappy-Hi-Spy', and 'Tap the Icebox' (Brewster, 1953, pp. 48–50). 'Tap the Icebox' is said to have been played by Chicago children 'for generations' (*Chicago Daily News*, 4 April 1961). 'Poke the Icebox' was played in Canada about 1925. In Melbourne, Australia, it is 'Tip the Finger'. In New Zealand, 'Tip the Finger' or 'Draw the Snake' (Sutton-Smith, 1959, pp. 67–8). And in West Berlin, according to Peesch, *Berliner Kinderspiel*, 1957, p. 42, it is 'Englisch Versteck', the children saying 'Ich mach das Fragezeichen, wer macht den Punkt?' ('I make the question-mark, who makes the point?')

The use, in some versions of the game, of the couplet

> North, South, East, West,
> Who's the king of the crow's nest?

shows how the game has evolved. This rhyme has been in juvenile employment for many years, and was formerly attached to the game in which one player, whose eyes were covered, sent each of his fellows to a different starting-place, prior to a race home (see under 'Hot Peas'). It will be appreciated that if the first part of the game is played on its own, and the player whose face is to the wall attempts to guess whose hand touched him, merely to put the other person in his place, the game becomes little different from 'Hot Cockles', popular in the Middle Ages (see pp. 293–4).

BLOCK

This form of hide-and-seek is both popular and exceedingly energetic: the seeker has not only to locate each player in hiding, but has to race him back to the starting-place. It is often this game that children are referring to when they speak of 'Hide-and-Seek', although local names abound, and the cacophonous 'Block', 'Blocky', or 'Block, One, Two, Three' is common in the south and north-east of England, and in Scotland. The game is usually played at night ('so that the person after you cannot see so good'), and the usual 'block', or starting-place, is a lamp-post, sometimes termed the 'blocking-post'. One person, 'the blocker', hides his eyes and counts to an agreed number while the rest scatter and hide. In Accrington a scarf is tied round his eyes, and when he has finished counting he ties it to the lamp-post to mark which one he started from. The players hide in an area around the block, if possible keeping it in view, for their aim is to reach the lamp-post while the blocker is out searching for them. If they can touch it and shout 'One, two, three, block home', they are free. Sometimes players will sneak up behind the blocker while he is still counting, in the hope of rushing the block the moment he has finished, but generally this is frowned upon. In Scarborough, if the counter hears them approaching, he warns 'No backs, no sides, no front'. In Caerleon he cries 'No behind the cat's tail'.

In this game it is not enough for the seeker to see a person, nor does it mean anything if he catches him. He can only put a player out of the game by racing back to the block, and shouting 'Block, one, two, three, I spy . . .', giving his name. (In Whalsay, Shetland, 'Block, one, two, three, you are not free'.) Sometimes, to make certain there has been no mistake, the blocker has also to shout the person's hiding-place, as 'Block, one, two, three, I spy *Alan* up a tree'. If his identification proves wrong it is called 'False Alarm' (in Ballingry, 'Burnt Spy'), and the blocker has to hide his eyes and start again. Occasionally the hiders change coats to confuse the blocker; but identification is generally difficult enough, especially if it is dark. Usually the game goes on until everyone has either been 'blocked' or has freed himself, and the first person blocked becomes the new blocker. But in some places, as in Forfar, they have a rule: 'Three free, all free.' If three people manage to reach the block and cry 'Free', those already blocked are at liberty to run off and hide again. In a few places, for example Welshpool, whenever a hider reaches the starting-place unseen he shouts 'Release, one, two, three', and those waiting at the starting-place are released to hide again (compare 'Buzz Off', hereafter). In Fulham, 'If everyone has been spied except one, and that person gets home, he says "Block, one, two, three, saved the lot", and they all go off again'. As a 13-year-old girl commented, 'This game hasn't an ending and goes on until the children are called in to bed'.

'Block' is the basic game in which attention is focused on the starting-place; and there are a score of local names for the lamp-post, or whatever serves as the 'home' or 'den'; for example, 'bay' in Spennymoor, 'billy' in York, 'blobbing place' in Annesley, and 'the bounce' in Jersey and Guernsey. In Wolstanton it is the 'bucking place', in Scarborough the 'carry post', in Bristol, sometimes, the 'cree', and at Welshpool the 'deno'. In Aberdeen and Forfar it is the 'dell' or 'dellie' (pronounced *dael* at Arbroath, as McBain remarked in 1887), and in Cumnock the 'dill' or 'dull' (cf. the *dule* or goal in the old game 'Barla-Breikis' described by Jamieson in 1808). In the southern part of Yorkshire—Pontefract, Barnsley, Ecclesfield—it is the 'dob' or 'dobby' ('the person races back and knocks the dobby three times'), while to the east of the county, and in Lindsey, it is the 'hob'. In Tunstall and Wolstanton it is the 'ducker' (as it also was in 1910), and in Manchester and around Knighton, where a stone usually marks the place, it is called the 'kick-stone', though they do not kick it. It is also known as the 'kig' (Stockton-on-Tees), 'fleaky post' (Leek), 'lurgy post' (Ipswich), 'mobbing post' (south Wales and Monmouthshire), 'relevo place' (Wilmslow), and 'rally' (Coventry). At Crickhowell in Brecon, 'You run to the tally post and tally the person'. At Holmfirth, 'You try to win the other person to the whipping den'. While at Oldham and Nelson children happily make for the 'whipping post', the game being commonly known in the north-west as 'Whip'.

Other names for the game, which also provide the call as the runner reaches home: 'Acky' or 'I-acky' in Warwickshire and Northamptonshire ('I acky *Freddie*, one, two, three'; 'First person "ackyed off" is "it" next time'), 'Hicky, One, Two, Three' (Chester), 'I-erkey' (Leicester), 'Hi-Lerky' (Newton Abbot), 'Erkie' (Plymouth), 'Urkey' or 'Murkey' (Helston), and 'Ookey' (The Lizard). It is 'Forty Forty' in Ipswich and district, Ramsey in Huntingdonshire, Cranborne in Dorset, and in south-east London generally, although in Walworth 'Fifty-Two Bunker' (seeker counts to fifty-two; shouts 'I see *Janet*, fifty-two bunker'). It is usually 'Mob' or 'Mob Mob' from Aberystwyth to

Bristol. At Lydeard St. Lawrence in Somerset it is 'Mop Mop'. It is 'Om Pom' or 'Pom Pom' in Norwich, Great Staughton in Huntingdonshire, Hayes in Middlesex, Weymouth, Chichester, Liss, and Fulham, S.W. 6 ('Pom, pom, *Ernie*, one, two, three, i-o-key'). Idiosyncratically, it is known as 'Billy, One, Two, Three' (York), 'Cocoa-Beanie' (Edinburgh), 'False Alarm' (Ponders End), 'Free' (Orkney and Shetland), 'Hidy-bo' or 'Hidy-bo-Seek' (Brinsley and Annesley), 'My Bounce, One, Two, Three' (St. Helier, Jersey), 'Tackie' (Letham, Angus), and 'Whip Out' (Windermere).

 ** The following names were familiar in the past: 'Billy Rush' (York, *c.* 1910), 'Block Block' (Gainsborough, *c.* 1910), 'One, Two, Three, Block' (Hull, 1890s), 'Bucky Bean' (Glastonbury, 1920s), 'Bunky-Bean Bam-Bye' (presumably this game, North Devon, 1867, see *EDD*), 'Eci' (pronounced *Ekki*, T. Hudson-Williams, *Caernarvonshire*, 1952, p. 65), 'Forty' (London, 1910 and 1916, probably this game), 'Gilty Galty' (Huddersfield, 1810, seeker counted to forty as today, after reciting 'Gilty galty, four and forty, Two tens make twenty'—*Almondbury and Huddersfield Glossary*, 1883), 'Hacky', 'Hi-acky', 'I Hacky' (Midlands, *c.* 1915), 'Ackee'(Somerset, 1920s), 'Jacky' (Warwickshire, *c.* 1890), 'I-erkee' (Oxford, *c.* 1910), 'I-urkey' (Earl Shilton, Leicestershire, *c.* 1910), 'Key Hoy' (Argyll, 1906), 'Lurky' (Nottinghamshire, 1902), 'Lerky' (? this game, D. H. Lawrence, *Sons and Lovers*, 1913, ch. iv), 'Mop-and-Hide-Away' (Cornwall, 1880), 'Mopan-Heedy' (Devon, 1889), 'Moppy-Heedy' (Cornwall, *c.* 1900), 'Point' (*London Street Games*, 1916, p. 20), 'Squat' (Sundon, Bedfordshire, *c.* 1900), 'Whip' (apparently this game, but see p. 164, Lancashire and Potteries, *c.* 1900).

It is hardly necessary to confirm that the game is international. When a child, today, in Canada or the United States speaks of 'Hide and Seek' it is this racing-home game that he ordinarily means. This is also so in Australia, although children there sometimes know the game by one of the English dialect names, e.g. 'I-ackey' in Queensland. In Germany and Austria, likewise, the ordinary game of 'Versteck' is this racing-home variety; and according to Jeanette Hills, *Das Kinderspielbild von Pieter Bruegel*, 1957, p. 36, the game appears in Veit Conrad Schwarz's *Bilderbuch*, 1550, and is there called 'Ekkete Eck'. Brewster (1953) collected Armenian, Greek, Rumanian, and Hungarian analogues, the one from present-day Greece being remarkable in that the seeker commonly counts to forty before starting his search: the same number that custom dictates children should count when playing this game in southeast London.

BUZZ OFF

This game, played mostly in Scotland and the north country, is no more than a variation of 'Block' or 'Hide-and-Seek', but is carefully differentiated by the children, and is never, it appears, referred to as 'Hide-and-Seek'. It has the one additional rule that any hider who succeeds in racing back to the starting-point without being blocked exclaims 'Buzz off' and frees everyone there. Further, on hearing the cry 'Buzz off' the seeker must return to the starting-point, hide his eyes, and count all over again to give those freed time to rehide, a rule that places considerable strain on his patience. As a 13-year-old girl in Aberdeen commented: 'I am not very fond of "Buzz Off", neither are the other people in my area. We think that it is unfair as the "man" [the seeker] may be the "man" for ten times at least.' In a few places the seeker is given the marginal concession that each time he has to count again he counts twenty-five less than he did the time before; and in New Cumnock he only counts to twenty the third time, and announces that this time 'Last in's het', that is to say whoever reaches the den last will be seeker next time, no matter what happens.

Names: 'Buzz Off' (Aberdeen, Forfar, Inverness, New Cumnock, St. Andrews, Spennymoor, York, and Knighton), 'Twenty Buzz Off' (Kirkcaldy), 'Bazooka' (Vale and Castel, Guernsey), 'Rescue' (St. Helier, Jersey).

COME TO COVENTRY

Here the seeker is at almost greater disadvantage than in 'Buzz Off', for a hider can rescue a prisoner merely by getting in sight of the den. However, the seeker has only to see and recognize a hider to make him a captive. He shouts 'Back to camp' or 'Come to Coventry' and the player's name, and the person must come out of his hiding-place and wait at the den or starting-point. The hiders can release the prisoner by creeping close enough to be seen and waving a hand. When a person at the den sees someone waving he is free to sneak away, provided, of course, that the seeker does not see him and call him back. It is important in this game that a good site is chosen for the den: if it is too exposed the seeker's task becomes very difficult; if it is too hidden-away it may be too easy. On the whole, as an 11-year-old remarked, 'It is tough for the person who is "on it", he or she does a lot of running about'. And there is a further point that is important, as another child commented: 'In this game you must not cheat, and go and hide without somebody waving to you.'

Names: 'Back to Camp' (Edinburgh), 'Coventry' (Ipswich), 'Come to

Coventry' (Bristol), 'Go to Coventry' (Bury St. Edmunds), 'Flashie' (Forfar, waving torches), 'One, Two, Three, Hide-and-Seek' (Hounslow).
'Wavy Wavy' (Colwyn Bay), 'Wave Hiding' (Triangle near Halifax, *c*,
1930).

** In the United States 'Beckon' and 'Sheep in My Pen' (Brewster,
1953, p. 42).

WHIP

Girls are the chief players of this curious game, which commences like the
previous games with the players hiding while the seeker counts to an
agreed number at the home or den. But thereafter the game embraces
elements of other games, for when the seeker catches sight of someone,
whether in hiding or attempting to reach home (merely to see the person
and recognize her is enough), she names the person, and calls 'Stop',
'Whip', or 'Whipit'. The player must immediately stop wherever she is,
and wait there until everyone else has either reached home or been similarly stopped. The seeker then estimates the minimum number of special
steps, such as 'fairy feet' or 'giant strides', that she thinks each person will
need to reach the starting-point ('The seeker is always amazingly fair in
this', comments one observer), and if the person can touch the home in
the prescribed number of steps she is free; while the first person to fail to
reach the home becomes the next seeker. The attraction of this game,
which is distinctly slow-moving, is not immediately apparent, but seems
to lie in its deliberateness and precision, in the variety of stages, and in the
opportunity it gives of watching other players. 'It is great fun to play, we
play it nearly every night', a 10-year-old assured us.

Names: 'Whip' (Accrington, Bramford near Ipswich, and Oxford),
'Whipit' (Scarborough), 'Wave Me' or 'Whip' (Doncaster).

** Alfred Easther, *Almondbury and Huddersfield Glossary*, 1883, giving
'Whip' as the local name for 'Hoop' or 'Hoop Hide' suggests it comes from
the local pronunciation, *hooip*.

TIN CAN TOMMY

'Tin Can Tommy' is probably the game that most commonly disturbs the
evening repose of the back streets. Its chief requirement, other than
energy, is a good-sized tin can; and sometimes the children's way of
determining who shall be the first seeker is to have everyone find an empty
tin (there is a scramble through the dustbins), and whoever is last back
with one acquires the uncoveted role. The best tin is then chosen, placed

in a chalk circle in the road, or on a manhole cover, and one player, perhaps the strongest, kicks it or throws it as far as he can down the street.
This is the signal for everyone to run off and hide, while the seeker (also
known as the 'canner', 'denner', 'den-keeper', 'hound', or, in Plymouth,
the 'slave'), has to retrieve the tin, walk backwards as he returns, replace it
in the circle, and has in addition, sometimes, to walk round the tin ten
times, or count to 300 'in fives', before he commences his search. When the
seeker sees someone he has to race back to the tin, place his foot on it or
rattle it on the ground, and shout 'Tin Can Tommy, one, two, three', and
call out the name of the person he has seen. This player is then obliged to
come out of his hiding-place and stand by the tin; while the other players,
hearing the cry, know that someone has been caught.[1] When the seeker
continues his search, any hider who thinks he will be unobserved can rush
to the tin and free whoever is standing there by kicking the tin out of the
circle. The seeker has no power either to hold people captive or to put
them out of the game when the tin is not in position. Thus the seeker has
two conflicting tasks, for he has to leave the tin to find further hiders, yet
repeatedly race back to see that his captives are not being released. Further,
should he make a mistake when he sees someone, and call out a wrong
name (and sometimes players will change coats or jerseys and deliberately
show their backview or just an arm to mislead him), there is, as in 'Block',
a jubilant cry of 'False Alarm' (or 'Blin' spy' in Langholm, 'Sly Fox' in
Stoke-on-Trent, 'Double D, double D-motion' in Bristol), and 'everybody comes out of their hiding-places, and the person who was "on it" has
to be on again'. In New Cumnock they jeer:

> Hard up, kick the can,
> *Archie Gibson*'s goat a man
> If ye want tae ken his name
> His name is *Ian Scott*.

Indeed the seeker's role is not an enviable one, and as the game progresses,
and he has perhaps acquired a number of captives, he becomes increasingly
unwilling to move far from the tin, while the impatience of the captives and
those still in hiding grows in proportion. In Scotland the captives taunt
'Go oot, go oot, ye lazy hen', urging him to give them a chance to be rescued:

> Leave the den, ye dirty hen
> An' look for a' yer chickens.

[1] The seeker's cry and actions vary with local custom. In Langholm he places his foot on the
tin, screaming 'Bob-e-tee-bob!' In Wolverhampton he raps the tin on the ground, calling out
'Tin can nerky, one, two, three'. In Glastonbury he 'daps' the tin three times, crying 'Dap, dap,
dap'. In Dundee he 'dunts' it ten times, but always counting in threes, 'One-two-three, one-
two-three, one-two-three, one'.

Plate III. TIN CAN TOMMY

If there are many players (and the more players the longer the game), the seeker's task is nearly impossible, and they sometimes make it a rule that if the captives have been released three times the next person 'caught' shall become seeker, and the game start again. Yet juvenile enthusiasm for the game is unflagging. Even a 14-year-old boy said, 'I spend hours playing this game. I love it, and so do all my pals'. There are, nevertheless, clearly two opinions about its virtues. 'This is a game which the people where I live don't like us playing very much', admitted a 13-year-old girl. 'They say it is too noisy, and the mothers say it wears out our shoes.' 'The truth is, it is a perfectly evil game guaranteed to put me in a bad temper', commented a headmistress. And a 15-year-old, attempting to defend it, remarked innocently, 'The only inconvenience it causes is that when the tin is being kicked about it has a tendency to wake up the neighbours' babies'.

Names: 'Tin Can Tommy', which is the basic name in London, is widely distributed, e.g. Glastonbury, Ipswich, Wolstanton. 'Kick the Can', the usual name in Scotland and the Isles, is also not uncommon in Dublin, Liverpool, Manchester, and much of Wales. Other names: 'Bobby, Kick the Tin' (Swansea), 'Can Can' (Tetchill), 'I-erky Kick the Can' (Lydney), 'I-o-kay' (Welshpool), 'Kick Can Copper' (Camberwell), 'Kick Out Can' (Featherstone near Pontefract), 'Kick the Bucket' (Hexham and Plymouth), 'Kick the Cog' (Spennymoor),[1] 'Kick the Tin' (occasional but widespread, e.g. Bishop Auckland, Rhondda, and Guernsey), 'Kicky-Off-Choff-Choff' (Spennymoor, alternative name), 'Kit Can and Hop It' (occasional, Wigan), 'Kit the Can' (Crickhowell, Breconshire, and Meifod, Montgomeryshire), 'Maggie, Kick the Can' (occasional, Spennymoor), 'Om Pom Rattle Tin' (Liss), 'Pom Pom' (Berry Hill, Gloucestershire), 'Rin Tin Tin' (usual name Norwich, not uncommon Ipswich, Bristol, Swansea), 'Tap the Tin' (Glastonbury), 'Throw Out Can' (usual name, Wigan), 'Tick Tock Tony' (Stornoway, Isle of Lewis, alternative name to the usual 'Kick the Can'), 'Tin-a-Lerky' (Annesley, Nottinghamshire), 'Tin Can Alley' (Croydon and St. Ives, Cornwall), 'Tin Can Annie' (Knighton), 'Tin Can Bosher' (Laverstock), 'Tin Can Copper' (Clapham and Millwall), 'Tin Can Leaky' (Caistor and Lincoln), 'Tin Can Lizzie' (Four Crosses, Montgomeryshire), 'Tin Can Lurky' (Leicester and Windermere), 'Tin Can Nurky' (Barrow-in-Furness and Wolverhampton), 'Tin Can Squash' (Holmfirth), 'Tin Can Topper' (Stoke

---

[1] *Cog* is an old northern term for a hollow, wooden vessel for holding milk or other liquid. Thus Aphra Behn, *Widow Ranter*, 1690, I. i, 'Come, Jack, I'll give thee a cogue of brandy for old acquaintance' (*O.E.D.*).

Newington), 'Tin Can Whippet' (Stoke-on-Trent), 'Tin in t' Ring' (Burnley and Rossendale), 'Tin Leaky' (Lincoln), 'Tin Pot Monkey' (St. Helier, Jersey), 'Tin Tam Tommy' or 'Tin Tan Tommy' (common Fulham, West Ham, and Camberwell, also Devon and Cornwall), 'Tin Ton Talley' (Henstridge, Somerset), 'Tin Tong Tommy' (Alton), 'Tin Whip' (Workington) 'Tinny' (Accrington, Bacup, and Colne), 'Tip the Copper' (Gower Peninsula), 'Whip the Can' (Liverpool).

'You cannot really play this game at school because it is awkward to get a tin' (Girl, 10), but in some places they regularly use a stick, stone, half-brick, or ball, and the game is known as 'Ball Out' (Kingerby, Lincolnshire), 'Chuck the Stick' (Langholm), 'Kick Ball Fly' (Cleethorpes and Grimsby), 'Kick Ball Kick' (Scarborough), and 'Kick Ball Lurky' (Sleaford).

** The game seems to have been well known in city streets before the First World War. In districts of London such as Canning Town it was called 'Kick Can Bobby', 'Kick Can Copper', and 'Kick Can Policeman'; in the Clifton district of Swinton 'Kick Can', and in the North Country and across the Border it was 'Kick the Block' (played with stone or block of wood). Other names: 'Ecky' (*Warwickshire Word-Book*, 1896), 'Fly Whip' (Gomme, vol. ii, 1898, p. 438), 'Foot in the Bucket' (Belfast, *c.* 1905), 'I-er-kee' (Lydney, *c.* 1870), 'I Spy, Tin Can' (*Folk-Lore*, vol. xvii, 1906, p. 97, Argyllshire), 'Kick the Bucket' (? this game, *Suffolk Words*, 1823, p. 238), 'Kick the Bucket' or 'Kick the Tinnie' (listed *Notes and Queries*, 5 June 1909, Stromness), 'Kicky Tin Spy-Ho' (Easingwold, *c.* 1930), 'Kickstone' (*Journal of Lakeland Dialect Society*, 1945, p. 7, Cross Fell, *c.* 1885), 'Lerky' (Nottinghamshire, before 1898, *EDD*), 'Mount the Tin' (Gomme, vol. i, 1894, p. 401, Beddgelert, Caernarvonshire), 'New Squat' (Gomme, vol. i, 1894, pp. 412–13, Earls Heaton, Yorkshire), 'Nurky' (Windermere, *c.* 1900), 'Old Tin Can' (Wrecclesham, Surrey, *c.* 1895), 'Releaser' ('played with a block of wood, a ball, or an empty tin', *More Organised Games*, *c.* 1905, p. 129), 'Squat' (Leeds and Midgley near Halifax, *c.* 1895, with stone), 'Tin Can Squat' (J. B. Priestley, Bradford, *c.* 1905).

In Canada and the United States generally 'Kick the Can' or 'Kick the Tin' (Newell, 1883, p. 160, describes it played in New York with a stick, and called 'Yards Off'; Brewster, 1953, 'Throw the Wicket' in Illinois). In Australia in 1930s, 'Kick the Block' and 'Kick the Tin'; 'I-Acky' in Sydney, *c.* 1890. In *Games of New Zealand Children*, 1959, p. 58, 'Kick the Tin', 'Kick the Boot', 'Kick the Block', and 'Homaiacky'. In Italy it is 'Barattolo'. In France it is 'La boîte', played with a ball. In Berlin it is

'Stäbchenversteck', played with two sticks placed together, and 'Ball-versteck' or 'Russisch Versteck', played with a ball (Peesch, *Das Berliner Kinderspiel*, 1957, p. 41). In Antwerp, *c.* 1900, it was 'Buske Stamp' (*Radio Times*, 21 March 1958, p. 54).

§ *Equal Number of Players Hiding and Seeking*

OUTS

When hide-and-seek is played between two teams, with equal numbers hiding and seeking, it is generally known in the London area as 'Outs', 'Outings', or 'Runouts', the 'outs' or hiders having no home, but keeping hid or moving around until caught. A feature of this game, usually played at night, is the indisputableness with which a person has to be caught. Thus at Blaenavon in Monmouthshire a player who has been caught has to be held while the catcher counts ten. In Aberdeen the catcher has to jump over his captive's back. In Welshpool, where the game is generally known as 'Buckum' (alternative name 'Find Them and Catch Them'), 'when you tick them you have to say "Buckum" '. At Dovenby in Cum-berland a person is not considered caught until he has been patted on the head three times, as also at Enfield ('You have him three times on the head'), although at nearby Ponders End the captive submits only when his captor has been able to 'pull his hand off his hair'. (Was he, originally, protecting his head from being tapped?) One reason for emphasizing the capture may be that those caught are out of the game, and have to wait until the rest have been caught and a new game can begin; or because, as at Enfield, they are made to change sides and join the seekers. Indeed, at Ponders End, when a hand has been separated from a head, the captive calls out 'Ripe Bananas', to warn those still in hiding that they should change their hiding-places, for he is now obliged to tell where he last saw them.

Other names: 'Chasey' (Dovenby), 'Night Chase' (Grimsby), 'Ripe Bananas' (Ponders End), 'Scouting' (Golspie), 'Spotlight on Sally' (Cwm-bran, the seekers being armed with torches), 'Topsy Turvey' (Aberdeen). At Mousehole, near Penzance, the game is known as 'Coosing' ('One side hides and the other side tries to coose them out'); in Bristol 'Bunk and Chase' ('We tossed to see which team bunked first and which chased'). Also 'Runouts on Bikes' (Herne Hill) and 'Cycle' (Grimsby, 'When you have them you must put your hand on them or the cycle').

⁎⁎ The two sides in a seeking game were referred to as 'Ins' and 'Outs' by Jamieson in his description of 'Hy Spy' (*Scottish Dictionary*, *Supple-*

ment, 1825). Compare *Folk-Lore Journal*, vol. v, 1887, p. 60, 'Buckey-How' (Cornwall); *Notes and Queries*, 11th ser., vol. i, 1910, p. 483, 'Inners and Outers'; *London Street Games*, 1916, p. 20, 'Inner and Outer'.

KISS CHASE

Sometimes the game of 'Outs' becomes a contest between the sexes. 'We play boys versus girls', says a 10-year-old girl in Camberwell. 'The boys get the longest outs because they hide on roofs where the girls can't get. The girls won't go very far away because they are frightened of the dark.' In Pontypool, where the game is also known as 'Outs', those caught 'are either kissed or head and tailed'. But ordinarily, if the game is to involve kissing, the name gives warning: it is 'Kiss Chase' or 'Kiss Catch', or in Norwich 'Kiss Cats' (according to every child asked), in Swansea 'Kiss Touch', in Langholm 'Catchie Kissie', in Liverpool 'Catch the Girl, Kiss the Girl', in Langham, Rutland, 'Hide-and-Seek Kiss', in Croydon 'K.C.'. They play when school is over, sometimes having a special place for the game, 'in the woods', in a park, or where there is shrubland, somewhere 'away from the watchful eyes of parents'. It is played in the street only after dark. 'We go round the backs very late at night and ask the girls if they want to play', reported a 15-year-old. 'First you give the girls a chance to get out of sight then the boys try to catch them, and when you catch one girl you kiss her as a reward.' In Swansea when a boy has caught and kissed a girl he lets her go and chases again, and the boys see how many kisses they can get. More often when the boys have caught the girls, the girls go after the boys. Sometimes there is more hiding in the game than running, sometimes it is mostly chasing about. There are also degrees in their acquiescence to the 'reward' or penalty for being caught. A 10-year-old girl declared that the boys do not always have their way: 'We struggle and whack . . . and run away like wild horses.' But an older girl said: 'According to who's chasing you, sometimes you run fast, sometimes you hardly run at all.' Some boys speak of 'having to kiss' the girl they have caught as if it was a dull duty in an otherwise enjoyable game. In general, however, from 12 upwards, both sexes show a certain willingness for this part of the game; and it is certainly much played, sometimes in sophisticated forms. In Monmouthshire, for instance, the game is sometimes 'Kiss, Hug, or Both' ('We usually say "Both" '), and on the Firth of Clyde, where it is called 'C.C.K.' (Catch, Cuddle, Kiss), a 12-year-old girl records:

'If a girl is caught by a boy, then she has to leave the game with that boy and kiss him. Therefore that couple cannot resume to play the game that day. Then the game goes on until everybody is caught and they all start kissing the girls.'

'Different letters', remarks a 15-year-old boy 'can be added to the C.C. to describe different variations of the game.'

In some parts of England (chiefly, it seems, in the west and north), a further and less-pleasing choice is offered, and the game becomes 'Kiss Torture', 'Kiss, Cuddle, or Torture', or 'Kiss, Kick, or Torture'. 'Some girls who are tough have kicks and torchers', writes a 12-year-old girl, 'but the older girls seem to enjoy being kissed. It is quite a nice game if you like that sort of thing.' And occasionally, but not often, the gentler alternatives disappear; the game is 'Tiggy Torture' or 'Kick, Prick, or Torture'. 'If the person wants a kick you kick (with the knee) in accordance with their age, i.e. 13 years, 13 kicks. If she wants a prick you prick her in the hand with a pin. If she wants torture you twist her hand or something like that. When all the team have been caught three times their team is on it' (Girl, 13).

** One ancestor of 'Kiss Chase', and indeed of 'C.C.K.', is the game of 'Stacks' recorded by Gomme (vol. ii, 1898, pp. 211–12) as formerly played in farm rickyards after harvest time in Lanarkshire. At the end of the game each lad tried to catch 'the lass he liked best, and some lads, for the fun of the thing, would try and get a particular girl first, her wishes and will not being considered in the matter; and it seemed to be an un-written law among them for the lass to "gang wi' the lad that catched her first" '. Indeed in the eighteenth century and earlier, as is well known, the chief pleasure if not purpose of certain roisterous games was the opportunities they afforded adult society for promiscuous, and even selective kissing. The author of *Round about our Coal-Fire*, 1731, for instance, directed that in the game of 'Hoop and Hide':

'The Parties have the Liberty of hiding where they will, in any Part of the House; and if it should prove to be in Bed, and if they even then happen to be caught, the Dispute ends in Kissing, &c.'

HUNTS

This is straightforward hide-and-seek between two teams, as in 'Outs', except that the aim of those in hiding is to get back to the starting-place without being caught. Should they succeed in this, or should an agreed number of the hiding side, say three players, be successful in this, the side has the privilege of going out to hide again. The game, which is now little played compared with 'Relievo' hereafter, is known as 'Chasing' (Knighton), 'Hunts' (Helensburgh), and 'Yelly Yelly' (Bacup).

⁎⁎ In *The Boy's Treasury*, 1844, p. 64, the game is called 'I Spy I', the seekers being declared the winners if they catch a specified number of hiders; in *Games and Sports for Young Boys*, 1859, pp. 2–3, it is 'High Barbaree'; in Gomme, 1898, 'Save All'.

HUNT THE KEG

This game, which has been reported in the present day only from St. Andrews ('Hunt the Keg') and Golspie ('Smooglie Gigglie'), is played in much the same way as 'Hunts', with the hiding side creeping around trying to get back to the starting-place without being caught; but one of the hiders carries 'the keg', and it is on his success in making his way back to the starting-place that the course of the game depends. The keg is a stone or penknife or anything easily held in the hand. The side that hides are smugglers; the side that goes after them are coastguards, and the coast-guards do not know which smuggler has the keg. If a smuggler is caught he is ordered to 'Deliver the keg'. If he has not got it he is merely taken into custody at the 'coastguard station' (some hut or den); but if he has the keg he has to hand it over, and the coastguards become the smugglers. The smugglers, however, may muster what ingenuity they possess to outwit the coastguards. Some of them as they approach home may let themselves be seen, and deliberately lure the coastguards after them, and perhaps even allow themselves to be caught, to distract attention from the player with the keg. It is of no account how many smugglers are taken prisoner, provided that the player with the keg is not amongst them. Even if the keg-bearer is the only player to reach home, his side has won and goes out again.

⁎⁎ 'Smuggle the Geg' (or Keg, Gag, Gage, Gig, or Giggie) seems to have been one of the usual round of boyish games in Scotland in the nine-teenth century. In *The Scottish Dictionary, Supplement*, 1825, Jamieson described boys in Glasgow playing it:

'The *outs* get the *gegg*, which is anything deposited, as a key, a penknife, &c. Having received this, they conceal themselves, and raise the cry, "Smugglers". On this they are pursued by the *ins*; and if the *gegg*, for the name is transferred to the person who holds the deposit, be taken, they exchange situations, the *outs* becoming *ins*, and the *ins*—*outs*.'

Other references: James Ogg, *Willie Waly*, 1873, p. 75 (Aberdeen); W. B. Nicholson, *Golspie*, 1897, p. 121; Gomme, *Traditional Games*, vol. ii, 1898, pp. 205–7; Maclagan, *Games of Argyleshire*, 1901, pp. 89–90;

English Dialect Dictionary, vol. v, 1904, p. 562; *Notes and Queries*, 10th ser., vol. xi, 1909, p. 445 (Stromness); *Buchan Observer*, 23 April 1929. In *Juvenile Games for the Four Seasons*, *c.* 1820, pp. 81–2, a game is referred to called 'The Wand' which appears to be identical. In Caputh and Dunkeld in the 1930s the game was called 'Gig or No Gig'.

The name 'Smuggle the Geg' is also sometimes given to a Scots indoor game played on the lines of 'Hunt the Slipper'.

RELIEVO

In Scotland, Wales, and the northern half of England, 'Relievo' is the principal seeking game with two sides. Leaders are chosen to pick the sides; a den, usually a portion of the pavement, is marked out with chalk, boundaries are agreed on (the greater the number of players the wider the boundaries); and the leaders toss to see which side has first 'outs'. The hiding side run off and scatter ('Favourite places where my pal and I hide are in tool sheds, dog kennels, up trees, behind dustbins, and lying in flower beds'); the seeking side wait until they hear a call, or until they have counted 'five-hunder in fives', and when they set off they leave one person in charge of the den. When a hider is caught he is taken to the den, and it becomes the object of his own side to try and release him. However, when a seeker finds someone he must catch him properly in the prescribed manner, for the hider need not submit until he has been ritually taken. In Annesley he has to be touched on the head or 'bobbed and tailed'. In York, similarly, 'the catchers catch the "off" by putting one hand on the off's head and one on his bottom'. In Spennymoor the hand must be kept on the head long enough to say 'Tally ho!' or 'Fliggy, one, two, three'. In Swansea he must be slapped on the back three times, as also in Penrith where the captor must shout 'One, two, three, rallio!'—'If the captive manages to tear himself away before this is said he is not caught'.[1] In Forfar, Helensburgh, and Rossendale, the captive has to be held for the count of seven before he can be marched to the den. At Langholm 'When you are caught you struggle your way out, but if they count ten on you, you are caught'. Similarly in Grimsby a person is 'not properly caught until "Two, four, six, eight, ten, ree-leave-i-o" is shouted'. In Wolstanton the hider will not submit until his captor has kept his grip on him while crying:

> Rallio, Rallio, one, two, three,
> You are the jolly man for me.

---

[1] Likewise in Texas, when children are playing 'Release', 'the capture is made by touching a player three times on the back' (Paul G. Brewster, *American Nonsinging Games*, 1953, p. 60).

And in Liverpool, where the leading of the prisoner to the den is some-thing of a triumphal ceremony, the prisoner is not allowed to try and break away, unless, that is, his captor has forgotten, in his excitement, to utter the formula 'Nockey-no-twist-no-breaks'.

The actions necessary for releasing a prisoner are less formal but not easier to perform, since the prisoners are guarded by the den-keeper, who has only to touch the one attempting the rescue to make him also a captive. However, the den-keeper is under certain restrictions. He may have to keep out of the den, or keep more than three yards from it, or keep one foot in it and one foot out. 'If he puts two feet in the den all those caught are free' (Liverpool). Commonly it is enough for the 'releaser' to shout 'Relievo' or 'Rallio' or perhaps 'Bish-bash' as he rushes through the den, although he may also have to touch the person on the head three times, and in Bishop Auckland he 'must spit in the bay [the den] and shout "Tally-ho" '. Very often he may manage to release a prisoner but get caught himself. It is up to the prisoners to keep awake while they are in the den, and sometimes, when there are several of them, they will form a chain from the den to make it easier to be rescued. At Accrington when the last hider is caught the prisoners chant 'Last man, last man', and the last man is allowed to struggle while he is being put in the den, and if he manages to pull one of his captors in with him, all the captives are free to run off and hide again.

Like most other seeking games 'Relievo' is 'best played at night in a dimly lit street where there are plenty of places to hide', and the game is mostly played by boys, or so the boys say, 'because it is a little rough for girls'. Nevertheless there are few games in which some girls will not join, and in several accounts we have been given, the girls make a side to play the boys.

General names: 'Leavo', 'Leavio', 'Rallio', 'Realio', 'Release', 'Releaso', 'Relievo', 'Relievio', 'Rileo', 'Tally-ho'. Local names: 'Bedlam' (New Cumnock), 'Bish Bash' (Knighton and Langholm), 'Chasies' (Forfar), 'Fliggie' (Spennymoor), 'Free Me' (Cumnock), 'Lallio' (Bootle), 'Le-oh' (Barrow-in-Furness), 'Li-vo' (Stromness), 'Mile a Minute' (Helensburgh), 'Offers and Catchers' (Kirkwall, Orkney), 'Relvo' (Scarborough), 'Rolio' (Tunstall), 'Rowlies' (Lossiemouth), 'Run, Sheepie, Run' (Cumnock), 'Sally-o' (Seaforth and Wrexham), 'Sides' (Penkhull), 'Skiely' (Wigan), 'Tiggin In and Tiggin Out' (Rossendale).

The game is sometimes given a dramatic form as 'Spitfires and Gliders' (York, 1961). In Rossendale, one of the many places where the game develops into 'Cops and Robbers', the den becomes jail, and the seekers dress up as policemen, wearing their caps back to front and their coats cape-wise. Occasionally the catching side is limited to only one or two players, and the game is then called 'Corner Tig' (Langholm), 'Tig and Release' (Knighton), 'Tiggin Out Hide-and-Seek' (Rossendale), 'Stag in Den' (Welshpool), 'Sticker in the Den' (Blaenavon), 'Bully Horn' (yet another game with this name, Ballingry, Fife), 'Daddy Grandshire' (the 'daddy' or seeker has a stick, Newbridge, Monmouthshire), and 'Sun, Moon, and Stars' ('Sun' chases 'stars' and only 'moon' can release them, Aberdeen). At Spennymoor, when played on bikes, it is called 'Bike Tally-ho'.

The game is scarcely known in London and the south, except in a play-ground or confined-space version, usually known as 'Release', in which the 'runners' or 'releasers' instead of hiding have a base like the seekers, and the emphasis is on the releasing—indeed the catchers may start with a volunteer prisoner. Other names: 'Hide and Seek Releaso' (Alton) and 'One, Two, Three, Post He' (Croydon).

** The manner of playing the game in the north of England seems to have altered little over the past hundred years. Alfred Easther, headmaster of King James's Grammar School, Almondbury, 1849–c. 1873, who knew the game as 'Stocks', described it fully, adding the detail that a boy's captor had to count ten while he held him (*Almondbury Glossary*, 1883). Clough Robinson, who knew the game as 'Bed-o!' and 'Bed-Stocks', said that the captor had to count 'Two, four, six, eight, ten', and spit over the captive's head (*Dialect of Leeds*, 1862). And S. O. Addy, born 1848, educated Sheffield Collegiate School, confirmed that in 'Bedlams' or 'Relievo', 'the tenter' who had charge of the den had always to 'stand with one foot in the den and the other on the road' (*Sheffield Glossary*, 1888). Other old names: 'Alla-Least' (Cwmavon, Glamorgan, c. 1905), 'Delievo' (Sheffield, 1888 and c. 1934), 'Inamon' (Forfar, c. 1910). With two chasers only: 'Shepherd in the Box' (Bearpark near Durham, c. 1925). London version: 'Box Release' (Wandsworth, c. 1912).

GEE

This game has been reported only from Meir, Stoke-on-Trent. Two sides are chosen, and having agreed on a starting-place, usually a lamp-post, the hiding side goes off and hides, to be followed in due course by the seekers. When one of the seekers sees one of the hiders he shouts 'Gee!' All the *seekers* then rush to the lamp-post and the hider who was seen chases them. If the seekers get back to the lamp-post without being caught, the hider becomes a captive. But if the hider manages to touch one or more of the seekers before they reach the lamp-post, he frees that number of hiders who have already been caught, and they run off and hide again.

⁎ 'Gee' is a survival of an old form of 'I Spy' (described in *Every Boy's Book*, 1856, p. 4) and of 'Spy-Ann' as Mactaggart knew it:

'A game of hide and seek, with this difference, that when those are found who are hid, the finder cries *spyann*; and if the one discovered can catch the discoverer, he has a ride upon his back to the *dools*.'

'*Gallovidian Encyclopedia*', 1824, p. 435

Other names for it have been 'Spy for Ridings' and 'I Spy the Devil's Eye' (*Notes and Queries*, 7th ser., vol. x, 1890, pp. 186, 331); and 'I Spy Charlie across the Sea' (*Folk-Lore*, vol. xvii, 1906, p. 98).

5

Hunting Games

Squeak, whistle, holly
Or the dogs shan't folly.
Traditional call, South Molton

In hunting games there are no boundaries (Boy Scouts aptly style them 'Wide Games'), and the feature of games without boundaries is that those being pursued must give some indication of the direction they have taken, which they do, according to the game being played, by shouting, by showing themselves, by showing a light, by leaving a trail, or by providing a guide. A secondary characteristic is that both pursuers and pursued operate as teams under leaders, and ordinarily move in a pack. In hunting games those who are pursued may, and often do, make use of cover; but the game is not over if they are seen, named, or even touched; the quarry has to be effectively captured, sometimes, as in Seeking Games, by ritual action on the body.

HARE AND HOUNDS

The general opinion is that the more players there are in 'Hare and Hounds' the better—'even twenty' says one enthusiast. Sometimes only one or two boys set off as the hares (or 'foxes'), and the rest count to 300 or 400, or 'fifty *very* slowly', to give them a decent start. At other times the players divide into two equal groups of hunters and hunted. 'Hare and Hounds' is essentially a country game, not infrequently played 'on the edge of the dark', and 'to have a good game you want an hour at least because you might run for miles'. 'We go all across the country, round the school, over the rivers, across bogs.' The hares must keep together, have a definite course in mind, and are generally supposed to circle back to the starting-place after a certain length of time. The hounds must catch up with them before they are home. If one of the chasers views a hare he shouts 'Tally-ho', to put the others on the right course. Otherwise the hounds rely on calls. Every now and again they cry out for guidance, their shrill voices sounding across valley and field:

Hollo, hollo, hollo,
The hounds can't follow.
                            Knighton and Lydney

Hollo, hollo, the dogs won't follow,
Whistle or shout or else come out.
                            Thirsk

Robinson Crusoe give us a call,
A peeweep whistle or nothing at all.
                            Luncarty

Faintly in the distance—or startlingly close sometimes if it is dusk—comes the answering 'peeweep' or 'hollo'. 'If the call is not answered,' explains a 10-year-old, 'the hounds are allowed to give up after calling three times.' Usually, however, 'the hounds don't come back until they've caught all the foxes'.

In Radnorshire 'when the hounds catch a fox they tap him on the head three times'. In Burton Salmon, near Pontefract, 'he must be patted three times on the back'. On the Isle of Lewis a player must be held while he is named and his captor says 'One, two, three, caught'. In other places 'you have to catch him in a way that he can't get away. This game is very rough so you have to watch that you don't get caught'. 'If the hounds catch the hares the hounds give them any punishment they like.' 'If we catch the fox,' says a 9-year-old at Dovenby, 'we pretend to eat him. If a boy sulks we just leave him and if he joins again we let him.' 'It is a very sporting game for boys,' added a small male participant, 'but a shade too rough for girls.' The girls deny this: 'We like playing it,' they say, 'because it is very exciting.'

Usual names: 'Hare and Hounds', 'Hounds and Hares', 'Hunts', 'Hollo', and in Wales, 'Fox and Hounds'. Other names: 'Fox Off' (traditional in Yorkshire), 'Robinson and Crusoe' (Luncarty and Stromness), 'Stag Hunting' (Helston), and 'Whistle or a Cry' (Eassie, Angus).

⁎ The names have had little cause to alter through the centuries. Randle Holme of Chester (born 1627) knew the game as 'Hare and Hound' or 'Hunting of the Hare', so did Edward Moor (born in Suffolk 1771), and William Howitt (born in Derbyshire 1792). George Mogridge (born near Birmingham 1787) knew it as 'Stag Chase'. Thomas Miller (born Gainsborough 1807) called it 'Stag Out'. John Clare (born 1793) refers to 'Fox-and-Hounds' (*Village Minstrel*, ii. 37); and Strutt states specifically in 1801 that 'Hunt the Hare' was the same pastime as 'Hunt the Fox' under a different denomination. Whether it is also the pursuit to which Hamlet refers, 'Hide fox, and all after' (IV. ii), is impossible to say (see p. 152), but

the game seems to have been played before Shakespeare's day. In *The Longer thou livest, the more Foole thou art*, published in 1569, Moros, the simpleton, boasts:

> Also, when we play and hunt the fox
> I out run all the boyes in the schoole.

It may be added that calls in the nineteenth century include:

> Hoot and holloa
> Or my dogs shall not follow!
> *George Mogridge, 'Sunny Seasons of Boyhood',*
> *1859, p. 106*

And the following, apparently from Somerset, the game being 'Pee-wip' or 'Pee-wit':

> Whoop, whoop, and hollow,
> Good dogs won't follow,
> Without the hare cries 'pee wit'.
> *J. O. Halliwell, 'Nursery Rhymes', 1844, p. 104*

Froissart records that in his childhood at Valenciennes, about 1345, he played 'au chace-lievre' ('L'Espinette amoureuse', l. 233).

JACK, JACK, SHINE A LIGHT

'Jack, Jack, Shine a Light' is one of the most played of after-dark games. In some places only one or two people are selected to go off with the light, while the others give them 'the most part of five minutes to get clear away'. In other places the players divide into two teams, and one team sets off with a number of lights. The lights are usually torches, but several children speak lovingly of candles fixed in jam jars, carried inside the jacket until the light has to be shown. Country children are sure the game is best played in the country 'where there is plenty of space and woods to run about in and hide'; town dwellers extol the pleasure of playing in a man-made maze:

'It is played on dark nights, down passages and on bombed buildings. There are lots of places you can hide, on shed roofs, in yards, in doorways, in trees, laying on the roofs of cars, inside lorries, laying on the ground in dark corners, and sitting on high walls.'   *Boy, 13, Grimsby*

Usually those who are hunted pick a place from which it is easy to get away, for if they hear a call they must show their light, and point it in the direction of the pursuers. In some places 'Jack' whistles or calls back instead of showing a light. At Allerton Bywater, near Pontefract, the searchers sing out:

> Jack, Jack, shine a light,
> Aren't you playing out tonight?

and the other team taunt their pursuers by chanting the same words back at them. In most places those who are caught join sides with the catchers. 'But if anyone proves tiresome by shouting out of turn or making some kind of noise to give the hiders away,' says a Stoke-on-Trent girl, 'he or she is tied up and left until the hiders are found.' Those not caught eventually make their way back to the den, and if no one is there, they shout out that they are 'Home!' 'There is generally a time limit of an hour, but,' says a 12-year-old ambitiously, 'it may last all night.'

Alternative names: 'Jack, shine your lamp' (Newcastle upon Tyne), 'Jack, shine your lantern' (Accrington and Tunstall), 'Jack, shine the maggie' (Spennymoor—the traditional name in County Durham), 'Jack, Jack, shiny eye' (Ponders End), 'Shine a licht, Jock' (Langholm),[1] 'Shine a light' (Grimsby), 'Shine the lamp' (Oxford), 'Shine, Jack, shine' (Pontypool), 'Flash the light, Dickie' (Caerleon), 'Dick, shine the torch' (Lydney), 'Dicky, Dicky, show your light' (Liss, and traditional in south-east Hampshire for fifty years), 'Nicky, Nicky, show a light' (Brockenhurst), 'Mickey Mike, show your light' (Brighton), 'Mickey-me-light, show your light' (the name current in Alton for the past sixty years), and 'Midnight Hide-and-Seek' (Sheffield).

** Boys have probably enlivened the darkness with a game such as this for as long as man has been able to carry a fire box. The game 'Chase Fire', listed by Randle Holme as one of the 'recreations and sports . . . used by our countrey Boys and Girls' (*Academie of Armory*, iii, 1688, xvi. 91), was probably this game, although possibly it was 'Will o' the Wisp' hereafter. 'Hunt the dark lanthorn' played at Eton in 1766, subsequently known at both Eton and Harrow as 'Jack-o'-Lantern', was certainly this game; the great point of the sport at both schools being to entice the pursuers into some pool or muddy ditch by showing the light exactly in a line on the other side (W. L. Collins, *Public Schools*, 1867, p. 312). A description of 'Jack! Jack! Show a light' occurs in *The Boy's Treasury of Sports*, 1844, pp. 63–4, the detail being added that 'the hiding party is provided with a flint and steel'. In 1870 when 'Dicky, Dicky, show a light' was described in *The Modern Playmate*, p. 10, the hider or hiders were armed with

---

[1] Thus, too, at the turn of the century: 'On nights when, as boys, we used to thread its dim streets playing "Jock, Shine the Light" . . . [Langholm] had an indubitable magic of its own.' —Hugh MacDiarmid, *The Listener*, 17 August 1967, p. 204.

'a policeman's dark lantern', and Dicky, when hard pressed, might effect his escape 'by turning sharp upon his pursuers and blazing his bull's-eye in their faces'. In *Cassell's Book of Sports and Pastimes*, 1881, p. 267, the game is recommended under the names 'Sam, Sam, show a light', and 'Nicky Night, show a light'.

WILL O' THE WISP

This game varies from 'Jack, Jack, Shine a Light' in that the hunters do not have to call for guidance, but are lured along by the 'Wills', who flash their torches in one place, and run quickly to another, endeavouring all the while to keep just out of reach of their pursuers. This is no easy business for the 'Wills', who need to be the best runners amongst the players and to know their territory well; in fact a single good runner on his own usually makes the most effective Will o' the Wisp. The game does not appear to be played often, and has been reported only from Alton and Billericay.

TRACKING

In country places tracking can take place only in the daytime; but in towns, where chalk arrows gleam under the street lights, tracking is generally a night game. The children divide into two equal gangs. The 'escapers' set off, equipped with pieces of chalk, and tend to be given a longish start because 'laying arrows' every ten or fifteen yards takes time and inevitably slows down their pace. 'In the country,' says a young informant, 'the arrows may be drawn on buildings, stones, trees, telegraph poles, fences, and any place where they show up. They may also be formed of three sticks or a number of stones in the shape of an arrow.' In Langholm they sometimes make an arrow by placing a small stone in front of a large one. In Manchester a refined method of laying a trail is to squeeze a bad orange, making a track of juice and pips. False trails are made by drawing the arrows back to front, or pointing them 'over walls or shelters', or getting one of the party to lay a trail down a side street. If he does this he puts a cross at the end of the trail X to show it is false. The real trail usually leads to a hiding-place. When the gang have chosen their hiding place they write HOME somewhere nearby, with an arrow pointing in its direction, or make an (F) sign, meaning 'find', or draw the sign ←↑→↓ to show they are hidden somewhere roundabout and the trackers must now search without further help. The hiders then crouch together waiting for the trackers to come upon them. As the trackers follow their trail they strike out the arrows, sometimes with a different coloured chalk, so that the trail will not cause confusion in the next game when the trackers become the hiders.

Names: 'Arrows', 'Arrow Chase', 'Follow the Arrows', 'Arrow Tracking', 'Find the lost Sheep' (Pontypool), 'Cat and Mouse' (Caerleon and High Green near Sheffield), 'Hares and Hounds', and, since snails leave trails, 'Chase the Snail' (Wilmslow). A variation, in which the hiding party lie in wait and hope to jump out on the trackers is known as 'Tracking and Ambush' (Bristol).

⁎ In Edwardian London tracking was known as 'Chalk Corners' or 'Chalk Chase'.

PAPER CHASE

Paper chases have generally been school-organized affairs. Even in their heyday, after being extolled in *Tom Brown's School Days* (1857), the amount of preparation they entailed: the collecting of enough newspapers, the tearing of them into innumerable small pieces, the packing of the soft piles of fragments into satchels or bags—fun though it was—was scarcely a procedure which could be undertaken spontaneously. Today, when the unfisting of satchelfuls of litter, however minutely fragmented, is generally looked upon with disfavour, the chasers are bequeathed the dreary sport, as a 10-year-old Welsh girl reports, of 'picking up the paper as they run so as not to litter the countryside'. The paper chase, it seems, will very shortly be a sport of the past. 'Anyhow there is too much litter lying about nowadays to make chases practicable', commented one cynic.

⁎ Dickens referred to 'paper chases' by name in *Household Words*, vol. xiii, 1856, p. 28; and they were popular enough at the time to be played on horseback by British officers on the plateau before Sebastopol. Sometimes, however, the chase was called 'Hare and Hounds' (e.g. in *Every Boy's Book*, 1856, pp. 10–12, referring to the 1830s), and it is occasionally so named to this day in books for the young.

STALKING

The attraction of 'Stalking' ('Gang Stalking', 'Shadowing', 'Indians') is that the hunters can become the hunted. The aim of the game is to follow the other gang and touch them unseen from behind. However, if the side which went out first becomes aware that it is being followed, it can turn aside into a gateway or clump of bushes, wait until the followers have gone past, and become the stalkers of those who suppose themselves to be stalking. Thus, in this game, unlike the majority of seeking and hunting games, those who follow rather than those who are followed are the ones who most strive to remain unseen.

Clearly the success of the game much depends on where it is played; and it is no coincidence that the time and place most often mentioned for the game is at night round 'the backs', that is to say, down the back alleys and through people's back gardens and back yards.

Other names: 'Polecat' (Street, Somerset), 'Wolves in the Dark' (Accrington).

HOIST THE GREEN FLAG

This curious game, in which the leader of the hiding party traces a cryptic map on the ground showing the whereabouts of the party he has just hidden, has been reported only from Scotland and overseas. Some twenty players gather together at a den where there is 'soft earth or maybe sand'. They choose two leaders who pick sides, and toss to see which side shall go off to hide. The leader of the hiding side takes his team to a place where they can all hide together, but does not necessarily lead them there by a direct route. He may set off in the opposite direction, deliberately lead them into a garden and out again, and unnecessarily climb over a wall. When he has his team well hidden he returns to the den, and draws a map of the route that was taken to reach the hiding-place. Overtly he appears as helpful as possible with the directions: he is under obligation to mark every turning that was taken, every gate that was entered, every wall that was climbed over. However, he uses only straight lines on his map, and the lines do not necessarily point in the direction he took; and, in Aberdeen, he further confuses his account with the word *chocolate*: 'We went through a chocolate garden, over a chocolate fence, round a chocolate corner, through a chocolate door, and flopped under the sky.' The seeking side study the map, decide where the hiding-place may be, and set off with the leader of the hiding side, who usually walks just behind them. This leader is allowed to communicate with his own side, shouting out news and instructions, but in a prearranged colour code, thus 'Green' may mean 'danger, stay quiet'; 'Blue'—'they've gone the wrong way'; 'Red'—'come out'. The crisis in the game occurs when the seeking side catches sight of the hiders, or when the seekers are about to catch sight of them, or when the two sides are far apart but the seekers are so much astray that they are as far from the den, or further, than are the hiders. In the first situation the leader of the seekers shouts to his side to run; in the other two the leader of the hiders gives the order to his side with some prearranged signal (as 'Red' or 'Rotten onions'), and they run for the den. The seekers must either try to catch the hiders or reach the den first, where (in Forfar) 'the captain of the side which is back first lifts something up above his head and the whole side call "Hoist the green flag" '; or, in most places, he—or whoever arrives first—rubs out the map. The side which manages to obliterate the map are acknowledged the winners, and have the right to go out and hide next time.

Names: 'Hoist the Green Flag' (Aberdeen and Forfar), 'Hoist the Sails' (Toronto), 'Flag' (Kinlochleven), 'Run, Sheep, Run' (Cumnock and Edmonton, Alberta, and commonly in the United States), 'Sheep Lie Low' (Glasgow), and 'Scout' (Isle of Lewis).

*** The game appears to be well known in the United States. J. H. Bancroft, *Games*, 1909, pp. 6–7 and 170–1, gives it as 'Run, Sheep, Run' in Minnesota, and 'Oyster Sale' (an obvious corruption of 'Hoist the Sail') in New York. W. L. McAtee, 'Indiana in the Nineties', *Midwest Folklore*, 1951, p. 245, describes it under the name 'Go, Sheep, Go'. And Brewster, *American Nonsinging Games*, 1953, pp. 40–1, who reports an almost identical game called 'Figs and Raisins' played in Greece, knew it as 'Run, Good Sheep, Run', in Indiana.

6

Racing Games

Now and again in the playground the urgent concern of the younger children is to organize races. Boys of nine and ten are to be seen striding about, being almost belligerent in their anxiety that nobody shall start running before they do; and it is clear to the observer that something more is astir than recreation, their pride in themselves is at stake. They arrange to run from one end of the playground to the other, or the length of the netball court and back, and the starter, who may be 'someone who is not a good runner', ordinarily says 'One, two, three, go', or 'On your marks, get set, go', or 'One to be ready, two to be steady, and three to be off'. But in Oxford children sometimes say 'One for the money, two for the prize, three to be ready, and four to be off', or 'Ready, steady, get your knives and forks ready, go', or, for some reason, 'Ready, steady, paddle, go'. In Newcastle upon Tyne they say, 'Ready, set, fire, go'. In Canonbie, 'Ready, steady, fire (or fire-engine), go'. In Birmingham, 'One to be ready, two to be steady, three to be perfect, four to be off'. And in Aberdeen, 'One to be ready, two to be steady, three to be balanced, and four to go'. Sometimes the starters cannot resist making themselves unpopular by saying, not 'Ready, steady, go', but 'Ready, steady, grapefruit', or 'Ready steady, galoshes', and everyone groans and has to start again. In Helensburgh they use the traditional rhyming start 'Scotch horses, Scotch horses', and in Swansea 'Bell horses, bell horses':

> Bell horses, bell horses, what time of day?
> One o'clock, two o'clock, three and away.[1]

The winner of a race, says a north London boy, will exult: 'Yah! beechyer', 'Lardy', 'Excelsior', 'Good ole me', 'Lapped yer', 'I'm easy best'. But he (or she) who has not done so well may console himself with a little chant:

[1] See *Oxford Dictionary of Nursery Rhymes*, 1951, pp. 69–70, where the rhyme is shown to have been current in George III's day. Cf. also J. O. Halliwell, *Nursery Rhymes*, 1844, p. 124: 'The following is used by schoolboys, when two are starting to run a race. One to make ready, And two to prepare; God bless the rider, And away goes the mare.'

First is fussiest,	First for fuzzy,	First is the worst,
Second is muckiest,	Second for ugly,	Second is the next
Third is luckiest.	Third for lucky.	Last is the luckiest.
Oxford and elsewhere	*Aberystwyth*	*Lydney*

In Lydney they have, too, the happy saying, 'Last gets luck, finds a shilling in the brook'.

TYPES OF RACES

It is our impression that children do not really enjoy competitive athletics. The only running-race that comes to them naturally is the one that follows the challenge 'Last one there is a sissy!' (or, as Samuel Rowlands reported in 1600, 'Beshrow him that's last at yonder stile'). The races they have when they are on their own are noticeably ones in which their respective running abilities are not too finely matched. They like races in which ordinary running is made impossible by the nature of the course; races in which they crawl beneath parked cars, edge their way along parapets, and clamber over the roofs of garages. One group of 10-year-olds were not much exaggerating when they named their course 'Devil's Death Ride'. They hold races in which other skills than running are required, such as 'Pat-Ball Races' and 'Spitting Races':

'You spit as far as you can, run to where your spit lands and spit again, continuing like this to the winning post.'

And they have races in which the competitors have to adopt laborious methods of progression, as hopping, crawling, rolling, skipping, running backwards, running sideways, or blindfold, or with both legs tied together, or one leg tied to a partner (the 'Three-legged Race'), or running or jumping with feet inside a sack (in the eighteenth century sack-races were popular at fairs), or on stilts (quite a number of children have stilts), or with tin cans tied under their feet (called 'Whip Tin Can' in Accrington), or walking on their hands while someone holds their feet ('Wheelbarrow Race'), or riding on people's backs ('Horse Race', 'Piggyback Race'), or being carried—at no great speed—by a combination of two people, or three, in a 'Camel Race' or 'Chariot Race' (see p. 218).

There are also the races which children markedly prefer, which they engage in when they have means of moving faster than their feet will carry them, races on roller-skates, ice-skates, bicycles, or home-made trolleys.

And there are the race-games in which progress is so severely controlled, as in 'May I?', or in which the start is so elaborate, as in 'Drop Handkerchief', or in which the restarts are so numerous, as in 'Peep Behind the

Curtain', that they scarcely appear to be races, although fundamentally this is what they are, as will be seen by the descriptions that follow. Likewise chases are included in this chapter where the chase can equally well be a race: the chase being to a fixed point or over a fixed course, as in a steeplechase, or the chaser having to follow the precise course taken by the one he chases. Indeed, in these games, race and chase are interchangeable, and children will chase each other in one version of the game and race each other in another.

It is to be noticed, however, that despite assiduous cultivation in schooltime, relay and team races do not thrive when children are playing on their own.

HESITATION STARTS

There is one style of race, or manner of starting a race, much adopted in the backstreets, in which the competitors are likely to hesitate before they run. Their hesitation is only momentary, but where space is minimal, and the race is merely across the road, each fraction of a second counts.

The most common of these races is 'Odds and Evens', in which the starter, who is on the far side of the road, calls out either 'Odd' or 'Even', and a number. If description and number agree, for example 'Odd seven', the players run across the road and back again, and the last one back is out. But should he call 'Even seven', no one must move, and anyone doing so is out. Similarly in Newcastle the game is 'Shops'. If the starter says 'I went to the butcher to buy some bread' nobody must move. 'You run only if you can buy the thing at the shop and the last one back to the wall is out.' Likewise in Millwall in 1952 there was a game known as 'King George'. The players stood ready to run with one foot in the gutter and one on the pavement, and the caller took up his position at the far wall and would say, perhaps, 'King George wears medals!' This being true the crowd would rush across the road, touch the wall, and race back. But if the caller said 'King George wears petticoats!' anyone who believed this and ran, or anyone, more likely, who began to run because he was over-eager, had run himself out of the game. In Scotland, where a game with the unlikely sounding name of 'Eatables and Drinkables' is nevertheless popular, one pavement is designated 'eatables' and the other 'drinkables'. The player who is starter, or 'mannie', stands in the middle of the road, and the players line up on one of the pavements ready to run:

'If the players are all at the Eatables side the mannie has to shout out something to drink and they all run over and the first one to say 'over' is mannie.

But just to confuse the people he can shout something to eat, like bread or biscuits, and if you move you are out.'

An 11-year-old lad in Orkney points out, however, that the person in the middle is not to be permitted to call out 'Water (pause) biscuits'.

Other names: 'Fruit and Vegetables' (Spennymoor) and 'Juicy Juicy' (Cumnock).

§ *Races in which the Progress of Those Taking Part is Dependent on Their Fulfilling a Condition or Possessing a Particular Qualification*

MAY I?

Some children say that this game is, as it certainly appears to be, a silly game. Nevertheless vast numbers of bright children (girls rather than boys) are amused by it, and it is played throughout Britain. One child stands on one side of the road, and the rest line up facing her on the other side. The object of the game is to be first across the road and able to touch the person in front. However, the players can move forward only one at a time according to the instructions they receive individually from the one in front, who employs a terminology more or less peculiar to this game, thus: 'Jean, take one giant and three babies', 'Pauline, do a lamp-post', 'Phil, five pigeon steps'. The player addressed asks 'May I?' and advances as instructed, then waits where she is until her turn comes again. However, should a player forget to ask 'May I?' and advance before receiving permission, she has to go back to the starting-line. The names of the movements which, say the young, 'do not need describing, for we know just what to do', are as follows, and are general throughout Britain unless a particular locality is specified.

Baby Step. A small heel-to-toe step, also known as a 'dolly step' or a 'fairy step'.

Bag o' Tatties. A jump landing heavily on the ground. Langholm.

Banana Slip. A slide forward with one foot as far as possible and the other foot drawn up after it.

Barrel. A spin-around, moving forward at the same time. Cf. 'umbrella'.

Black Pudding. An order to go back to the beginning again, that is fairly widely understood, e.g. at Brightlingsea, Grimsby, and Hounslow; but in some places the equivalent order is 'rotten egg'.

Bob Jump. A big jump from a crouching position. Also known as a 'frog jump' and, in Alton, as a 'chair'.

Box of Chocolates. Five jumps. A 'big box' is ten jumps. Aberdeen.

Bucket. Player steps through own linked hands. Also known as a 'Jackdaw'.

Bunny Rabbit. A hop with both feet together.

Cabbage. A step forward taken in a crouching position, with arms folded round body 'like a cabbage'. At Ellesmere in Shropshire a 'Cauliflower' is similar but with hands placed on the head.

Caterpillar. A movement forward lying face down on the ground, drawing feet up beneath the body and pushing forward. 'You graze your knees horribly doing it.' Petersfield.

Crocodile. 'Person lies flat on the road with his feet touching the kerb and his or her hands out as far as they can go.' Person stands up at the point reached. Aberdeen. Cf. 'Lamp-post'.

Cup and Saucer. One 'bob jump' forward, and one jump with legs apart.

Cushion. A jump while sitting on haunches, hands behind back. Aberdeen.

Dolly Tub. The twirl-around step more often known as an 'umbrella'. In the north children remain familiar with the cross-headed stick turned in a wash-tub, and this step is called a 'dolly tub' as far south as Chester.

Ghost Walk. Progression sideways with feet together, pivoting alternately on heels and toes. Sussex.

Giant Walk or *Giant Stride*. As large a step as possible.

Knock. Player is pushed, and stands where he lands. Aberdeen.

Lamp-post. Player stretches out on ground, and then stands on the spot reached by his finger-tips. In Peterborough the player steps forward with arms upright and body as stiff as possible.

London and Back. Player runs to the person in front, then back to the starting-place, and forward again until told to stop. London suburbs. Known elsewhere as 'train' (q.v.), 'running bucket' (Langholm), 'running river' (Berry Hill), 'trip to fairyland' (Workington). The player in front may have his back turned so that he does not know where the runner is when he says stop.

Long Needle. Running until told to stop. Market Rasen.

Lotus Walk. Walking on knees. Wispers School, near Chichester.

Needle. A heel-to-toe step. North-west England.

Newspaper. Three steps forward. Doncaster.

Pigeon Step. A step the length of the foot, or, in some places, the width of a foot.

Pin. A step the width of a foot. North-west England. In Guernsey a step on the points of the toes.

Poker. Occasional alternative to 'lamp-post'. In Aberdeen a jump while standing stiff.

Policeman's Walk. Both the caller and the walker shut eyes. The walker steps forward until the caller shouts 'stop'.

Posh Lady. A high step, holding head in air. Edinburgh.

Postage Stamp. A number of ordinary steps according to the value of the stamp quoted. Kirkcaldy.

Scissors. A jump forward with feet apart, and a further jump landing with feet together. At Finedon, Northamptonshire, however, and doubtless elsewhere, this is known as 'Open and Shut Bible'.

Soldier, *Chocolate Soldier*, or *Wooden Soldier*. A step forward with the body held stiff in supposedly military fashion.

Spit in the Bottle. Player spits as far as he can and moves forward to where the spit lands. Liverpool. In Workington 'Spitfire'. In Ellesmere 'Cuckoo Spit'. More often called 'Watering-can' (q.v.).

Squashed Tomato. Both caller and called run towards each other, with arms crossed in front of them. The one advancing remains at the spot where they squash into each other. The caller returns to his place in front.

Tablecloth. Two large steps forward. Barrow-in-Furness.

Train. Player runs round the person in front, back to the start, forward to the person in front, and so on until told to stop. This is sometimes called 'fire engine'. Alternatively he may be told to do 'slow train' (the same course but walking), 'blind train' (walking with eyes shut), or he may have eyes shut and be led, which is 'ghost train'. See also 'London and back'.

Trip round the Moon. Player shuts eyes and is taken to wherever person in front wishes. Norwich.

Umbrella. A twirl-around on one foot with arms extended, a step forward being taken with the disengaged foot as the turn is completed.

Waterbottle. A 'hot waterbottle' is running until told to stop. A 'cold waterbottle' is walking until told to stop, the person in front having his eyes shut. Edinburgh.

Watering-can. Player spits as far as possible and stands where the spit lands. (Spitting races are sometimes called 'watering-can races'.)

Wheelbarrow. One player whose feet are held off the ground, as for a wheelbarrow race, reaches forward as far as he can, and both players go to the spot touched.

Since the movements of each player are under the control of the person in front, the game gives, or appears to give, little scope for initiative, and much scope for favouritism; but this seems to trouble adult observers more than it does the children. In some places the person in front (the 'leader', 'caller', 'outer', or 'on it') has her back turned to the rest, and the racers are allotted numbers, or the names of the days of the week, or of the months of the year, so that the leader does not know whom she is instructing, but this is not usual. The fact is that the players do not care greatly about who wins, but like watching their companions having to do absurd movements, such as 'lamp-posts' and 'watering-cans'. In Scotland, when the one in front has been touched, the game sometimes concludes—as games commonly did in the past—with all the players rushing back to the starting-line, the one who is 'het' chasing them. The player who next goes in front is then not the first to have got across the road, but the one who is caught on the way back.

Names: 'May I?' is the usual name, but sometimes the game is known as 'Steps', 'All Sorts', 'Walk to London', 'Variety', or 'Mother, May I?' It is also quite often known by the name of one of the steps, for instance, 'Fairy Footsteps', 'Giant Steps' (in Inverness, 'Giant Spangs'), 'Banana Slides', 'Scissors', 'Cups and Saucers', 'Squashed Tomato', 'Black Pudding', 'Pins and Needles' (the usual name in the north-west), 'Box of Chocolates', 'Chocolate Boxes', 'Umbrellas', 'Trip to Fairyland', and 'Pigeon Walk'. When players are known by the days of the week the game is called 'Days'.

⁎ The history of this game is obscure. It has certainly been popular in Britain for the past forty or fifty years, known in Hanley, *c.* 1915, as 'Lankeys and Strides', in Kilmington, Devon, 1922, as 'Granfer Long-legs', but not known to Gomme, 1894–8. It is played in Canada, the United States, and Australia, but Sutton-Smith, *Games of New Zealand Children*, 1959, p. 49, who describes it under 'Steps and Strides', gives a slightly different game under 'May I?' in which the contestants them-selves suggest the steps they should take, for example, 'May I take two banana skins, please?' Since it is not unusual for a component of one game to become attached to another game, it is understandable that new games arise. The present game is certainly an improvement on the one called 'Judge and Jury' described by Mrs. Child in *The Girl's Own Book*, 1832, p. 40. In this game a player, when called upon, had to jump up and spin a plate, yet was not to make any movement without first asking leave: 'May I get up?' 'May I walk?' 'May I stoop?' 'May I pick up the plate?' The game 'May I?', as played today in Britain, is, however, well known in southern Europe. It is played throughout Italy, usually under the name 'Regina reginella', the children asking 'Queen, Queen, how many steps must I take to reach your beautiful castle?' and the queen allots to each so many 'lion's steps', or 'tiger's steps', or 'ant's steps' (Lumbroso, *Giochi*, 1967, pp. 291–3). It is well known in Austria under the names 'Kaiser, wieviel Schritte darf ich machen?' or 'Vater, wie weit darf ich reisen?' (Kampmüller, *Oberösterreichische Kinderspiele*, 1965, pp. 199–200). It is played in Yugoslavia under the name 'Gospodična, koliko je ura' (Brewster, *American Nonsinging Games*, 1953, p. 164). And an observer tells us he saw the game played in Israel, the child in front being called 'Abba', and the steps being given in the same cryptic style, for example, 'ten elephants' and 'eight ants'.

AUNTS AND UNCLES

In this game, as in 'May I?', one player stands alone on one side of the road or playground, the rest line up facing him on the other side, hoping to get across. The player on his own calls out a name such as 'Uncle Leslie', and any competitor having an uncle of this name has the right to take one pace forward. Other names, such as 'Aunt Pat' and 'Uncle Henry', are then called in as quick succession as the caller is capable of thinking of them, and whoever has a relative so named takes a step forward, or if he has two Uncle Henrys or two Aunt Pats, two steps forward. It is clearly an advantage in this game for a player to have, or to suppose that he has, an infinite number of blood relations. Indeed the relatives may be of any kind. As one boy observed, 'This game, although it is called "Aunts and Uncles" includes mothers, fathers, grandfathers and grandmothers', and it is, of course, not improbable that the same name will be called more than once. Whoever crosses the road first wins, and becomes the next 'caller' or 'relationer'.

The game is so slow moving it is remarkable that it should be popular in all parts of Britain, even though chiefly with younger children. Their excuse that the game is 'quite interesting really' scarcely disguises the fact that the 'interest' is less in the action of the game, than in the information it discloses about the number and nomenclature of other people's relatives. Thus, as in 'May I?', the fun lies more in watching other people's progress than in making progress oneself. In consequence little variation in the way the game is played has been noticed between one place and another, except that in Kirkcaldy a player having a relation who has been named, is then told, as in 'May I?', what kind of step forward he should take, for example, a 'baby step' or an 'umbrella'.

The game is sometimes known as 'Relations'.

LETTERS

'Letters' is much like 'Aunts and Uncles' except that the person in front, instead of calling the names of possible relations, calls out letters of the alphabet. Contestants take a step forward, or in some places a jump, each time a letter is called which comes in their name; and if the letter comes twice in their name they take two steps or jumps. In some places two steps are also allowed if the letter is a person's initial; and an Inverness boy says that they then 'take a spang, that is a lang jump'. Thus the game is simpler, but quicker-moving than 'Aunts and Uncles'; and often seems to be played of an evening by no more than four or five friends having quiet fun together 'in our road'. Occasionally when the person in front is reached there is a chase back to the starting-line.

The game is also known as 'Names', 'Letters in Your Name', and 'Alphabet'.

COLOURS

This game is similar to 'Letters', but more personal. The child in front calls out a colour, and those who have it 'on their body or on their clothes' take a step forward. When the caller is touched there is usually a chase back to the starting-line, and if the person who was in front does not manage to catch someone he, or more often she, has to 'call' again. Although the game is scarcely more than a poor relation of 'Farmer, Farmer, may we cross your Golden River?' (q.v.) it has a wide distribution, being played in places as far apart as Perth, Spennymoor, and Welshpool. It is also reported from Australia and New Zealand.

EGGS, BACON, MARMALADE, AND BREAD

In this variant of 'Colours', reported only from south-west Manchester, the caller is concerned not with what is on the body, but within it. 'This game is played by two or more persons. One person stands at one side of the road while the rest line up at the other side of the road. The person who is on says, "What did you have for breakfast, did you have porridge? If so take two steps." This keeps on until the person who is asking the questions is took, and they all run back to the other side of the road. If the questioner ticks a player running back he is on' (Boy, c. 12).

PEEP BEHIND THE CURTAIN

'Peep Behind the Curtain' is the delight of 8-year-olds, for success in it comes not from athletic ability, but from luck and cunning. One player stands on the far side of the road, or somewhere twenty to thirty feet away, with his back turned to the rest, and usually facing a wall. The rest try to be first to sneak up on him and touch his back without his seeing them move, although he is allowed to turn round often and suddenly to try and spot them moving. Usually he is supposed to count ten under his breath before turning round, or to mutter a formula. But he says the words quickly or slowly as he likes, and the other players do not know when he is going to turn round. In some places there is a standard formula for him to say. In the south-west it is generally 'L-O-N-D-O-N spells London', in Monmouthshire 'On the way to London—one, two, three', in Welshpool 'Cat's in the cupboard and can't catch me', at Hoyland, near Barnsley, 'Crack, crack, the biscuit tin', and in Penrith 'One, two, three, four, five jam tarts'.[1] As the person in front turns round, the players

---

[1] An informant recalls that in her day in Bath, c. 1905, it was customary to say 'Sunlight Soap is the best in the world'. It seems that advertisers were as able then, as now, to have their slogans adopted by children. Indeed, in the period before the First World War, commercial advertisers had a number of amusing devices for attracting juvenile attention which are unknown today.

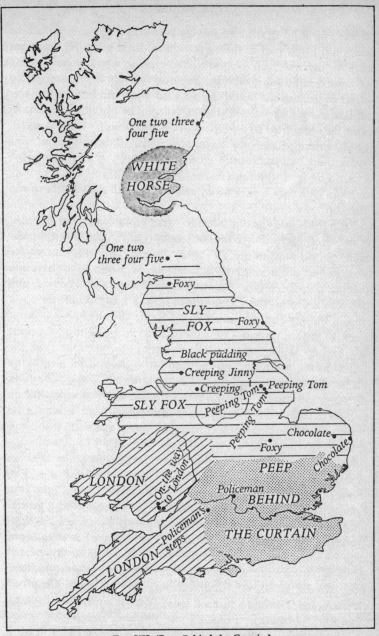

FIG. VII. 'Peep Behind the Curtain.'
Simplified map showing distribution of principal names.

'freeze', for if he sees anyone moving even a hand, let alone a foot, he sends that player back to the starting-line to begin again. He then turns back to the wall, and everyone resumes their advance. Some move cautiously, as if treading on eggs; others dart forward hoping to reach the player in front in one bold dash, but usually are unable to stop moving quickly enough when he next turns round. The player who is first to reach the one in front takes his place in the next game, for this is a game in which, for some reason, the coveted role is that of the player who turns round. 'Sometimes there's an argument who is to be on it first,' remarked a Caerleon boy, 'but we settle it after a while. I don't know why everybody wants to hide their eyes. I would prefer to be with the other boys who creep up.'

'Peep Behind the Curtain' is a quiet game, probably the most-played of 'quiet' street games. But in the north country and in Scotland it is tradi-tional for the game to end with a race back to the starting-line. In the West Riding, and consequently in Guernsey (see p. 66 n.), the player who succeeds in touching the one at the wall shouts 'Black Pudding', and everyone rushes back to the starting-line. On the Isle of Bute the cry is 'Queen Victoria' (and the game is so called); at Annesley in Nottingham-shire the cry and name is 'Queenie'. In other places (see map) the cry is 'Sly fox' or 'White horse'. In some places (e.g. Spennymoor) all the players have to touch the wall, not the player in front, and it is only after the last of them has touched the wall that they run back and are chased. In Aberdeen and Kirkcaldy a player seen moving is sent back only a certain number of paces rather than back to the starting-line; and, as is often the case, the Scottish practice is also that of the United States (cf. 'Red Light' in Brewster's *American Nonsinging Games*, 1953, p. 35).

Names: There are more than thirty names for this game, sometimes two or more being current in one place. (The distribution map, Fig. VII, shows only the predominant names.) 'Peep Behind the Curtain' prevails in London and the south-east. The name 'London' does not become general until children are a hundred miles from the metropolis, e.g. at Gloucester, Worcester, and Birmingham. The name 'Grandmother's Footsteps' seems to be known mostly in private schools. Other names include: 'Black Peter' (Golspie), 'Black Pudding' (Halifax, since *c.* 1900, and Guernsey), 'Bull's Eye' (Wickenby), 'Carrot, Carrot, Neep, Neep' (Perth), 'Chocolate' (Norwich and Yarmouth), 'Creep Mouse' (Bristol, Wells, New Radnor), 'Creeping' (Wilmslow), 'Creeping Jinny' (Wigan), 'Creeping Up' (usual name in Australia and New Zealand), 'Crystal Palace' (St. Helier, Jersey), 'Cuckoo' (Troutbeck, nr. Ambleside), 'England, Ireland, Scotland, Wales'

(Helensburgh), 'Fairy Footsteps' (Manchester), 'Five Jam Tarts' (Penrith), 'Foxy' (Carlisle, Scarborough, Peterborough), 'Giant's Treasure' (Helston),[1] 'Granny' (Lockerbie), 'King's Palace' (Castel, Guernsey), 'On the way to London' (Caerleon, Cwmbran, Pontypool), 'One, Two, Three' (Scalloway), 'One, Two, Three, Four, Five' (Aberdeen and Langholm), 'Peeping Tom' (Market Rasen), 'Piggy Behind the Curtain' (Wilmslow), 'Policeman' (Oxford), 'Policeman's Steps' (Wootton Bassett), 'Red Light' (Liverpool, Blackburn, Spennymoor, Peterborough, Helensburgh, and Edmonton, Alberta), 'Statues' (Stamford, Cumnock, Forfar, Kirkcaldy, but in other places 'Statues' is a different game, see pp. 245-7), 'White Horse' (Edinburgh, and eastern Scotland).

⁑ At the beginning of the century the game was generally known as 'Steps', e.g. at Eastbourne, c. 1910, Victoria, N.S.W., c. 1915, and in Bancroft's *Games for the Playground*, New York, 1909, pp. 188-9. In Austria it is 'Ochs am Berg', 'Der Hase läuft über das Feld', or 'Küche-Zimmer-Kabinett' (Kampmüller, *Oberösterreichische Kinderspiele*, 1965, p. 199). In Germany it is 'Eins zwei drei—sauer Hering!' (Peesch, *Berliner Kinderspiel*, 1957, p. 32). In Italy and Sardinia it is usually 'Uno, due, tre, stella', though interestingly, at Forenza in southern Italy it is 'Per le vie di Roma' (Lumbroso, *Giochi*, 1967, pp. 289-91).

§ *Games in which only Two Competitors Run Against each Other at a Time, One of them Generally being Instrumental in the Selection of the Other*

BLACK MAGIC

Several racing and chasing games have fused, under names such as 'Black Magic' (Hackney and Oxford), 'Black and White' (West Ham and Stockport), 'Flip Flop' (Ipswich), 'Pit-a-Pat' (Titchmarsh), 'Tip Tap' (Walworth and Cwmbran), 'Tip Tap Toe' (Sheffield), and 'Tit for Tat' (Swansea), owing to the singular manner in which the person who is 'on' selects whom he will run against. The players line up, and hold their hands out in front of them, with their palms all facing upwards or all facing downwards. The person who is 'on' or 'dipper' goes along the line tapping or smacking the top of each hand, but smacks the underside of the

---

[1] 'Giant's Treasure' is more often the name of a related Scout game in which the person in front, who is blindfold or has his back turned, guards a knife, bottle, or other object, which the players try to snatch from him without being heard. This game also goes under the names: 'Blind Knife', 'Blind Pirate', 'Sleeping Pirate', 'Sneaking Pirates', 'Blue Peter', 'Stalking', 'Creeping', 'Crack-a-nut', 'Farmer's Apples', 'Stealing the Honey Pot', 'The Bear and the Honey', 'Doggie and the Bone', and 'Princess in the Tower'.

hand of the person he elects to run against. Alternatively, he walks along the line repeating some hocus-pocus such as 'Black, black, black, black . . . *magic*', 'Black, white, black, white . . . *red*', 'Black, white, black, white . . . *whisky*', or 'Tip, tip, tip . . . *tap*', 'Flip, flip, flip . . . *flop*', or 'Tip, tap, toe, my first go, if I touch you, you must, must, must . . . *go!*' Thereupon the pair race to the opposite wall and back, or to a selected point, or, sometimes, the one who is 'on' is chased to a certain point by the player he has chosen, and, as a 10-year-old explained, 'If he "has" the dipper he is now "it", but if he doesn't the dipper remains the dipper'.

The game 'Kerb or Wall' is also sometimes started in this manner.

KERB OR WALL

This game is essentially a street game. One player is made 'on' or 'out' by the ordinary process of dipping (e.g. 'Ickle ockle, chocolate bottle, ickle ockle out; if you want a titty bottle please shout out'). The player then discovers his rival by lining the rest up against a wall on one side of the road, and 'hitting' each of their hands in turn as he repeats the particular rhyme associated with the game in his locality, for instance:

Peter Pan said to Paul,
Who d'ya like the best of all—
Kerb stone, or the solid
　　　brick wall?
Said Peter Pan to St. Paul
Who lives at the bottom of the
　　　garden wall.
> *Walworth version. Several others current in London*

Two little dicky birds
Sat upon a wall,
One named Peter
The other named Paul.
Paul said to Peter,
Peter said to Paul
Let's have a game
At 'Kerb and Wall'.
> *Birmingham. Cf. 'Oxford Dictionary of Nursery Rhymes', pp. 147–8*

Bim, bam, boo, and a wheezey anna,
My black cat can play the piano,
One, two, three, kick him up a tree,
Kerb or wall?
> *Stockport. Versions in various stages of decomposition throughout north country*

Kob or wall or lucyanna,
Jack and Jill went up the ladder.
Which would you rather have
Kob or the old brick wall?
> *Newcastle upon Tyne*

The player whose hand is struck last has to choose 'kerb or wall'. If he chooses 'kerb' he runs to the edge of the near kerb, back to the wall where he started, then across the road to the far wall and back to the home wall again. If he chooses 'wall' he runs first to the far wall, then to the home wall, then to the near kerb and back. The person who dipped covers the same ground, only he does it the opposite way round. If 'kerb' was chosen he does 'wall'; if 'wall' he does 'kerb'. The dipper always gives the other person the choice of 'kerb or wall'. Whoever wins becomes 'on' or 'dipper' next time. If there is a tie they may be made to run again, or to go out of hearing and each choose a similar name, as 'candy rock' and 'candy stick', and the bearer of the name selected by the rest is declared winner.

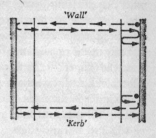

This game is fast, rough, and sometimes conducted in reckless fashion, especially when played at night, if the older children join in, for the kerb can become a trip-stone sending players sprawling on the pavement, and the runners crash against the wall in their effort to reach it first. Nevertheless—or perhaps in consequence—there seem to be few places where the game is not played.

Names: Usually 'Kerb or Wall' or 'Kerb and Wall', but also 'Crib and Wall' (Stornoway), 'Kerby Wall' (Helston), 'Big Black Wall' (Hounslow), 'Brick, Stone, Kerb, Wall' (Birmingham), and there are three names emanating from the rhymes: 'Peter Pan' (Peckham and elsewhere, not uncommon), 'Ombombay' (Oxford and Spennymoor), and 'Ina Vana Vea' (Wigan).

TIME

Two players, chosen by conventional means, go off a little way and agree on a time of day such as 'Half-past six'. They come back to the others, who have lined up, and one of them goes along the line saying 'Half-past what?' and receiving a suggestion from each person in turn. The player who chances to give the right time races, or sometimes chases, the other time-fixer. Usually the race is to an agreed spot and back again. In Barrow-in-Furness the running does not begin until the one who is taking the guesses shouts 'Clock's right'. In Scalloway, Shetland, the one who is standing out has his back to the others and arms outstretched. The player who guesses the correct time runs up to him and touches his arm, and this is the signal for the race to begin. In Aberdeen the time-fixer, whose part in the game is little more than that of referee (although not looked upon as such), takes the two runners to the middle of the road, stands with one on either side of him, and declares:

> My father had a fiddle
> And he broke it through the middle.

When he says 'middle' the two players run round whichever is the nearest lamp-post on their side of the road, then cross the road and run round the other player's lamp-post and back to the person in the middle. The player who wins becomes the referee time-fixer in the next game, and the referee becomes the runner time-fixer.

The game is usually known as 'Time' or 'Timey', although in Swansea it is 'Old Timer', in Barrow-in-Furness 'Clock Time', and in Guernsey 'One o'clock'. In a variant called 'O'Clock', played in Swansea, each player in the line is given a number, while the player who is 'on it' is out of hearing. He then calls out a number not knowing which player this means he will have to run against.

** The game was current in Fraserburgh in the nineteenth century under the name 'Time' (Gomme, *Traditional Games*, vol. ii, 1898, pp. 293–4), although the game called 'American Times' played in Fraserburgh today is purely a guessing game (see pp. 279–80). In Argyllshire it was known as 'Time' or 'Polly in the Ring' (*Folk-Lore*, vol. xvii, 1906, pp. 102–3).

DROP HANDKERCHIEF

They say 'Drop Handkerchief' is best played with twelve people or more. One of them is chosen to be out and the rest join hands and make a big circle. They then loose hands and either remain standing or sit on the ground. The one who is out has a handkerchief or 'something that won't make a noise when it is dropped'. He—or, more usually, she—goes round the outside of the circle singing—perhaps to the tune of *The Jolly Miller* or *Bobby Bingo*:

> I sent a letter to my love
> And on the way I dropped it;
> One of you has picked it up
> And put it in your pocket.

Sometimes everybody sings. Then the one who is running round outside the circle drops the handkerchief quietly behind one of the players and races on. The player who has had the handkerchief dropped behind her must pick it up and race round the circle in the opposite direction, trying to get back to her place before the dropper reaches it. Usually the dropper is first. She can see if she is going to win by the point where they meet on the far side of the circle. Sometimes there is jostling and pushing when they meet, as the one who is losing hopes by force or craft to make up lost ground. The player who arrives back last has to go round the circle with the handkerchief in the next round.

Less often now than formerly the game is a chase not a race, and the player who has the handkerchief dropped behind her runs after the dropper in the same direction around the circle.

In many places in Britain the game is still embellished with actions or imagery from the past, which are sometimes also to be found in versions of the game on the Continent. In Govan, south-west Glasgow, the children crouch down in a ring singing:

> I sent a letter to my lad (*pronounced* '*lod*')
> And on the way I dropped it,
> I dropped it once, I dropped it twice,
> I dropped it three times over:
> Over, over, in amongst the clover,
> Over, over, in amongst the clover.

Here, as of old, the player who is pursued may run in any direction he wishes in and out of the circle, while he who chases 'must go the very same road as the other person, or he is called "a rotten", and he must sit in the middle' (Girl, *c*. 10). In Spalding, Lincolnshire, the one who loses the race for the empty place goes into the centre and is made to suck her thumb until someone comes to replace her. In Aberystwyth, and also at Frodsham in Cheshire, if a girl does not notice that the handkerchief has been dropped behind her she must stand in the centre and, similarly, place her thumb in her mouth. In the neighbourhood of Welshpool, where the game is sometimes known as 'Dummy', the girl who drops the handkerchief aspires to run the full circle before the player notices that the handkerchief has been dropped behind her. If she succeeds in this she shouts 'Dummy', and sends the booby into the centre of the circle to suck her thumb. But should the player who had the handkerchief dropped behind her be a fast runner, and catch the dropper, the catcher shouts 'Dummy', and sends the dropper into the centre to suck her thumb. And so the game continues 'until nearly everyone is in the centre sucking their thumbs'. In Radnorshire, and some other parts of Wales, the procedure is different. When the girl who is outside the ring has finished singing 'I sent a letter to my love', she holds the handkerchief up and calls out 'All see it?' The children in the ring reply 'We see it'. Then the runner says 'You'll never see it again until you find it'. Whereupon the children in the ring shut their eyes, and the runner drops the handkerchief behind one of them, calling out 'All people look behind their back', and the race begins. Across the border, at Upton Magna in Shropshire, the same ritual is adopted, and is expressed in song:

I wrote a letter to my love
And on the way I lost it,
Some of you have picked it up
And put it in your pocket.

Now is the time to close your eyes,
Close your eyes, close your eyes;
Now is the time to close your eyes
And see (*sic*) if you've got the letter.

Now is the time to look behind,
Look behind, look behind;
Now is the time to look behind
And see if you've got the letter.

Similar words have been reported from County Durham. Other songs, sung to various tunes, among them *A tisket, a tasket* and *Yankee Doodle*, are traditional in other places.

Sent a letter to my love
And on the way I dropped it,
I dropped it once, I dropped it
  twice,
I dropped it three times over.
Shut your eyes, look at the skies,
Guess where the letter lies.
                                    New Cumnock

Lucy Locket lost her pocket,
Someone must have picked it up
And put it in her basket.
Please drop it, drop it,
  drop it. . . .   *Birmingham*[1]

Who goes round my house at night?
Nothing but dirty Tommy.
Who stole all my chickens away
And left me only one?
So it's rise up and run, run, run.
I sent a letter to my love
And on the way I dropped it,
I dropped it once,
I dropped it twice,
I dropped it three times over.
Over, over, in amongst the
  clover,
Not yet, not yet, not yet. . . .
                                    Paisley

At Arncliffe, in the West Riding, the handkerchief is not a letter but a pigeon, and the one who is outside the circle sings:

I've got a pigeon in my pocket,
I won't tell you where I got it.

In most places, even though the players keep their eyes open, they are not allowed to watch the handkerchief-dropper, nor to look behind them until she has passed. However, the race is more exciting when a player is left in no doubt that it is her turn to run. In Stirling, for instance, and also in Flotta, the player who is about to drop the handkerchief goes round

---

[1] In Accrington, where the game itself is called 'Lucy Locket', the song is exactly as in *The Oxford Dictionary of Nursery Rhymes*, p. 279.

touching each person's head saying 'Not you, not you', until she reaches the player she wishes to compete against, when she says 'but you!' In Swansea, after the player has dropped the handkerchief she continues her circuit, touching each person on the head, saying 'It wasn't you, it wasn't you, it wasn't you . . . it was *you*!' Only then do the two players race round in opposite directions. And in Blackburn, Lancashire, the girl carrying the handkerchief sings:

> I sent a letter to my lad
> And by the way I dropped it,
> I dropped it, I dropped it,
> An old man came and picked it up
> And put it in his pocket, pocket.
> The doggie won't bite you, nor you, nor you, nor you . . .

Not only does this song hold an echo of past times, but so does the action, for if the one behind whom the handkerchief is dropped succeeds in catching his challenger—or tempter—he has the right, as of old, to a kiss.

⁂ A miniature history of deportment could be based on a study of this old and much-loved game. 'I sent a Letter to my Love' is depicted in *A Little Pretty Pocket-Book*, 1744 (earliest surviving edition 1760), where a youth is shown dropping a letter, not a handkerchief, behind one of the maidens in the ring. 'Dropping the Letter' is named, but not described, in *Suffolk Words*, 1823, p. 238, as a Suffolk game of about 1780. 'Dropping the 'Kerchief' is described in the 1829 edition of *The Boy's Own Book*, p. 38, played as in Glasgow today: 'the pursuer is bound to follow precisely the course of the pursued'; and this rule appears, too, in *The Gallovidian Encyclopedia*, 1824, where the game is titled 'Allicomgreenzie', and said to be 'played by young girls at country schools'. The first mention by name of 'Kiss in the Ring' is in *Sports and Pastimes of the English People*, 1801, p. 285, where Strutt says, in a description of 'Cat after Mouse':

'When this game is played by an equal number of boys and girls, a boy must touch a girl, and a girl a boy, and when either of them be caught they go into the middle of the ring and salute each other; hence is derived the name of "Kiss in the Ring".'

Throughout the nineteenth century 'Kiss in the Ring' was a favourite game at Christmas time and midsummer, at rustic weddings, bank-holiday outings, fairs, and flower shows, and it was played by 'grown lads and lassies' as well as by children. Pleasing references to the game in these settings occur in Hone's *Every-Day Book*, vol. i, 1826, col. 692; *Sketches by Boz*, 1835, ch. xii; Chambers's *Popular Rhymes*, 1869, pp. 129–30 (played in eighteenth-century Dumfriesshire under the name 'Pease and Groats'); *Punch*, 16 August 1890, p. 78; and in Flora Thompson's *Candleford Green*, 1943, p. 62; as well as in several of the Opie manuscripts. There are couples who have celebrated their golden weddings (see e.g. *Hereford Times*, 7 October 1960, p. 11, and *Hants and Sussex News*, 7 June 1961, p. 1), who first met when they 'saluted each other' while playing this merry game, in the days when a kiss between young people could still be a light-hearted courtesy.

In *The Girl's Own Book*, 1832, p. 45, the American authoress Mrs. Lydia Maria Child describes the game under the name 'Hunt the Squirrel', as also does Newell in *Games of American Children*, 1883, pp. 168–9. In England this name has not been common, although mentioned by Horace Walpole in a letter, 8 October 1742, and known to London elementary schoolchildren in 1910. This personation of a small animal that bites, which occurs today in the Blackburn game, was quite general in the nineteenth century (see Gomme's *Traditional Games*, vol. i, 1894, pp. 109–12, 305–10, and *The English Dialect Dictionary* under 'Hitch-Hatch'). One such version, not uncommon around 1900, opened with lines which belong to the eighteenth century:

> I had a little moppet, I put it in my pocket,
> And I fed it on corn and hay;
> And it won't bite you, and it won't bite you . . .
>   but it *will* bite you.

(Cf. *Oxford Dictionary of Nursery Rhymes*, p. 313.) The game itself, or something like it, considerably antedates the eighteenth century. It appears to be referred to in a Reichenau glossary, compiled before 1300, where, as transcribed by F. J. Mone, *Anzeiger*, 1839, p. 395, an entry reads:

'Circulatorius ludus est puerorum in circulo sedentium, post quorum tergum discurrit puer unus, portans aliquid in manu, quod ponit retro aliquem sedentium ignorantem, vulgariter dicitur: Gurtulli, trag ich dich!'

The closest parallels to 'Drop Handkerchief', however, are found in the Romance languages, 'Le Jeu du mouchoir' in France, 'Pan para todo el año' in Spain, 'Ovo marzo' in Trieste, and 'Bacio nel cerchio' in Sardinia. In Germany a more robust circle game is usual, in which the person pursued is hit rather than kissed; and it may well be that the medieval monks at Reichenau were describing the start of the closely related game 'Der Fuchs geht rum', an account of which appears hereafter under 'Whackem'.

BUMP-ON-THE-BACK

This is 'Drop Handkerchief' without the handkerchief. The player who goes round the outside of the circle taps or pushes the back of the person he wishes to run against, and the race is on. When Brownies or Wolf Cubs play the game it is the practice, sometimes, for the two who are racing round the circle to stop when they meet on the far side, and for them to shake hands and say 'Good morning', before they race on; or even for them to say 'Good morning' as they shake their right hands, 'Good afternoon' as they shake their left hands, 'Good evening' as they shake their right hands again, and 'Good night' as they shake their left hands again. This is usually done at speed although both parties are being held up for the same length of time. When these formalities are customary the game is usually known as 'Good Morning'.

⁎ Gomme, *Traditional Games*, 1894–8, gives the names 'French Jackie' (Keith), 'French Tag' (Forest of Dean), 'Gap' (Barnes), 'Push in the Wash Tub' (Crockham Hill, Kent), and 'Tap-Back' (Bitterne, Hampshire). Correspondents have known the game as 'Pushing in the Buttertub' (Eastbourne, 1914), and 'Filling the Gap' (Honiton, 1922).

In Germany in the nineteenth century the game was called 'Ringschlagen' or 'Komm mit', or, in Königsberg, 'Guten Morgen, Herr Fischer', this being the greeting with which the players addressed each other on the far side of the circle (F. M. Böhme, *Deutsches Kinderspiel*, 1897, p. 589). The game is popular in Italy, being played under a number of names such as 'Pugno', 'Chi tardi arriva male alloggia', and 'Giorni della Settimana' (Lumbroso, *Giochi*, 1967, pp. 114–15, 117–19).

WHACKEM

This ring game, which is similar to 'Drop Handkerchief' but more virile, is a favourite with boys in Acocks Green, south-east Birmingham. The players stand in a circle with their eyes closed and their hands behind their backs, while one of their number runs round the outside of the circle with a short piece of rope. As he runs he places the rope in somebody's hands, and the boy who receives it instantly opens his eyes and belabours his neighbour to the right of him. This neighbour, though taken by surprise, must set off around the circle as fast as he can, for he is subject to as many further blows as his pursuer can inflict upon him until he has completed the circuit and returned to the safety of his place. The player with the rope then continues round the circle, and places the rope in another boy's hands.

The game is also played in gymnasiums, and by Wolf Cubs, usually under the name 'Beat the Bear'.

⁎⁎ 'Whackem' may be compared with the games 'Whacko' and 'Daddy Whacker' (qq.v.); and is perhaps allied to the game played by Irish girls, known in County Kerry as 'Burning' (Irish Folklore Commission, MS. vol. 470, 1938). In this game the player going round the outside is said to be 'burning', and she strikes the person she means to burn, which is the signal for the chase to begin. The notion that the one who runs round the ring is hot or burning is not apparently confined to County Kerry. J. O. Halliwell, *Popular Rhymes and Nursery Tales*, 1849, p. 113, describes a game known as 'Drop-Cap' in which the child selecting who shall chase him makes his progression round the circle chanting:

> My hand burns hot, hot, hot,
> And whoever I love best, I'll drop this at his foot!

Halliwell also describes (p. 130) the game of 'Drop-Glove' (a game listed by Randle Holme in 1688) in which the player whose part it is to carry the glove cries 'It burns, it scalds!' as he drops it behind a player. Likewise in the game 'Tartan Boeth', a version of 'Drop Handkerchief' played at Beddgelert in north Wales (Gomme, vol. i, 1894, p. 112), the child with the handkerchief says 'Tartan boeth, oh mae'n llosgi, boeth iawn' ('Hot tart, oh it burns, very hot'), and drops the handkerchief at the words 'very hot'. In the game 'Black Doggie', a further version of 'Drop Handkerchief', played at Rosehearty in Aberdeenshire (Gomme, vol. ii, 1898, p. 407), any player in the ring who did not notice that the handkerchief had been dropped behind him, while the dropper completed a circuit, was declared to be 'burnt', and had thereafter to kneel down and be out of the game. Similarly in Argyllshire, in the game 'Drop the Napkin', if a girl was unaware that the handkerchief had been dropped behind her, or did not follow exactly the course of the one she was pursuing, the other children raised the cry 'Ye're burnt' or, quaintly, 'You burned a hole in your porridge' (Maclagan, *Games of Argyleshire*, 1901, pp. 213–14, and *Folk-Lore*, vol. xvii, 1906, p. 102). And in Hungary, where a comparable game is called 'I Carry Fire' (Brewster, 1953, p. 91), the players recite while the handkerchief is being carried round:

> I carry fire; don't see it.
> If you see it, don't tell anyone.

It seems probable that this calefaction by a secretly deposited article, whether cap, glove, or handkerchief, is connected in some way with the childish contention that a hidden object—such as the thimble in 'Hunt the Thimble'—is hot, and that he who approaches is 'getting warmer' or 'burning'.

It appears from Böhme that the usual way boys in Germany played the game 'Fuchs geht rum' in the nineteenth century was little different from the way the boys play 'Whackem' at Acocks Green today. The players stood in a circle, with their hands behind their backs, and none must look round. A player called 'the fox', who was armed with a knotted handkerchief (variously termed *Plumpsack*, *Klumpsack*, *Plumser*, *Knötel*, or *Tagel*), skirted round the outside of the circle saying:

> Seht euch nicht um,
> Der Fuchs geht rum.

He secretly placed the *Plumpsack* in someone's hands, and that person struck his right-hand neighbour with it, and chased him round the circle back to his place (*Deutsches Kinderspiel*, 1897, pp. 556–7). Further, J. C. F. Gutsmuths, in his pioneer work *Spiele für die Jugend*, 1796 (1802, pp. 232–4) exactly describes the game, under the name 'Das böse Ding', the player who goes round the circle with a knotted handkerchief singing:

> There goes a wicked thing around
> That will properly sting you:
> If anyone looks behind him now
> He'll get it on the neck.

However, Gutsmuths considered that the original method of play was much the same as the way many children continued to play it in Böhme's day, particularly in northern Germany. The players sat in a circle facing inwards, and the player making the circuit behind them intoned 'Die Gans, die Gans, die legt ein Ei':

> The goose, the goose, it lays an egg,
> And when it falls, it falls in two.

He let the *Plumpsack* fall behind one of the players without him knowing. If the person noticed it, he picked it up and chased the dropper round the circle; but if he failed to notice it and the dropper came round the circle again and picked it up, he had the privilege of chasing the inattentive player round the circle. In fact the game was initially the same as 'Drop Handkerchief', but the chaser commonly used the knotted handkerchief (or rope, or belt) to strike the person he was chasing whenever he came within range. In Holland, Böhme added, the game was known as 'De Vlugt of Sackjagen', and while the dropper went round the circle his words were 'Cop, cop heeft ghelecht' (the little hen has laid). In present-day Austria, where the game is called 'Der Plumpsack geht um', the association with the poultry yard is maintained. A player who fails to hit the one who has dropped the knotted handkerchief behind him is obliged to go into the centre of the ring and is jeered at for being a 'rotten egg' (Kampmüller, *Oberösterreichische Kinderspiele*, 1965, pp. 140–1). In Switzerland a version of the game is actually called 'Faul Ei'; and the unsuccessful player is similarly relegated to the centre of the circle, and called a rotten egg (cf. the Govan version of 'Drop Handkerchief'). In Trieste, too, as already reported, a version of 'Drop Handkerchief' is called 'Ovo marzo', the player who fails to notice that the handkerchief has been dropped behind him being dubbed a 'March egg' (Lumbroso, *Giochi*, 1967, p. 118). And in Berlin, where the inattentive player is struck three blows if he fails to run, the striker chants as he delivers them, 'Eins, zwei, drei, ins faule Ei!' (Peesch, *Das Berliner Kinderspiel der Gegenwart*, 1957, p. 15).

There can be little doubt that the game 'Die Gans, die Gans, die legt ein Ei' is descended from the ancient Greek game 'Schœnophilinda', described by Pollux (ix. 115), in which, it appears, one player endeavoured to drop a short rope, without being seen, behind one of the company who were squatting in a circle. If he succeeded the squatter was belaboured round the circle; if detected, the dropper was chased round the circle. And it may be noted that this was also the opinion in Elizabethan times. Adrianus Junius in his *Nomenclator*, Antwerp, 1567, named the equivalent game 'Cop cop heeft geleyt'; and when John Higins prepared his English edition in 1585 he named the equivalent game in his day 'Clowte, clowte, to beare about', and gave, as an alternative name, 'My hen hath layd'. This name has not been found subsequently in Britain, but girls in Scotland still seem to have been echoing it 300 years later in a nonsense song they sang while playing 'Drop the Napkin':

> Drip, drop the napkin,
> My hen's laying,
> My pot's boiling,
> Cheese and bread and currant-bun,
> Who's to get the napkin?

(R. C. Maclagan, *Games of Argyleshire*, 1901, p. 213.)

STONEY

In 'Stoney' which several girls of eight and nine describe as 'my best game', the others do not know who has been chosen to run, nor when she

is going to start. The attraction of this game for small girls is its secretive-
ness. They stand in a row—perhaps there are only four or five of them—
with one girl, 'the outer', who has a pebble or cherry stone, standing in
front. They hold out their hands with palms together, 'as if praying', but
leave a small hole at the top. The girl with the stone holds her hands in
similar fashion and proceeds along the line placing her hands on top of
theirs. She drops the stone into one person's hands, and neither of them
makes any sign that she has done so. When the person who has the stone
thinks the others are not looking, she dashes across the road and back, or
runs to an agreed point, having to complete the course without any of the
others managing to touch her. If she succeeds she is the 'outer' next time.

This game, popular in London, and sometimes called 'Cherry Stone',
has been reported only from the Home Counties, and from Berkshire,
Oxfordshire, and Devon.

§ *Games in which the Players Start Running from Different Places*

PUSS IN THE CORNER

The fun of 'Puss in the Corner' is that the players themselves negotiate
when they are going to run; its disadvantage is that it is normally for five
players, no more and no less. Four of the players stand at four points:
lamp-posts, drain covers, or, if indoors, the four corners of a room. The
fifth stands in the middle, or where he likes. Those at the corners call to
each other: 'Puss, Puss, come here', 'Puss, Puss, come and get some milk',
'Puss, Puss, come to my corner', and change places when they think the
one in the middle is not looking. The player in the middle has to try and
reach one of the corners when it is empty; and if he is successful the player
he has raced goes into the middle. Any two corners may change places,
and often all four run at once. They keep switching about, even to opposite
corners, until one player becomes confused, dashes to the wrong corner,
and makes it easy for the middle player to slip into an empty corner. This
game is popular. As one youngster remarked, 'I like it because it makes you
feel happy and gay and it is very funny'.

Names: 'Puss in the Corner' and 'Puss, Puss' (both common), 'Poor
Pussy' (Wigan), 'Puss in Four Corners' (Wolstanton), 'Fox and Chickens'
(Market Rasen), 'Bear's in his Den' (Golspie). The name 'Corners' is not
uncommon, especially when the players use the corners of a netball court.
In Aberdeen the game is known as 'Poles', in Forfar 'Polecat', the players
taking their positions at the poles of the washing-lines. Sometimes small

circles are chalked on the ground for the corners, and they draw five or six, or however many are required, so that more than five people can play.

A variant game in which more than five children can take part is played in West Ham and called 'Bad Penny'. Each player has a partner, except 'Bad Penny'. They stand in a line with their partners, and 'Bad Penny' just in front of them. At a given word one from each pair races to a certain point and back again, their partner marking their place. 'Bad Penny' runs with them, and tries to get back to one of the partners first, whereon the player he has displaced becomes 'Bad Penny'.

*** 'Puss in the Corner' was one of the games Randle Holme listed in his *Academie of Armory*, 1688, III. xvi, § 91. It has frequently been alluded to since, e.g. by William King, *Useful Transactions in Philosophy*, no. i, 1709, p. 43; John Arbuthnot, *Memoirs of Martin Scriblerus*, 1714 (Pope's *Works*, vol. vi, 1757, p. 115); in *Round about our Coal Fire*, 1731, p. 9; *The Craftsman*, 4 February 1738, p. 1; by Dorothy Kilner, *The Village School*, vol. ii, *c*. 1785, p. 42; in *The Happy Family* [1786], p. 18; *David Copperfield*, 1850, ch. vii; and by Keats in a letter to Georgiana, *c*. April, 1819:

'You may perhaps have a game at Puss in the Corner—Ladies are warranted to play at this game though they have not whiskers.'

Descriptions of the game appear in Strutt's *Sports and Pastimes*, 1801, p. 285; *Juvenile Games for the Four Seasons*, *c*. 1820, p. 36; and in most other nineteenth-century collections, the player in the middle (unlike today) usually being nominated the 'puss'. Thus a writer in *The Boy's Own Paper*, 12 November 1887, p. 103, recalled:

'With us we had "any number of players", and a puss for every five of them. I have played with twenty-eight corners, and then there were six pussies in the centre.'

In the West Country the game was called 'Catch-Corner' (*EDD*). In Furness and Westmorland it was 'Chitty Puss'. A Cartmel Fell schoolboy said, about 1935, that the players shouted to each other 'Chi-chi-chi-chi, come to my den'—'Chi-chi-chi-chi' being the lakeland manner of calling a cat.

In the United States, where the game is often called 'Pussy wants a Corner', the player in the middle is the cat, and proceeds to each corner in turn pleading 'Pussy wants a corner', only to be told 'Go and see my neighbour'. When the cat goes to the next corner, the others move. In Austria, likewise, where the game is called 'Schneider, leih ma d'Schar', the reply is 'Go to my neighbour' (Otto Kampmüller, *Oberösterreichische Kinderspiele*, 1965, pp. 151–2). In Germany, where the game was recorded in 1851 and called 'Die Schere leihen', the player in the middle went round begging 'Tailor, lend me your shears', and was told in reply 'Da läuft sie leer' (E. Meier, *Deutsche Kinder-Spiele aus Schwaben*, p. 111). Brewster states that the game is played in exactly the same way in Hungary, the one in the middle saying 'My sponsor-woman, where are the scissors?' and the player at the corner replying 'I have lent them to my neighbour'. This is also the practice in Czechoslovakia, where the one in the middle says 'Godmother Anne, lend me a sieve', and the player in the corner replies, 'I have lent it to my neighbour' (*American Nonsinging Games*, 1953, pp. 96–7). He also states that in Greece, where the game is very popular, the player without a corner begs 'Light my candle', and gets for reply 'Go to another corner'. In Sicily, similarly, it appears that the player who is *sotto* used to approach each corner on the pretence of having a candle to light (G. Pitrè, *Giuochi fanciulleschi siciliani*, 1883, pp. 272–3). In Sweden and Switzerland, too, the player asks if he may 'borrow fire', and it appears that the game is related to the child-stealing drama 'Mother, the Cake is Burning' (q.v.), especially as in Germany there was a similar game called 'Kinderverkaufens' (Meier, p. 382). In Italy, however, the game is played without these formalities, and known prosaically as 'Quattro Angoli' or 'Quattro Cantoni' (M. M. Lumbroso, *Giochi*, 1967, pp. 295–8). In France, too, it is generally played as in England, and has long been popular as is apparent, for instance, from Lancret's picture 'Le Jeu de Quatre Coins'. A form of the game is also common in Japan, where the player in the middle is termed 'oni' or devil.

'Puss in the Corner' is also closely allied to the party game known variously as 'General Post', 'Move All', 'Stations', and 'King, King, Come Along', or, as in *Rob Roy*, 1817, ch. xxxi, 'Change Seats, the King's Coming'.

HOT PEAS

In this game the players race on different courses which are assigned to them by chance. 'Everybody assembles together at a certain gate or street lamp,' writes a 13-year-old. 'Whoever has to be out turns his back to the rest of the players and faces the gate or lamp. The rest of the players put their hands on top of each other on this person's back. One by one they pull their hands away and the person who has his back turned tells each one to go to some certain place, as it might be "Go to the bus stop", or "Go two gates down", or "Go across the road". When everybody has gone where they have been told to go, the person who was out turns round.' He shouts 'Hot potatoes', 'Hot soup', 'Hot mackerel', and they must not move; but when he shouts 'Hot peas' they race back to the gate or street lamp, and 'the last one back goes under an arch and the people thump him or her on the back, and he is "het" in the next game' (Glasgow).

This race-game is also known as 'Black Pudding' in Glasgow: the caller shouting 'different colours of puddings before he calls "Black pudding" '. In Cumnock it is 'Hot Peas and Vinegar' ('Hot peas and chips . . . Hot peas and sauce . . . Hot peas and *vinegar*'), and the same name and formula is reported from Enfield, the last player home being made to go 'through the mill'. In Wolstanton the game and call is 'Hurry Home!', in Perth 'Sheep, Sheep, Come Home', and in Accrington, where the game is played under the name of 'Ralliho', an additional player is employed, and while the first has his face to a wall, he points at the others in turn, chanting:

> North, South, East, West,
> The wind blows the robin's nest,
> Where shall this one go to?

These words not only neatly join the game to its history, but help to show the evolution of the still more popular pastime 'I Draw a Snake upon Your Back' (q.v.).

*** Robert Chambers's description of the game in *Popular Rhymes of Scotland*, 1869, pp. 122–3, shows how little it has altered in the past hundred years:

'One boy stands with his eyes bandaged and his hands against a wall, with his head resting upon them. Another stands beside him repeating a rhyme, whilst the others come one by one and lay their hands upon his back, or jump upon it:

> Hickety, bickety, pease scone,
> Where shall this poor Scotchman gang?
> Will he gang east, or will he gang west;
> Or will he gang to the craw's nest?

When he has sent them all to different places, he turns round and calls: "Hickety, bickety!" till they have all rushed back to the place, the last in returning being obliged to take his place, when the game goes on as before.'

Presumably the boys' sport named 'Pirley Pease-weep', mentioned in *Blackwood's Edinburgh Magazine*, August 1821, pp. 36–7, was also this game, for it had the words:

> Scotsman, Scotsman, lo!
> Where shall this poor Scotsman go?
> Send him east, or send him west,
> Send him to the craw's nest.

And Caleb, in *The Bride of Lammermoor*, 1819, ch. xxvi, refers to 'the bairns' rhyme—

> Some gaed east, and some gaed west,
> And some gaed to the craw's nest.'

Other references: W. H. Patterson, *Antrim and Down Glossary*, 1880, under 'Hurly-burly' (Patterson subsequently described the game and called it 'Capball'—Gomme MSS., 30 January 1915). Northall, *English Folk-Rhymes*, 1892, pp. 401–2, ' 'Otmillo', i.e. 'Hot Mill' or 'Through the Mill' (Warwickshire). J. Inglis, *Oor Ain Folk*, 1894, p. 110, 'Het Rows and Butter Baiks' (Angus). E. W. B. Nicholson, *Golspie*, 1897, p. 118, 'Cabbage-stock'. *English Dialect Dictionary*, vols. i and ii, 1897–1900, 'Burn the Biscuit' (north country), and 'Eettie ottie for a tottie, where shall this boy go?' (Aberdeen, 1853). Maclagan in his *Games of Argyleshire*, 1901, pp. 215–16, gives 'Hickety, Bickety'; and 'Huggry, Huggry, Piece, Piece', in which the last back was asked if he would have 'wind' or 'rain'; if he chose 'wind' he was thoroughly fanned with their bonnets, if 'rain' he was spat upon.

Correspondents played the game under the names: 'Hurly Burly, Pim Bo Lock' (Midgley, Halifax, *c.* 1900); 'Ickery, Ickery, I Cuckoo' (Alfriston, Sussex, *c.* 1910); and 'Tally-ho Dogs' (Millom, Cumberland, *c.* 1925).

Compare some of the games, but not the conjectures, given by Gomme under 'Hot Cockles' (vol. i, 1894, p. 229; vol. ii, 1898, pp. 429–30). It is possible, however, that the names 'Hot Peas' and 'Hot Cockles' are not coincidental.

The game seems to have been well known in Germany in the nineteenth century under the name 'Salzhäring'. One player knelt down and put his hand behind his back. The other players came up behind him in turn and touched his hand asking:

> 'Tik, Tak, wo schall de Mann hen?'

Each player was sent to a different place and waited there until the one who sent them called out 'Solten Hering, solten Hering, solten Hering!' They then ran back and, as in Britain today, the last back was punished 'with blows and shoves', and became the kneeling player in the next game (H. Smidt, *Wiegenlieder, Ammenreime und Kinderstubenscherze* (Bremen) 1859, p. 60, cited by Böhme, 1897, pp. 581–2).

7

Duelling Games

THE games that follow are those in which two children place themselves in direct conflict with each other, yet scrupulously observe the conventions of the encounter. Whether the test be of their courage when steering directly at each other on bicycles, or of the simple ability to choose a more resilient stem of ribgrass, the naïvety of their conduct is generally such that it does honour to Rousseau. Thus in 'Slappies' a boy will continue to accept punishment from a faster mover, blaming only his own slowness for the pain; and in the 'autumnal jousting with horse chestnuts', as Richard Church has observed, a boy's word is always accepted by his fellows when he states that his conker is a 'fiver' or a 'tenner'.

Obviously the most interesting duels are the ones children engage in when away from supervision; and we omit here the contests, or 'partner activities' (to use P.E. jargon) which have now become domesticated in school gymnasiums, such as 'Chinese Boxing' (trying to force opponent to hit himself), 'Chinese Wrestling' (wrestling with one hand), 'Japanese Wrestling' (trying to get an opponent off a mat), 'Chinese Tug' (standing back-to-back and attempting to pull opponent under legs), 'Cock Fighting' (squatting on the floor, or on hunkers, with hands clasped under the knees),[1] 'Knee Boxing' (fighting only with the knees), 'Pulling the Cow to Market' or 'Obstinate Calf' (attempting to pull opponent along by clasping hands behind the back of his head), and 'Uprooting the Slipper' (attempting to remove opponent's shoe while keeping on one's own). Away from school children feel these contests are dull sport compared with a prohibited game like 'Split the Kipper'.

§ *Contests Mainly Requiring Strength*

ELBOWS

When a boy boasts of his superior strength he may be challenged to 'Elbows'. A table is cleared and the contestants sit facing each other,

[1] Several styles of combat are known as 'Cock Fighting' (see pp. 214–15, 218–19); in fact the terms 'cock' and 'cock fight' are recurrent in children's speech, continuing testimony to the place the 'royal diversion' once held in British life. It has been prohibited since 1849.

placing their right elbows on the table and clasping the other's hand, so that the two forearms are upright against each other. The boy who succeeds in forcing back the other boy's forearm on to the table is the victor; and if a lighted candle is placed on one side of the table and a knife point fixed upwards on the other, it wonderfully increases the interest of the contest.

¸ One instance of this combat occurred at Mulligan's drinking-place in James Joyce's story 'Counterparts' in *Dubliners*, 1914; another was between the 'tough' of the class and the new master in John Townsend's *The Young Devils*, 1958, p. 56; a third was fancifully depicted between Khrushchev and President Kennedy in a French cartoon reproduced in *The Sunday Times*, 20 August 1961, p. 2.

KNIFING

The two adversaries find sticks of the same length and holding them in their right hands, take their opponent's wrist in their left hand, and attempt to stab each other, while avoiding being stabbed themselves. Alternatively they have one long and strong stick or pole, of which each takes an end single-handed, and tries to stab or jab his opponent with the other end. ' "Knifing" is a game I like playing,' says a 12-year-old. 'We kneel down and start fighting and we see who can get the other's throat first. But there are certain rules, which are, no kicking, and keep your free hand behind you.'

DIVIE DAGGER

A 14-year-old boy writes: 'Divee Dagger is another game I learned at Langholm. We played it at the age of eleven when we played on the hillside. We find a flat stretch of ground of about twenty yards long and two teams are picked. Someone then makes a wooden "dagger" and places it upright in the ground. The first people from the teams come out and stand at the same distance from the "dagger" and on the cry "Now" they both race for the "dagger". One of them gets the "dagger" and one of them does not. The two people then fight with the dagger until one of them is hit with the point and killed (in pretence). Then the second people of each team fight until each person has had a turn, and then the team with the most people alive is the winner.'

LIFTING

The rivals sit on the ground opposite each other, feet to feet, and either hold each other's hands, or both take hold of a short stout stick, held crosswise between them. They then pull against each other as hard as they can, to discover which is the weaker, a fact which is not made more palatable to the loser by being so apparent, for he finds himself raised from the ground while the other remains seated.

** This trial of strength is clearly depicted in the illuminated manuscript 'The Romance of Alexander', 1344 (Bodley MS. 264, fol. 100), and in a French Book of Hours of the fifteenth century (Douce MS. 276, fol. 39ᵛ). It is also shown in *Les Jeux et plaisirs de l'enfance*, 1657, under the title 'Le covrt baston', and this was the name of one of Gargantua's games (1534). In more recent times it has been a popular entertainment in Scotland. It was described by Jamieson in 1808, under the title of 'Sweir-tree', and was also known as 'Drawing the Sweirtree'—the 'sweirtree' or 'lazy tree' being the stick with which the contestants sought to draw each other off the ground. Mactaggart, *Gallovidian Encyclopedia*, 1824, p. 26, quotes an old man of nearly ninety boasting: 'I hae seen the day I wad hae pulled ony o'm aff their doups at the sweertree.' Jamieson later reported (1825) that in Tweeddale when persons grasped each other's hands, without using a stick, the sport was called 'Sweir-drauchts'. In Ireland, where the handle of a spade or pitchfork might be used, it was known as 'Sweel Draughts' (O'Súilleabháin, *Irish Wake Amusements*, 1967, pp. 39–40). In Indiana it was 'Pulling Swag' (*American Nonsinging Games*, 1953, p. 175, where Brewster gives several references to northern Europe). And amongst the Turks it is or was 'Quvvet' meaning *strength* (*Folk-Lore Society Jubilee Congress*, 1930, p. 143).

· COCK FIGHTING

The two cocks who are to fight each other are often chosen with the rhyme:

> Hop, hop, hop to the butcher's shop,
> I dare not stay no longer,
> For if I do my mother will say
> I've been with the boys down yonder.

Once they have been selected they have to stay on one leg and keep their arms folded. They hop towards each other, and butt, barge, or 'dunt' their opponent, trying to knock him over, or at least force him to put his second foot on the ground. Whoever unbalances the other, without of course putting both his own feet on the ground, is the victor, and takes on someone else.

This contest is sometimes known as a 'Shoulder Fight'. In Scotland it is 'Hopping Davy' or 'Hoppy Dig' ('Humphy Dick' says an Aberdonian); in Dublin 'Hopping Cock'; and amongst Wolf Cubs 'King of the Ring'. Without barging it becomes 'Catch Leg'.

A similar game is called 'Bumpers' or 'Bumper Cars'. In this the antagonists must keep their arms folded, but have both feet on the ground, and the victor is he who knocks the other player over. The feature of this combat is the initial charge and collision, which in itself is liable to prove decisive, one party or the other usually being sent sprawling (Banbury and St. Peter Port).

⁂ The earliest reference found to a fight between two hopping boys is in *Games and Sports for Young Boys*, 1859, p. 4, where, like today, it is called a 'Cock Fight'. In Forfar, *c.* 1910, it was 'Hockey Cockey Fechtie'.

EGGY PEGGY

This is a duel between two hoppers, exactly as in 'Cock Fighting' but with a set ritual for determining who shall be the second combatant. One person is named 'Eggy Peggy', and keeps out of hearing, while each of the other children (the players are usually girls) chooses a colour and stands in line. Eggy Peggy, who is supposed to have a bad leg, hops up to them and says: 'Eggy Peggy has broke her leggy'.

Children: 'What on?'

Eggy Peggy: 'A barbed wire gate'.

Children: 'What do you want?'

Eggy Peggy: 'A pair of stockings'.

Children: 'What colour?'

Eggy Peggy names a colour. If there is a child who has chosen this colour she hops out, and the duel begins. Whoever wins is the next Eggy Peggy.

Names: 'Eggy Peggy' (Oxford and Lydeard St. Lawrence, Somerset); 'Heggy Peggy' (Bristol); 'Heckety Peckety' or 'Hippety Skippety' (Swansea). Compare the games 'Little Black Doggie' and 'Limpety Lil'.

In Scotland and the north country the game is often known as 'Cigarettes'. Each of the players, except the 'caller', chooses a brand of cigarettes, for instance, Bristol, Players, Senior Service, Batchelor, Woodbine, Passing Cloud, Piccadilly, Capstan, etc. The 'caller' names a brand, and if anyone has chosen it he hops forward and the barging begins. In Liverpool the game is played in the same way but using the names of 'Comics', and is sometimes so called; or it is known as 'Fighting Cocks', and the winner who remains standing triumphant in the middle of the road on one leg is called 'king cock'.

DANCE, FIGHT, OR WINDMILL

This entertainment has been reported only from Aberdeen. The players are counted out, and the one who is 'out' stands in front and may challenge whom he wishes to 'Dance, Fight, or Windmill'. The player challenged then has the privilege of choosing the form of the contest. If he chooses 'Dance', both contestants must hop on one leg, and keep turning round all the time until one or other of them is dizzy and collapses. If the player chooses 'Fight', they both fold their arms, approach each other hopping, and have a 'cock fight'. And if he chooses 'Windmill', the one who challenges whirls his arms, and the other must twice run under the flailing arms without being hit.

BRANCH BOY

An 11-year-old in St. Peter Port, Guernsey, reports:

'Two boys climb on to the branch of a tree, clinging on with their hands and facing each other. Then they pull and push with their legs to get one another off the branch. Very often they both fall to the ground, but if one wins he shouts "Branch boy!" and jumps to the ground and sits on the other, who objects and fights. If he does not object to being sat on he is tied to the tree.'

** This duel is, of course, merely a make-shift version of the old fairground sport in which two contestants perched themselves astride a cross-pole, and each being armed with a bag of flour, attempted the not very difficult task of knocking the other to the ground, and the more difficult feat of remaining on the cross-pole themselves.

BUCKING BRONCO

In this exercise the contestants are not on equal terms: one boy allows another to mount his back, and then devotes his energy to bouncing him off. However, when a horse succeeds in throwing his rider, the rider becomes the bronco, and it is the other's turn to be discomfited. Occasionally the mount is made up of two boys, one bending down and holding the waist of a boy in front, as in a pantomime horse. The amusement is also known as 'Shaking Horse' (Alton), and 'Donkey' (south-east London).

** It always seems to be necessary to the enjoyment of this sport that the person underneath be thought of as other than a playfellow. As long ago as 1824 Mactaggart described a frolic which the Irish 'seem to enjoy' at their wakes, called 'Riding Father Doud', and Father Doud turns out to hold no more elevated office than that of the present-day bronco (*Gallovidian Encyclopedia*, pp. 320–1).

PIGGYBACK FIGHTS

Two small boys mount the backs of two 'well-built sturdy lads', usually by taking a good run and leaping on. The challengers, thus mounted, face each other some ten yards apart and charge. The object of the engagement, carefully explained by our informants, 'is to make the rider and horse part company', 'to knock your opponent down', 'to pull the rider off by any possible means or to upset both rider and horse'. 'The rule is that only the person on the back is allowed to fight, the others can only barge.' Sometimes the riders sit on the bigger boys' shoulders instead of their backs, and are then, in the metropolitan area, known as 'flying angels'. 'I prefer the shoulder fights because they are rougher and more falls are seen,' states a north-country boy. 'The shoulder fights usually end up in a flaring temper,' says another, 'because one man can be pulled off and hurt, and he may get up and hit the other man.'[1]

Sometimes more than two pairs fight at once. A 12-year-old Oxford boy writes:

'The craze at our school is piggy-back fighting. Every playtime all the boys from our school collect on the field and find a partner bigger than themselves and mount him. . . . To have good fun you need about twenty boys, ten mounts and ten horses. Sometimes more than one boy gets hurt and sometimes the injury is quite serious. Some boys' injuries are a cut leg, a broken tooth, a cut ear. Once or twice a boy's jumper gets torn like mine did last week all round the neck. Some nights all the boys collect on the local park to play.'

Sometimes the boys use sticks as swords, and think of themselves as knights. At Knighton, in Radnorshire, where they play 'Knights on Horseback', the one who 'stays on longest' gets carried shoulder high by the rest. They have a king, and carry the winner before the king, and set him down, and the king ruffles the winner's hair. Then the king presents him with a bunch of dandelions or other wild flowers. 'It is a very good game although rather rough.'

In Edinburgh, and doubtless in other places, they practise jousting on their two-footed steeds. Each rider has a buff ('which is a pole with a muff on one end'), and 'the object of the game is to jolt the other person out of the saddle'. 'The only rules are that you can't aim for the head and you can't knock the horse.'[2]

---

[1] However inappropriate the term 'flying angels' in this context, it is traditional. See James Greenwood, *Odd People in Odd Places*, 1883, p. 45 and Norman Douglas, *London Street Games*, 1916, p. 19, ' "Horse Soldiers" (also called "Flying Angels") is rather rough'. In Aberdeen to ride on someone's shoulders is termed 'cocksie coosie', hence the game 'Cocksie Coosie Fight'.

[2] It is scarcely necessary to confirm that little boys have jousted in play ever since the days

Sometimes the combats are conducted on other types of mounts.

Camel Fighting. The camel is made by one boy standing upright, and another boy bending down behind him holding on to his hands, or, occasionally, clutching his waist. The rider clambers on to the second boy's back: 'Then you charge.'[1] This formation is also known as a 'Donkey' and, in Dublin, as a 'Milk Car'.

Elephants. Two people stand upright, side by side, with their arms behind their backs holding each other's hands, right hand in right, left hand in left, and a third person bends down and grasps their hands. The rider mounts the third person's back.

Chariots. Sometimes made with the same formation as 'Elephants', the two people in front acting as horses; more often made by two people standing beside each other linking their outer hands, and the charioteer using their inner hands as stirrups, and keeping his balance with his hands on their heads.

Battering Rams. Two boys hold a third boy horizontally on their right shoulders with his feet forward, and try their human weapon against other players mounted piggyback.

Rumblin' Rhinos. Three boys are held horizontally on the shoulders of eight others, four linked together in a line in front, and four behind. A Liverpool boy says:

' "Rumblin' Rhinos" are used to smash up piggyback fights. The boys who are the crosspieces are chosen because they have got hard shoes, either leather, or rubber soles with studs on. "Rumblin' Rhinos" can have fights between themselves, but this usually ends up in somebody getting their teeth kicked in. "Rumblin' Rhinos" are known as "Icecream Carts" and "Tanks" in other places. Although this game is dangerous it is great fun, especially for the winners. The loser is the rhino first rumbled. The game is mostly played in the playgrounds of schools, and as people get hurt, the school usually ban it, although the kids still play it.'

Names: 'Piggyback Fighting' (rarely 'Pick-a-back') is also known as 'Horse Fighting', 'Horses and Riders', 'Jockeys', 'Donkey Fights', 'Cock

---

when their elders jousted in earnest. A representation of two small boys riding poles, by way of hobby-horses, and jousting with sticks, is reproduced by Strutt, plate xv, from a fourteenth-century manuscript book of prayers. An early sixteenth-century Flemish Calendar in the British Museum shows a group of boys likewise engaged, carrying whirligigs for lances (MS. Adds. 24098, fol. 23b). William Fitzstephen, in his account of the sports of Londoners in the twelfth century, refers to the lay-sons of citizens coming out into the fields on Sunday afternoons in Lent equipped with lances and shields, and adds that there were also present 'the younger sort with pikes from which the iron head has been taken off, and there they get up sham fights'.

[1] A duel between camel riders is shown in Bruegel's picture of children's street games, painted 1560. Here the two riders are shown holding either end of a loop of rope, and are attempting to pull each other across a line marked by two bricks.

Fights', 'Fighting Cocks', 'Collie-bag Fights' (Aberdeen), 'Collie-buckie Fighting' (Edinburgh), 'Hunch Cuddy Hunch' (Glasgow), 'Jousting', 'Charging', 'Tournaments', and (in Petersfield) 'Humpback Fighting'.

*** Piggyback riding and fighting is presumably universal. Robert Louis Stevenson in Tahiti, 1888, noted that 'the boys play horses exactly as we do in Europe'; and Samuel Marshak recalled bloody 'cavalry battles' in his youth at Ostrogozhsk in which 'the boys would charge into the fray mounted on their class-mates' (*At Life's Beginning*, 1964, pp. 126–7). Such play is certainly ancient. A sarcophagus in the Vatican Museum, Rome, shows boys engaged in a piggyback fight, although with one of the *horses* clasping the hand of the opposing rider and seemingly attempting to pull him down. (Since the rider is seated in the common classical style, merely clinging with his knees to the upstanding mount, his task should not have been difficult.) Medieval artists depict not merely piggyback riding, but duels between the riders exactly as today (see Rutland Psalter, *c.* 1250, fol. 70b; Stowe MS. 17, Hours of the Virgin, Flemish, *c.* 1290–1300, fol. 100ᵛ; Yates Thompson 8, Breviary, v I, French, *c.* 1290–1310, fol. 92; Luttrell Psalter, *c.* 1340, B.M. Adds. 42130, fol. 62; Bodley MS. 264, 'The Romance of Alexander', 1344, fol. 91; MS. Royal 2, B vii, Queen Mary's Psalter, *c.* 1320, fol. 161ᵛ). In *Les Trente-six figures*, 1587, piggyback is termed 'le jeu de sainct Chretofle'.

§ *Contests Requiring Nerve and Skill*

DANGER RIDE

Two boys mount their bicycles and circle round a patch of wasteland. They may do anything they can to separate their opponent from his bike, swerve in front of him, cut across him, and even aim at him, provided only that they do not actually touch him or his bicycle. At Barrow-in-Furness the two riders charge each other, sometimes down the length of a street, and just before they meet they have to apply their brakes and balance motionless. 'The first one to get off, gets scragged by the other lads.' The game becomes unpleasant when, in some places, it is called 'Chicken' and the two riders charge each other with no intention of stopping: he who swerves first being judged a chicken while the other is a cock (see also p. 270).

SPLIT THE KIPPER

In this contest, which, as one boy admitted, 'takes quite a lot of nerve', the two adversaries stand facing each other, a yard or so apart, with their feet together. The first boy throws a knife, preferably a sheath knife, so that it sticks in the ground not more than twelve inches to the left or right of one of his opponent's feet. The other boy, without moving his feet, plucks the knife out of the ground, and moves his nearest foot to the place where the knife went in. In this position he makes a return throw (most boys specify that the knife must be thrown by the blade), and the first boy, likewise, moves his foot to where the knife stuck in. However, should the point of the knife not stick into the ground, or should it stick in more than twelve inches away from the person's foot ('if more than a span', says a Durham lad; 'if over two knife lengths', says a Fife boy), the player does not have to move his foot, and the throw is lost. The object of the game is to force the opponent to stretch his legs so far apart that he cannot move them further, and gives in, or falls over while attempting the stretch. In this form 'it is a short game and suitable for short breaks at school'. But in many places, particularly in the north, a player is allowed to 'split' his opponent. If a person's legs are uncomfortably wide apart, and his opponent's likewise, or at least moderately open, he may attempt to throw the knife so that it sticks between his opponent's legs, and if he succeeds in this, may close his own. Usually 'splitting' is allowed only once, twice, or three times, otherwise the game can 'go on for ages'. Alternatively in Scotland, including the Isles, if the knife sticks between the opponent's legs, that person has to turn round, and thereafter throw less surely, and undoubtedly more dangerously, with his back to the other player.

This game, or ordeal, which is sometimes played with a dart, iron spike, or geometry dividers, has become popular only during the past decade. In our survey conducted in the early 1950s we heard of it only in the vicinity of Manchester, while in the early 1960s it was reported from virtually every contributing school, a number of boys describing it as their favourite sport. Further, the game now seems to be established, and to be developing its own oral lore, despite the fact that it has caused injuries (not to mention cut shoes and split trouser-seams), and has been banned in a number of schools. In Guernsey a doggerel rhyme or threat is now repeated by the contestants, and it will be noticed that it is similar in style to that used by conker players:

> Hick, hack, hoe,
> My first go.
> I'll split you yet
> And you'll forget
> That it's your go.[1]

[1] When we tried the game ourselves we were at first uneasy, and justifiably so, about where

In England and Wales the usual names for the game are 'Split the Kipper', 'Splits', and 'Stretch'. In Wigan it is 'Split Leg', in Enfield sometimes 'Slit Jack', and two West Ham boys call it 'Chinese Torture'. In north Lincolnshire it is often 'Black Foot', or occasionally 'Black Jack'. One report from Guernsey names it 'Watch Your Foot'. Throughout Scotland, where the game is rapidly supplanting the older knife-throwing games, it is 'Knifie'. Several informants state that they first learnt the game at Scouts.

In South Wales and East Anglia a variation is appropriately known as 'Nerve' or 'Chicken'. In this contest the players *start* with their legs apart and aim between each other's legs. The player closes his leg to where the knife stuck. 'It goes on like this until one of the boys gets scared and gives in', says an 11-year-old. 'He is then called "Chicken".'

TERRITORIES

This game, apparently well known in the first half of the century, now seems tame compared with 'Split the Kipper'. A circle is drawn on the ground, with a line through the middle, and one boy has one half and one the other. The boys throw in turn into each other's territory. When one of them succeeds in sticking a knife into the other's sector, a new line is drawn through the place where the knife fell, and in the direction that it points, and the old line is rubbed out. The other boy then has to throw his knife into the smaller segment in an 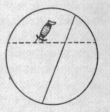 attempt to regain territory, while the first boy throws to diminish the other boy's land. Alternatively each boy stands in his own territory, and cuts off pieces of his opponent's territory, until he no longer has space to stand on—Manchester and Aberdeen.

⁎ Cf. Jamieson, *Scottish Dictionary, Supplement*, 1825, 'Cat-Beds'; Maclagan, *Games of Argyleshire*, 1901, p. 144, 'Sgrothan'; E. Meier, *Deutsche Kinder-Spiele aus Schwaben*, 1851, p. 394, 'Ackerles oder Kluvander'.

KNIFIE

This curious trial between two boys, which of them can first complete a prescribed series of feats with a pocket knife, is another game rapidly

the knife was going to land. But after some practice we found that our determination not to be humiliated by our opponent was such that we had developed an icy coolness: we were concentrating on the flight of the knife from the point of view of the game rather than of our own or our 'partner's' safety.

being supplanted by 'Split the Kipper', and in Scotland the new game has appropriated the name of the old. 'Knifie', so called in Scotland since the beginning of the century, is a game that has apparently held youthful attention for 400 years. It is here described, perhaps for the last time in Britain, by a 10-year-old boy in the Isle of Lewis:

' "Knifie" is a game for two people. First you do "Handy" in which you place the knife in the palm of your hand and try to make it stick in the ground by tossing it into the air. The same with "Backsy" except that you place it on the back of your hand. "Fisty" also has the same idea but it is your fist you place it on. Then you try to stick the knife into the ground six times in succession. Then try "Pitch and toss". Stick it in the ground, then try to hit it with the palm of your hand and try to toss it into the air, so that it will land blade first in the earth. First person to reach this point is the winner. This is a very good game, but it can be very dangerous.'

*** In former times it was the victor's privilege to drive a peg into the ground with as many blows of his knife-handle as the loser required additional throws to complete the game; and the vanquished, by way of penance, had to pull the peg out of the ground with his teeth. Hence the game was commonly known as 'Mumble-the-peg', 'Mumbletypeg', or 'Mumblepeg', a name-form which survives to this day among boys in parts of the United States, and in the outback of Australia, where the loser is made to pull a matchstick out of the ground.

'Mumble the pegge' is mentioned, along with 'scourge-top', 'Leape frog' and 'Nine holes' in the prologue to *Apollo Shrouing* by William Hawkins, a drama acted by the boys of the grammar school at Hadleigh, Suffolk (where Hawkins was master), on Shrove Tuesday, 1626. The youth Nehemiah, in Brome's *The New Academy*, 1659 (II. ii), who looks forward to playing games with a young wife, declares 'at Mumbledepeg I will so firk her'. (His mother tells him, 'When y'are married, you'll finde other pastime.') And a hundred years earlier, Moros in *The Longer thou Livest, the more Foole thou art*, written about 1569, speaking of his diversions at school, apparently refers to the sport when he says, 'You would laugh to see me mosel the pegge'.

As is to be expected with so old a diversion, it has a wide distribution. Sutton-Smith, *Games of New Zealand Children*, 1959, pp. 126–7, describes it under the names 'Stagknife', 'Stabknife', 'Jack-knife', 'Knifey', 'Momley Peg', and 'Bites'; Newell, *Games of American Children*, 1883, p. 189, gives it as 'Mumblety-peg'; Brewster, *American Nonsinging Games*, 1953, pp. 142–5, as 'Mumblepeg', and states that it is played in identical fashion in Hungary. The game has also been described to us by a 12-year-old Polish boy as played in Poland:

'You get a knife and set the point of the blade on your finger and toss it into the air so that it spins. If it stick in, you set the point on your foot and toss it into the air to see if it will stick, you then do the same on your nose and chin. The person who takes the least number of throws to stick it in wins.'

It is still played by boys in Trieste, where it is called 'Il Magnatappo' (Lumbroso, *Giochi*, 1967, p. 23). And it appears to be shown, being played with bricks, in Breugel's 'Kinderspiele', painted in 1560.

Our English correspondents recall playing the game with a sequence of up to eighteen tossings, including from nose, shoulder, and forehead. They knew it either as 'Stick Knife' or 'Stickie Knife'. *EDD* describes it under 'Spinny-diddl-um'.

§ *Contests Requiring Fortitude*

KNUCKLES

'Knuckles', in Scotland 'Knucklie', is a kind of conkers with the hands, played for no other reason, it seems, than to test the physical fibre of the players. One boy holds out a clenched fist and the second strikes the knuckles as hard as he can with his own. The first boy then strikes the other's knuckles, and so on, alternately hitting and being hit. 'As the game proceeds', says a 12-year-old, 'the bashing gets more fierce.' Nevertheless 'you mustn't pull your knuckles away, you've just got to take it. The first to cry is a baby'; or, as another youngster put it, rather more sedately, 'the loser is the one who retires earlier'. Yet some lads seem constitutionally unable to give in, and the game (so called) continues long after skin has been torn from their hands. Two 11-year-olds were observed in a playground taking turns at each other; both were in agony, yet they were found to be still at it ten minutes later, and neither had given in when, mercifully, the whistle sounded for the end of playtime. A 9-year-old on the outskirts of Gloucester told us, with a glint in his eye, that thereabouts the game was known as 'Knuckledusters'.

⁎ British children have no monopoly of this pastime. A fearsome bout of 'Knuckles' is described in Arthur Roth's autobiographical novel *The Shame of our Wounds*, 1962, p. 44, set in a New York Catholic Boys' Institute.

FLAT JACK

Reported only from Northumberland. One boy bends down and the other hits him on the back of the head. Then the other bends down and is hit likewise. If a boy dodges when his head is down, the one who is hitting has another turn. 'They take it in turns and the one who gives in or runs away is the loser.'

BOB AND SLAP

More widespread and slightly less primitive than 'Flat Jack' is the contest in which one person tries to duck his head between another's hands, held about two feet apart. The one who bobs his head between the hands may choose his moment to move; and this gives him a certain advantage, for surprisingly he very often slips through without having his head slapped, no matter whether moving upwards or downwards. The exercise is sometimes known as 'Slapping the Duck', especially in the United States.

** In nineteenth-century rural England this was a public-house sport, in Hampshire called 'Catch the Crow', the only difference being that one player was seated, with his knees apart, and his hands against his knees, while the second player knelt down and attempted to pass his head between them. It was also known as 'Quack'. 'George Forrest' (The Revd. J. G. Wood) described it in *Every Boy's Annual*, 1864, as 'one of the most mirth-provoking games imaginable', and added, 'I have seen ladies nearly ill through continued laughter'. Indeed, in Capri, today, the Tarantella dancers entertain guests in restaurants and hotels with the humours of the contest, which in the Neapolitan dialect is called 'U Scarrafone du Camp'. Urquhart in his list of English games (*Gargantua*, ch. xxii, 1653), names 'Bob and Hit' and 'Bobbing, or flirt on the nose', one or other of which is possibly this sport.

SLAPPIES

This is another test of a player's willingness to absorb pain. One player stretches out his hands palm upwards, and the other player, at first less fortunate, has to rest his hands flat on top of them, with the backs of his hands uppermost. The player whose hands are underneath attempts to slap one or both of the hands placed on his by suddenly withdrawing his hands and whipping them down on the backs of his opponent's. His opponent may attempt to move his hands out of the way, but not until the other has begun to move, or he forfeits a turn. If the one who is slapping misses the other's hands, he has to go on top; but if he succeeds he has another go. Although the person slapping has further to move, he has the advantage of being first to move (cf. 'Bob and Slap'), and in practice, when the players are of the same age, the chances of being hit or missed are about equal. On occasion, however, one player may gain ascendancy over the other, and, on a cold day, the result can be very painful. In Cumnock the game is known as 'Leathery', in Forfar as 'Puir Pussie'.

** This sport is also reported from Maryland (Howard MSS.), from Ireland, under the names 'Poor Little Pussy' or 'Poor Snipeen' (*Irish Wake Amusements*, 1967, pp. 46–7), and from Belgium under the name 'Leu Tchôde Main' (*Le Pays de saint Remacle*, no. 2, 1963, p. 144). A party game on the same principle, called 'Copenhagen', is given in *The Girl's Book of Diversions*, 1835, p. 10. A Cornish game called 'Scat' in which a player has to pick up a stick from the hand, with which to hit it, is described in *Folk-Lore Journal*, vol. v, 1887, p. 50.

STINGING

In 'Stinging' the punishment befalls the player who is unlucky rather than unskilful, or so it seems to him, for he has failed to guess the other player's mind. The two contestants flash their fingers at each other, usually making one of the three finger formations used in 'Odd Man Out' and 'Ick Ack Ock' (p. 26), thus, fingers clenched ('stone'), fingers flat out ('paper'), and two fingers stretched out and kept apart ('scissors'). In a number of places, however, for example, Edinburgh, Norwich, and Petersfield, they also make the signs 'rain', bunching their fingers and pointing them downwards, and 'fire', pointing their fingers upwards. 'Rain' puts out 'fire'. and rusts 'scissors', while 'fire' burns 'paper', and blackens 'stone'. More often the game is played merely with 'stone' beating 'scissors', 'scissors' beating 'paper', and 'paper' covering and thus beating 'stone'. Whoever wins each time wets his first and second fingers by licking them, pulls back his opponent's sleeve, and smacks his bare wrist good and hard, an operation which is more painful than those who have not experienced it might imagine. Play then resumes and continues until one boy or the other considers he has had enough. This game, like 'Slappies', is particularly prevalent in boarding schools. One informant tells us that on occasion it becomes an obsession in a classroom or dormitory, and some people's wrists, after a run of losses, would be 'absolutely bloody red'; yet they will not give in because they hope to start winning and make the other person suffer. When girls play the game they usually agree that a player must win three times before she has the privilege of giving a slap.

STAMPERS

The two rivals stand face-to-face, placing their hands on each other's shoulders. They may move about and dodge aside as they please, provided they keep their hands on the other's shoulders, but mostly they push each other, since the object of the exercise is to stamp as hard and as often as possible on the opponent's feet.

⁂ Formerly players used to kick each other's legs, but 'Shinning' or 'Cutlegs'—a rural, hobnail-booted sport of the nineteenth century—has not been reported in the present day.

§ *Duels by Proxy*

SOLDIERS

This gentlest form of combat, conducted with a stalk of ribwort plantain (*Plantago lanceolata*), has been a pastime with the young for more than 750 years. One player challenges another and they take turn about striking each other's stalk with their own until one of the plantain heads is whipped off, whereupon the victor crows his victory, and the loser feels no loss: another stalk is quickly produced—this is the pleasure of the game—and a new duel commences. During the flower's long summer season the game seems to be played everywhere, usually called 'Soldiers', although the plantains are also known as 'Blackmen', 'Cocks and Hens', 'Fighting Cocks', 'Hard Heads', 'Knights', and, in Perthshire, 'Carldoddies'.

⁂ In Scotland and the north country the game was long known as 'Kemps' (cf. the Old English *cẹmpa*, a warrior; Middle English *kempen*, to fight or contend with: the Norwegian *kjæmpe*, and Swedish *kämpa*, a plantain). Jamieson's description in the Supplement to his *Scottish Dictionary*, 1825, might well be of today:

'Two children, or young people, pull each a dozen of stalks of rib-grass; and try who, with his kemp, can decapitate the greatest number of those belonging to his opponent. He, who has one remaining, while all that belong to the other are gone, wins the game. . . . They also give the name "Soldiers" to these stalks.'

Jamieson also gives the name 'Carldoddy'; Moor, *Suffolk Words*, 1823, knew the game as 'Cocks'; Brockett, *North Country Words*, 1829, as 'Hard Heads' (Lancashire); Forby, *Vocabulary of East Anglia*, 1830, as 'Fighting-Cocks'; and W. H. Marshall, *Rural Economy of Norfolk*, 1787, knew ribwort as 'Cocksheads', presumably from the game, for it was certainly already well established. In the historical poem *Histoire de Guillaume le Maréchal*, written soon after 1219, the story is told how the boy William Marshal, later to become Earl of Pembroke and Regent of

England, but then not 10 years old, was detained as a hostage in the king's camp, while Stephen was besieging Newbury. One day the boy picked out the plantains (*les chevaliers*) from the cut grass strewn on the floor of the tent, and challenged the king, 'Beau sire chiers, volez joer as chevaliers?' The challenge being accepted, William laid half the 'knights' on the king's lap, and asked who was to have first stroke. 'You', said the king, holding out his knight, which the small boy promptly beheaded, greatly to his own delight. King Stephen (strictly in accordance with the rules of the game) then held out another plantain, but the game was interrupted (ll. 595–619). It matters not whether the story is apocryphal; as early as the thirteenth century a poet has shown himself to be familiar with the game.

Mrs. Craik, who knew the amusement as 'Cock-Fighting' (*Our Year*, 1860, p. 149) says that in her day children made the tough stems tougher still by twisting their necks round and round.

LOLLY STICKS

The game of 'Soldiers' is now also played with ice-lolly sticks, the sticks being, in urban surroundings, more available than are plantains. The stick, which is usually about four to four-and-a-half inches long, is held at both ends while the other player strikes it with his own stick held at one end. However, according to a young Mancunian, it is advisable first to examine the opponent's stick as 'some people try to cheat by playing with cardboard lollipop sticks which are a lot harder to break'. Others, it seems, practise one-up-manship by soaking their sticks 'so that they are waterlogged and won't break'. He continues, 'The lollipops, from which the lollipop sticks come, can be bought outside the school for the price of fourpence. They are bought from two or three ice-cream men who come outside the school at dinner time'. One informant states that at his school 'This craze was very popular until the masters ordered some boys to pick up all the pieces of stick in the school. Then it gradually faded away'. Another boy says 'This is a silly game and sometimes a sharp knock can cause blood on the hand'. Nevertheless, spasmodically, it has its adherents. In Liverpool it is called 'Chop Sticks', in Aberdeen, 'Icicle Sticks', in Sheffield, 'Foggy Plonks'.

CONKERS

For a brief spell in early autumn this game is as much part of the English scene as garden bonfires, and hounds cubbing at break of day. The boys are out searching for conkers, throwing sticks and stones up into the chestnut trees (the best conkers are believed to be at the top of the tree) and, with or without permission, invading people's gardens. They meet with little opposition. The youthful pleasure of prising a mahogany-smooth chestnut from its prickly casing is not easily forgotten; and when a vicar wrote to *The Times* complaining about the depredations of small enthusiasts, readers' sentiment was clearly against him.[1]

'Conkers', always so spelt, are also known as 'cheggies' in Langholm, 'hongkongs' in Grimsby, 'obbley-onkers' in Worcester, and 'cobs' in the area of Welshpool and Shrewsbury. A flat conker is popularly a 'cheeser' or 'cheese cutter'. Some boys bake their conkers for half an hour to harden them, or put them by the fire for a few days, or up the chimney. Some soak them in salt and water, or in a solution of soda; the majority prefer vinegar. An Edinburgh recipe is a teaspoonful of sugar and a little water in a jar of vinegar. A Putney boy puts them in vinegar for an hour, and then in water, explaining: 'If you did not put them in water the smell would keep on the conker and then people would not play you because they would think it was harder than theirs.' Others, more patient, put their chestnuts in a dark cupboard and leave them until next year. This makes them shrivelled and tough, easily recognizable as 'seasoners', 'yearsies', or 'second yearsers'. A boy with this year's conker, plump and shiny, sometimes called a 'straight conker', being straight from a tree, seldom cares to venture his new acquisition against a 'seasoner' and almost certain destruction.

No boys' game is more ruthless or carried out with more finesse than a conker fight. The conker is carefully selected. The hole through it is made with a meat skewer so as not to split the edges, and exactly through the middle. A strong piece of string, or a lace from a football boot, is procured which is long enough to be wound twice round the hand with about eight inches to hang down, and it is tightly knotted at the bottom to ensure the nut does not slip off and smash to pieces on an asphalt playground. It is a tenet of schoolboy faith that a conker is more likely to survive if it is the striker rather than the stricken, hence the conker-player's concern to have the first shot, which he secures by calling out 'First!' or, according to locality, 'My firsy' or 'Firsy jabs' (Bishop Auckland), 'First swipe' (York), 'First donks' (Shenfield), 'First hitsy' (Hornchurch), 'Bagsie first cracks' (Wigan), 'Iddley, iddley, ack,

---

[1] 'Vicarage grounds, chestnut trees in autumn, "conkers" and children—what more could a parish priest want?' Correspondent to *The Times*, 13 October 1962, p. 9. One 10-year-old in our survey delightfully confessed, 'My granny sends me all my conkers'.

my first smack' (Knighton), 'Hobily, hobily, honker, my first conker' (Lydney).

Conker Jeremy,	Iddy, iddy, onker,
My first blow,	My first conker,
Conker Jack,	Iddy, iddy, oh,
My first whack.	My first go.
Boy, 11, Cranford, Middlesex	*Boy, 13, Oxford*[1]
Ally, ally, onker,	Obbly, onker,
My first conker,	My first conker,
Quack, quack,	Obbly oh,
My first smack.	My first go.
Boy, 12, Thornton, Yorkshire	*Boy, 11, Gloucester*[2]

The other boy then holds up his conker, dangling on the string, at whatever height best suits his opponent, and keeps it still. The first boy sizes up to it, holding his own conker between his thumb and forefinger, or behind his first two fingers, as if they were a catapult, and pulls the conker loose with a swinging downward motion on to his opponent's nut. If he hits it the other player has his turn; if he misses he may be allowed two more tries. If his string tangles in the other boy's string there is immediate cry of 'strings'—or 'clinks' (Manchester), 'clinch' (Ferryhill), 'clenches' (Wigan), 'plugs' (Wolstanton), 'lugs' (Leek), 'tags' (Hainton), 'twits' (Cranborne), 'twitters' (Newcastle-under-Lyme), or 'tangles' (Bristol); and whoever cries first has an extra shot, or, in some places, two, three, or even six extra shots. Sometimes boys deliberately play for 'strings' so that they can claim extra shots, but this is not popular since the wrench the hand receives when the strings tangle can hurt and even cut the skin.

When one conker breaks another into pieces so that nothing remains on the string, the winning conker becomes a 'one-er', in Plymouth and Cornwall a 'one-kinger', in Sheffield a 'conker one', in Edinburgh, St. Andrews, Kinlochleven, Oban, and doubtless elsewhere in Scotland, a 'bull', 'booly one', or 'bully one', in Cumnock a 'bullyanna'. If it then breaks another person's conker it becomes a 'two-er', if a third a 'three-er', and so on. If one boy's conker is a 'tenner' and another boy's a 'fiver', whichever wins will become a 'sixteener'. A conker that is a 'sixteener',

---

[1] All children questioned in Oxford gave this formula, as did a correspondent recalling his Oxford schooldays in the 1880s.

[2] 'Isn't that rather a mouthful?' we asked. 'Oh no, you just say it quickly', and the boy gabbled it like one-o'clock. Compare the following which Mrs. Chamberlain records in *West Worcestershire Words*, 1882, p. 15, as 'written down for me by a National School boy'—'Hobley, hobley, Honcor, My first conkor. Hobley, Hobley ho, My first go. Hobley, hobley, ack, My first smack.'

and perhaps becoming battle-worn, is unlikely to be matched against anything less than a 'tenner'. It will not be worth the risk. In London, if a player drops his conker, or it is knocked out of his hand, or it slips off the string, the other player can shout 'stamps' and jump on it, and add its score to his own; but should its owner first cry 'no stamps' it cannot be counted as a victory, even if jumped upon and crushed. The worst disaster that can happen is that both conkers break at once, then both scores are lost.

When the conker craze is at its height 'there are pieces of conker flying in every direction, and we have to clean the yard up every day. The bins and wastepaper baskets are nearly full'. And the girls play too. One girl remarks that when boys are not very good at conkers they come to the girls' end of the playground 'because then they think they'll win, but sometimes the girls win'. When there is to be a match between two skilled players, each with a 'hundreder' or more, excitement flows through a junior school, bets may be laid, and the contest attracts as much attention as any sporting event in the school calendar. In 1952 the B.B.C. staged a contest on TV between a 460-er, a 1136-er, a 2385-er, and a 3367-er. The winner became a 7351-er. More recently a conker championship has been arranged annually at the village of Walton-on-Trent in Derbyshire. But happily the season is too short for much adult exploitation. Suddenly there are no more conkers to be had, and the game dies out. Those boys with a treasured handful set aside for next year hug their secret.

It is presumably just part of man's struggle with nature that one local authority (Lowestoft) has planted a commemorative horse-chestnut which will not bear chestnuts, so that it shall not be a temptation to the young; and that one toy manufacturer has attempted to popularize plastic conkers which, when broken, can be reassembled.

✷✷ The Horse Chestnut tree (*Aesculus hippocastanum*), introduced into England about the beginning of the seventeenth century, does not seem to have been common in the eighteenth century, and children do not appear to have played with horse chestnuts until the nineteenth century. Previously they played with cobnuts, and records of cobnuts as playthings go back to the fifteenth century; although it is difficult to tell the type of play. However, the name *cobnut* itself may come from *cob*, to strike, to top, to excel; and colloquially a cobnut was a large hazel nut that out-matched the others. Dorothy Osborne, in a letter written probably 29 January 1653, speaks of a friend who wears twenty seals 'strung upon a riban like the nutts boys play withall'; and from Hunter's

Hallamshire Glossary, 1829, p. 24, it is clear that this was how boys strung their nuts for a cobnut fight:

'Numerous hazel-nuts are strung like the beads of a rosary. The game is played by two persons, each of whom has one of these strings, and consists in each party striking alternately with one of the nuts on his own string, a nut of his adversary's. The field of combat is usually the crown of a hat. The object of each party is to crush the nuts of his opponent. A nut which has broken many of those of the adversary is a *cob-nut*.'

In the game of 'Cobs', 'Cobblers', 'Cock Haw', 'Conger', or 'Scabby', hazel nuts were hardened, strings which became 'twizzled' were cried at, and victories enumerated, virtually as with conkers today (see *Notes and Queries*, 7th ser., vol. ix, 1890, pp. 137–8). But cobnuts do not provide such a robust game as horse chestnuts. They are lighter, more difficult to obtain, and when obtained are better eaten.[1]

The Oxford English Dictionary (*Supplement*, 1928) suggests that the name 'conker' for a horse chestnut comes from an earlier game played with snail shells, sometimes called 'conkers', and that this stems from *conch*. There is, however, no record of boys calling snail shells 'conches', although they often called them 'conquerors'. Southey, in a letter to John May, 28 December 1821, recalled that in his schooldays at Corston, near Bristol, in 1782, there was a 'very odd amusement', greatly in vogue, in which snail shells were pressed against each other, point to point, until one of them was broken in:

'This is called conquering; and the shell which remained unhurt, acquired esteem and value in proportion to the number over which it had triumphed, an accurate account being kept. A great conqueror was prodigiously prized and coveted, so much so indeed, that two of this description would seldom have been brought to contest the palm, if both possessors had not been goaded to it by reproaches and taunts. The victor had the number of its opponents added to its own; thus when one conqueror of fifty conquered another which had been as often victorious, it became conqueror of an hundred and one.'[2]

---

[1] It is a fixed belief with schoolboys, and some adults too, that horse chestnuts are poisonous, and should on no account be put in the mouth, a superstition probably arising from their bitter taste. Formerly, however, they were considered beneficial to horses with the cough (hence their name?); and instructions for making a 'strictly edible and agreeable flour' for human consumption may be found in *Household Words*, 24 December 1881, p. 176. In Yorkshire, where chestnut trees are not as common as in southern England, boys continued to contest with hazel nuts anyway up to the Second World War, and possibly do so still.

[2] *The Life and Correspondence of Robert Southey*, vol. i, 1849, p. 55. Southey continues with a nice illustration of the schoolboy ethos, showing it to have been as particular in the eighteenth century as today. Southey, at this time a little fellow, came one day upon a boy whom he considered to be cheating. He 'had fallen in with a great number of young snails, so recently hatched that the shells were still transparent, and he was besmearing his fingers by crushing these poor creatures one after another against his conqueror, counting away with the greatest

This name is confirmed in *The Boy's Own Book*, 1829, pp. 10–11, where the game of 'The Conqueror' is described as being played with snail shells in the west of England, and is compared with a similar game of the same name in which one player threw a stone marble at another, hoping to split it:

'A strong marble will frequently break, or conquer, fifty or a hundred others; where this game is much played, a taw that has become the conqueror of a considerable number, is very much prized and the owner will not play it against any but those which have conquered a respectable quantity.'

This, in turn, appears to be little different from the game described by Ovid, in his poem 'Nux', where one boy, standing, is said to aim his nut at a nut on the ground, and must split it with a single blow.

The game of 'Conquerors' with snail shells, a parallel to the 'dumping' or 'jarping' of hard-boiled eggs that remains part of the Easter ritual in the north country (*Lore and Language of Schoolchildren*, pp. 252–3), was not peculiar to the west. One of John Clare's favourite pastimes in Northamptonshire was gathering 'pooty shells', threading them on a string, and playing 'what we called "cock fighting" by pressing the knibbs hard against each other till one broke'. See also 'Conkers' in *Holderness Glossary*, 1877; 'Fighting Cocks' in *Leicestershire Words*, 1881; 'Cocks and Hens' in *Northumberland Words*, 1892; and 'Cogger' in *Northamptonshire Words*, 1854. Amongst others, a correspondent to *The Athenaeum*, 28 January 1899, remembered duelling with snail shells when a schoolboy at Newport, Isle of Wight, in 1848, and he also 'often played conquerors with both horse-chestnuts and wallnut-shells', which is the earliest date we have for a contest with horse chestnuts, there being no mention in juvenile literature of playing 'Conqueror' with chestnuts until *Every Boy's Book*, 1856.

satisfaction at his work. He was a good-natured boy, so that I, who had been bred up to have a sense of humanity, ventured to express some compassion for the snails, and to suggest that he might as well count them and lay them aside unhurt. He hesitated, and seemed inclined to assent till it struck him as a point of honour, or of conscience, and then he resolutely said, no! that would not do, for he could not then fairly say he had conquered them'.

8

Exerting Games

'When you play "Jumping Jack" you keep jumping up and down. You go on jumping until you can't jump any more and fall on the ground. The last one who is jumping is the winner.'

Boy, 10, Alton

SOME games are little more than statements of vitality. They have few rules, they offer small scope for subtlety because the options open to the player are few, and they make correspondingly large demands on the player's forcefulness. They are games that are the pleasure of those whose talents lie in their limbs; and they are made bearable, very often, only by the pride that the young take in the practice of stoicism. Thus the game of 'Cat and Dog', played in Peterborough, is a simple trial of endurance: which of two boys can longest support the pain of bearing a third boy, who has jumped up between them, and exerts his weight on a shoulder of each, the first to give way being 'cat'. 'Grippy', at Langholm, amounts to no more than one person holding a button or coin in his fist, and the others trying to force his fingers open and gain the object. 'Steps' (otherwise known as 'Fire', 'Letters', or 'England, Scotland, Ireland, Wales') consists of jumping down a flight of steps in a given number of jumps, or taking one jump up or down on to a particular named step. And the game of 'Bundles', at Enfield and Fulham, seems to be merely an excuse for violence: 'One person is chosen and he is given ten seconds to run, then the rest run after him. When he is caught he is thrown on the ground and punched.'[1] In short, these are games in which the less courageous, or the more civilized, of the juvenile community are able to endure the part of spectator with some fortitude.

TUSSLES

Children's battles scarcely merit inclusion in a catalogue of games except that 'games' are what the children call them. It is true that some tussles

---

[1] The victim on these occasions does not attempt to resist. He curls up to protect himself, and takes what comes, consoling himself that the next person to be given ten seconds to run will be the one who caught him.

for high ground, with names such as 'Gain the Summit', 'Taking the Crown', 'Territories', 'Pull me Down', and 'King of the Castle' ('I'm the king of the castle, get down you dirty rascal'), retain a few of the conventions of a game, when a single player holds the keep.[1] But when one gang faces another the contestants are liable to be uninhibited, or like to think that they are, and, according to accounts received, this is so even when the battleground is the school steps:

'If one gang captures the "fortress", the other attacks it. Each side has a war cry, ours is "Aaarghrghrghaargh". The gang who has captured the fortress shouts out "Repel boarders". The attackers grab them by their legs and pull them down, sit on them, and pulverise them. Then the other gang comes down from the steps, and strives to rescue their friends, while my friend and I nip round the back and claim the steps. "Come on you miserable runts," we say, and one of the runts runs up the steps, only to be thrown over the side of the railing. By then the others from our gang are with us, dragging one or two hostages by the hair. These we rough up a bit. We give 'em one down the cake 'ole or punch up the bread basket, or give them the "knee-arm", and fling them at their friends when they try to "board" us' (Boy, 13).

Such activity, it is alleged, continues to the classroom door:

'When the bell rings there are cries of "Skunks", "Cheats", "Useless nits", and a final push or belt as we run into class. Hardly anybody gets away without some rip, bruise, cut, or black eye.'

Accounts like this are not the testimony of disinterested witnesses; and despite their almost monotonous conformity the adult reader is, of course, content to allow for a certain exaggeration. Then he may happen upon a detail that makes him glad nature has arranged he will not again be a schoolboy. Another 13-year-old writes:

'If you are lucky to get away without any affliction then your gang bashes you up because you have not fought well.'

CHAIN SWING

'Chain Swing' is almost as unpleasant as a fight, yet children, particularly young ones, seem unable to resist its thrill, and for a season it becomes a school craze, until banned, or until several players have been injured, and its attraction becomes less obvious. A number of children

---

1 '"King of the Castle" and "No Man Standing" are just red savagery set to rules', remarked Edwin Pugh in *Living London*, vol. iii, 1903, p. 267. 'King of the Castle'—in the United States generally known as 'King of the Mountain' or 'King of the Hill'—is included amongst the games in Bruegel's picture of children at play, painted in 1560. 'No Man Standing' or 'No Time for Standing' is a game in which every player tries to uproot every other by grasping at his ankles.

hold hands in a line, sometimes as many as twenty, often girls. They run forward together, and when they have gathered momentum, the child at one end, the leader, some say 'the bully of the playground', stops short and swings the chain round, which means that the ones at the far end have to go further and faster than the rest, and the endmost one is easily jerked off his feet, and dragged along the ground.

This sport also goes under the names 'Chain', 'Drags', 'Tally-ho', 'Whizz Bang', 'Stretching Snake', and, especially in London, 'Long Sausage'. In Canada and the United States it is 'Crack the Whip'.

** Compare 'Bulliheisle' in Jamieson's *Scottish Dictionary*, *Supplement*, 1825:

'A play amongst boys, in which all having joined hands in a line, a boy at one of the ends stands still, and the rest all wind round him. The sport especially consists in an attempt to *heeze* or throw the whole mass over on the ground.'

Gomme, vol. ii, 1898, reports the game under the name 'Port the Helm'.

TUG OF WAR

'Tug of War', whether between two people or two teams, is one of the more elemental struggles of the playground. 'When some boys start a tug of war with a girl's skipping rope, it usually ends up with the girls going to tell the teacher on playground duty and the boys get into trouble', reports a 9-year-old. However, a skipping rope is not the most favoured rope for tug of war, merely the most available. 'It is best to play with bull rope or your father's car rope because it does not snap so easily', asserts a 10-year-old. Generally two leaders pick sides; the sides take hold of the rope, with the heaviest person on each side at the end because he makes the best anchor man, and each side attempts to pull the other across a line marked on the ground, or, not infrequently, they pull until the other side lets go, for this is noticeably an activity in which short-trousered tempers quickly fray. 'Before long some people get sick and sit down', remarked another 10-year-old. 'The side that gets tired first usually loses. Some people are bad losers and start quarrelling, but others just say "Well it was only a game it doesn't matter".'

'Tug of War' is not as popular as it used to be, except among young children at the end of a singing game such as 'London Bridge is Falling Down', when the opposing leaders grip each other's hands, and the others line up behind them, each taking hold of the waist of the person in front. Only in the north, it seems, does tug of war continue to be an acceptable playground sport, and to be still called by the names of the opposing sides, as 'Rats and Rabbits', 'Soldiers and Sailors', and 'French and English'.

*** 'French and English' was the usual name in England throughout the nineteenth century, being given in, for example, *The Boy's Own Book*, 1828, and *The Boy's Handy Book*, 1863. The significance of the name was that the side which lost were Frenchmen, while the winners considered they had earned the right to be called true Englishmen. The editor of *The Boy's Handy Book* comments:

'The sport was exceedingly popular in our young days, when the belief held by most Britons concerning their neighbours on the other side of the channel, was still embodied in the beautiful lines (*not* by Tennyson) which ran thus:

> Two skinny Frenchmen, and one Portugee,
> One jolly Englishman thrash 'em all three!

And all schoolboys used to believe it, too.'

Maclagan in *Games of Argyleshire*, 1901, p. 132, speaks of 'Tug of War, as it is now called, what used to be called "French and English" '; and Chesterton in *All Things Considered*, 1908, pp. 35–6, recalls the game of 'tug-of-war between French and English'. In the eighteenth century the name seems to have been 'Pull Devil, Pull Baker', for this is the title of a picture by William Hamilton, of which Bartolozzi made an engraving. Yet earlier, in the sixteenth century, the name was apparently 'Sun and Moon'. John Higins in his *Nomenclator*, 1585 (and dictionaries following) defines the ancient Greek game *Dielcystinda* as 'a kinde of play wherein two companies of boyes holding hands all in a rowe, doe pull with hard holde one another, till one side be ouercome: it is called sunne and moone'. (Cf. Pollux, ix. 112.) It is probable that in the sixteenth century the opposing sides were formed by each player being asked privately whether he would be 'sun' or 'moon', in the way players today are invited to be an 'orange' or a 'lemon' in the game of 'Oranges and Lemons'; and it may be noted that a game of this kind, in which the players were asked whether they would be sun or moon to determine their side in a tug-of-war, was still being played in Alexander county, North Carolina, about 1928 (*North Carolina Folklore*, vol. i, 1952, p. 140).

ADDERS' NEST

A bucket, upturned box, tin can, boy's coat, stone, or stick is placed on the ground, and the players hold hands in a circle around it. At a given signal each player tries, without breaking the circle, to force one or other of his neighbours to touch or knock against the object in the centre, while taking good care himself to keep away from it. When a person touches it he has been poisoned and has to retire. The game recommences without him, and continues until only two players are left, which is the 'really exciting part of the game' since these two are usually the strongest players.

A Banbury girl says: 'There is a special way of holding hands, called "the hand grip". You bend your fingers and so does everyone else. Then you grip each other's fingers.' Sometimes, particularly when girls are playing, they first skip round as if unconcerned, and then all of a sudden begin pushing. In Enfield boys stand together in a ring on a large drain. One boy shouts out 'This drain is poisonous!' They leap off, still holding hands, and the struggle starts to force someone back on to it. In N.W. 2 four or five children have been observed standing on a manhole cover chanting:

> Five little sausages, frying in a pan,
> One went pop, and the others went bang!

On the word 'bang' they jumped off, and instantly set about ensuring that someone other than themselves should be shoved back on to it. In Accrington, when only one player is left, and he has been declared the winner, the rest of the players come back and try to get the winner 'out'. If this player gets the rest out again he is declared 'double winner'. 'This game can last a long time,' remarks a 10-year-old, 'and usually the winner is very exhausted.'

Other names: 'Poison', 'Poison Pot', 'This Drain is Poisonous' (Enfield), 'Smudger' (alternative name to 'Adders' Nest' in Croydon), and 'Herring on a Plate' (Forfar). The game has been taken up by the physical training instructors under such crummy names as 'Poison Circle Tag', 'Knock the Block', and 'Pull them in the Ring'.

*** In the nineteenth century the game was often played with a heap of boys' caps in the centre of the ring, hence the names 'Bonnety' in Keith and Nairn (Gomme, vol. i, 1894, p. 43), 'Dinging the Bonnets' (*Games of Argyleshire*, 1901, p. 1), and 'Chimney Pots' or 'Upsetting the Chimney' (correspondents with London childhoods, *c.* 1910–15).

BULL IN THE RING

The name aptly describes the game. The players link hands and enclose one boy who has volunteered or, more likely, been manhandled into the centre, and is now obliged to smash his way out, not using his hands, but charging wherever he thinks the link is weakest. He is allowed to butt, barge, turn-about, take by surprise, and generally display the characteristics of a baited bull, bellowing, looking ferocious, and treading on people's feet in an attempt to weaken their resolution; while those forming the ring 'shout and jeer' in return, but must merely use their bodies to stop the bull from getting out, 'not kick'. Formerly, when the bull escaped, he was chased by the members of the ring, and he who caught him became the next bull. This is still sometimes the rule in Golspie, where the game is also known as 'Bull in the Barn'. In Dovenby, Cumberland, where the game is called 'Farmer and Bull', there is a farmer who remains outside the ring, helping to support the circle whenever needed, and when the bull breaks through the bull has to catch the farmer to make him the new bull.

⁎ This game seems to be less played today than in the nineteenth century, when it was frequently recorded, e.g. in *Suffolk Words*, 1823, 'Bull in the Park'; *Youth's Own Book of Healthful Amusements*, 1845, 'Bull in the Ring'; *Nursery Fun*, 1868, 'Bull in the Barn'; *Cheshire Glossary*, 1877, 'Cry Notchil'; *Folk-Lore Journal*, vol. v, 1887, p. 50, 'Pig in the Middle and Can't Get Out'; *Traditional Games*, vol. i, 1894, 'Tod i' the Faul' ('Fox in the Fold'); and *Games of Argyleshire*, 1901, p. 239, 'Breaking through the Fence'.

A more ritualistic form of the game, known as 'Here I Bake and Here I Brew', is described in *The Girl's Own Book*, Boston, 1832. The girl within the ring went round saying:

> Here I bake, here I brew,
> Here I make my wedding-cake,
> Here I mean to break through.

A yet more formal version was called 'Garden Gate' or 'Have you the Key of the Garden Gate?' It was recorded by Gomme in 1894, and was still being played by children in Somerset in 1922 (Macmillan MSS.). The player in the middle sang:

> Open wide the garden gate,
> The garden gate, the garden gate;
> Open wide the garden gate,
> And let me through.

The players in the circle, forming a 'garden fence', danced round as they replied:

> Get the key of the garden gate,
> The garden gate, the garden gate;
> Get the key of the garden gate,
> And let yourself through.

The prisoner cried:

> I've lost the key of the garden gate,
> So what am I to do?

Still dancing, the others sang:

> Then you may stop, may stop all night,
> Within the gate,
> Until you're strong enough, you know,
> To break a way through.

Only after these time-taking preliminaries, which provide an example of how a basically simple game can be embroidered when it is popular (see Introduction, p. 9), does the player in the middle attempt to break out.

In another game on the same principle, 'The Wolf and the Lamb', played at Fraserburgh in 1892 (Gomme, vol. ii, p. 399), a wolf is outside the circle, a lamb within, and the wolf has to try and break *into* the circle. According to N. M. Penzer, editor of Basile's *Pentamerone*, 1634, the players in the Neapolitan game 'Rota de li cauce' similarly tried to keep one player from breaking into their circle, forming a wheel of kicks. However, from the text of the *Pentamerone* (the opening paragraph) it appears more likely that the player was inside the circle and attempting to get out.

RED ROVER

This game, which is a particular favourite with girls in Scotland, is played with two sides who face each other about five yards apart. The players on one side link hands (sometimes they grip each other's wrists), and advance and retreat chanting:

> Red Rover, Red Rover,
> Please send someone over.

In Cumnock they name the player they want:

> Red Rover, Red Rover,
> We call *Mary* over.

In Aberdeen they issue the challenge:

> Red Rover, Red Rover,
> We'll bowl *Mary* over.

The player named charges the opposing team, throwing her weight at a weak link, trying to break through (one child is reported to have had her arm broken in resisting the charge), and if she gets through she is free to return to her own side, and, in some places (e.g. Cumnock and Glasgow), she takes someone with her. If she fails she has to join the other side. The sides take turns in naming someone. 'This continues until there is one big line of people instead of two teams.' In England, where the game is only rarely played, they sometimes have a variant (e.g. at Crewe) in which there are two dens. One person in 'on' and the others link hands and run across to the other den. The person who is 'on' tries to break the chain, and if he succeeds the two players between whom he broke join him, until there is only one player left free. Occasionally the game is played with all the members of one side rushing together against the other side.

⁂ The history of 'Red Rover', which is also played in Canada, the United States, Australia, and New Zealand, is obscure. It is similar to the game 'Il Re di Spagna' played in Calabria, in which the King of Spain and his men oppose the King of France, and the King of Spain dispatches one of his men to break the chain formed by the French (Lumbroso, *Giochi*, 1967, pp. 177–8). It is also much the same as 'Der Kaiser schickt Soldaten aus' played in Austria (Kampmüller, *Kinderspiele*, 1965, p. 151), and the game called 'King' played in Czechoslovakia (described by Brewster, 1953, p. 170). No record of 'Red Rover' has been found in England or Scotland earlier than the Macmillan MSS., 1922, and it would seem to be an importation from the Continent were it not for a Scots game 'Jockie Rover' which possesses certain similar features. Walter Gregor told Gomme (*Traditional Games*, vol. ii, 1898, pp. 435–6) that in 'Jockie Rover', as played at Dyke, one player, the chaser, had a 'den', and before he emerged called out:

> Jockie Rover, three times over,
> If you do not look out, I'll gie you a blover.

He had then to run, keeping his hands clasped in front of him, and catch someone by crowning him on the head.

'When he catches one he unclasps his hands, and makes for the den along with the one caught. The players close in upon them, and beat them with their caps. The two now join hands, and before leaving the den repeat the same words, and give chase to catch another. When another is caught, the three run to the den, followed by the others pelting them.
During the time they are running to catch another player, every attempt is made by the others to break the band by rushing on two outstretched arms, either from before or from behind. Every time one is taken or the band broken, all already taken rush to the den, beaten by those not taken.'

It will be seen by reference to 'Warning' and 'Widdy' that this is a Scots version of the chasing game once general in England. In fact at

Tarry Croys, near Keith, it is still remembered that children used to chant:

> I warn ye once, I warn ye twice,
> I warn ye three times over;
> I warn all you Buckie wives
> To flee from Jack the Rover.

The game was also played by boys in Brooklyn about 1890 under the name of 'Red Lion' (*Journal of American Folklore*, vol. iv, p. 245), with the challenge:

> Red Lion, Red Lion,
> Come out of your den,
> Whoever you catch
> Will be one of your men.

And in addition the boys played a side-to-side catching game they called 'Red Rover' in which the Rover, in the middle of the street, called out a boy by name who had to run from one sidewalk to the other, and, if caught, had to stay and help catch the others (cf. 'British Bulldog', version 2). In Maryland, too, in the 1940s, children ran back and forth across an open space, being challenged by the chaser with the words:

> Red Rover, Red Rover,
> All I catch must come over.
> Red Rover, Red Rover,
> Let's everybody run over.

Those caught joined the chaser (Howard MSS.). This is not dissimilar from 'Red Rover' as played today at Crewe, in which the free players try to break the linked hands of those attempting to catch them. But it does not explain how 'Red Rover' has come to be played in its present form around the English-speaking world. It seems that at some period a game of breaking through people's hands such as 'Forcing the City Gates', described in Bancroft's *Games for the Playground*, 1909, pp. 89–90, and there taken from I. T. Headland's *The Chinese Boy and Girl*, 1901, was grafted on to the old game of 'Jockie Rover'. (Compare also 'King of the Barbarees' hereafter.)

KING OF THE BARBAREES

This rhythmic and ceremonious game is much like 'Red Rover' in that the objective is to break apart two linked hands, or two pairs of linked hands. It is played throughout Great Britain with little variation, and is here described by an 11-year-old Bristol girl:

'My favourite game is "The King of Barbarees". This is how you play it.

There is a King, Queen, Princess, Captain of the Guard, and some soldiers, also a Castle which consists of two children holding hands. The king tells the Captain of the Guard to march round the Castle singing,

> "Will you surrender, will you surrender,
> The King of the Barbarees?"

The Castle replies,

> "We won't surrender, we won't surrender,
> The King of the Barbarees."

Captain,   "I'll tell the king, I'll tell the king,
> The King of the Barbarees."

Castle,   "You can tell the king, you can tell the king,
> The King of the Barbarees."

The Captain goes back to the King and, stamping his foot, says,

> "They won't surrender, they won't surrender,
> The King of the Barbarees."

The King says, "Take two of my trusty soldiers."
The soldiers follow the Captain and the rhyme is repeated again,

> "Will you surrender, will you surrender,
> The King of the Barbarees?"

This goes on until all the soldiers have joined the ring, then the King says "Take my daughter". Next to go is the Queen, and last of all the King. The King says "We'll break down your gates," and after the rhyme has been said again with the King joining in, everybody makes a line with the King in front. He takes a run and jumps on the hands that are linked together and tries to break through them, while the two who are the Castle count to ten. If he does not break through, he goes back and one of the soldiers has a turn. They all jump on the Castle, one at a time, and try to break it down. If they do not succeed the Castle has won.'

In Scotland and Ireland the king is often referred to as 'King George', and sometimes the game is so named. Other names: 'King of the Barbarie' (Market Drayton), 'King of the Bambarines' (Cumnock), and 'Gates of Barbaroo' (Blackburn).

** Alice Gomme, *Traditional Games*, vol. i, 1894, pp. 18–21, gives only four versions, all from the south of England, which does not indicate that the game had the popularity or distribution in the nineteenth century that it attained between the wars, and still largely retains. Her earliest recording had been printed only the year before, 'The Tower of Barbaree', in Gurdon's *Suffolk Folk-Lore*, 1893, p. 63. Yet the game seems to have been long known on the Continent. A version called 'La tour, prends garde' is given in Marion Dumersan's *Chansons et rondes enfantines*, 1846, p. 37, which is little different from the English game. Two girls hold hands to form the tower, and the other players are the Duke of Bourbon, his son, and his soldiers. Before the tower is attacked a colonel and a captain are sent to the tower and sing: 'La tour prends garde de te laisser abattre.'

The tower replies: 'Nous n'avons garde, nous n'avons garde, de nous laisser abattre.'

The colonel says: 'J'irai me plaindre au duque de Bourbon.'

The tower: 'Va t'en te plaindre au duque de Bourbon.'

The colonel then complains to the Duke. The Duke gives him soldiers with which to attack the tower, and eventually comes himself. One by one they attempt to break through the girls' hands, and the player who succeeds in invading the tower is proclaimed Duke for the next game.

HONEY POTS

This is another game in which the pleasantries of feminine play-acting are the prelude to a sharp test of strength. One player is selected shopkeeper, one a customer, and the rest crouch in a row ('bop down' says a Suffolk girl), and clasp their hands under their knees, pretending to be honey pots. The customer arrives, inquires about the price of honey, and pretends to sample the various pots, touching each child on the head and licking her finger. She chooses one of the pots, and she and the shopkeeper proceed to weigh it, lifting the child by the arms, and swinging or shaking her, or otherwise attempting to break her grip, until they are satisfied that she is sound. She is then purchased. But if she 'breaks', the pot is rejected, and the player has to stand out. This continues until all the pots have either been bought or cast aside.

** The absurdity of this amusement is almost evidence in itself of its antiquity, and it has certainly been one of the joys of childhood for the past 150 years. 'Honey-pots' was described as 'a common game' among girls in Edinburgh in *Blackwood's Magazine*, August (part ii), 1821, p. 36; was listed amongst old Suffolk games by Edward Moor in 1823; and under the name 'Hinnie-pigs' was described by Mactaggart in his *Gallovidian Encyclopedia*, 1824, as a game for boys:

'The boys who try this sport sit down in rows, hands locked beneath their hams. Round comes one of them, the honey-merchant, who feels those who are sweet or sour, by lifting them by the arm-pits, and giving them three shakes;

if they stand these without the hands unlocking below, they are then sweet and saleable, fit for being office-bearers of other ploys.'

The popularity of the game through the years can be gauged by the number of times it has been described or mentioned, e.g. by the editor of *A Nosegay, for the Trouble of Culling*, 1813, who was clearly very familiar with it; by Hugh Miller, *My Schools and Schoolmasters*, 1852, ch. iii (played by the apprentice Francie 'though grown a tall lad'); by Henry Mayhew, *London Labour*, vol. i, 1851, p. 152 (played by a little water-cress seller aged eight); by Mrs. Valentine, *The Home Book*, 1867; by Alice Gomme, *Traditional Games*, vol. i, 1894 (many accounts); and by Norman Douglas, *London Street Games*, 1916. In some places before the First World War special words were chanted sing-song fashion while the honey pot was being tested, for instance, at Helmsley in Yorkshire:

> Is she rotten, is she sound,
> Is she worth a million pound?
> Toss her up and toss her down
> She is worth a million pound.

It is from the game that a crouched position has come to be known as a 'honey pot'.

The game is also popular amongst little girls in Italy, and is usually played almost exactly as in Britain under the names 'Laveggio', 'Pentole', and 'Pignatte'. The two children who test the pots go round patting the players' heads, and asking 'Sarà rotta, o accomodata?' 'Is it broken, or mended?' The child selected is then swung by her arms and if she holds she is a good pot, but if she does not hold she is a cracked pot, and is put in another place, and may eventually have to pay a forfeit (M. M. Lumbroso, *Giochi*, 1967, pp. 11–13). Signora Maroni Lumbroso also gives an account of the game 'Pignatte' as played at Forenza, near Potenza. Here one of the two players who are standing whispers a colour to each pot. When she has finished the other player approaches and taps her on the shoulder, saying 'tup, tup'. 'Who is it?' 'I am the Madonna.' 'What do you want?' 'I want a pot.' 'What kind?' The Madonna names a colour. If there is a pot of this colour the crouching player is tested, and according to whether she is found to be a good pot or a bad one she is deposited in paradise or hell. (Cf. 'Jams'.)

A similar game, 'Les Pots de Fleurs', was played in Paris in the nine-teenth century. A number of little girls crouched down, clasping their hands under their legs, and sticking their arms out like the handles of flower pots. Three further girls stood facing them who represented God

Almighty, the Virgin Mary, and the Devil. A fourth player was a flower-seller, and a fifth his helper. Each of the pots of flowers was given the name of a plant, and according to Rolland, *Jeux de l'enfance*, 1883, pp. 134–5, the following dialogue took place:

> Le bon Dieu: Pan! pan!
> Le marchand: Qui est-ce qui est là?
> Le bon Dieu: C'est le bon Dieu qui vient acheter un pot de fleurs.
> Le marchand: Laquelle voulez-vous?

God Almighty named a plant. The flower-seller and his helper lifted the plant up by the handles ('c'est-à-dire par les bras') and brought it to God Almighty. Then the Virgin Mary and the Devil made purchases in their turn. When all the pots had been sold, and successfully carried to their purchasers, those flower-pots acquired by 'le bon Dieu' and 'la Sainte-Vierge' made horns at those purchased by the Devil. It is possibly relevant that when 'Honey Pots' was played in York about 1910, and a girl's grip broke while she was being tested, she was (according to a correspondent) slapped on the bottom with cries of 'Fire and brimstone'. The game clearly has affinity with the latter part of 'Mother, the Cake is Burning' (q.v.).

STATUES

'Statues' is a contest to find who can best act and pose under somewhat trying conditions. The players, usually girls, line up at a wall, or on top of the wall if it is not too high, and hold out their right hands. One person, who is out in front and may be known as the 'puller' or 'twister' (in Scotland the 'birler' or 'burreller'), pulls each person from the wall in turn, either pulling as hard as she can, or giving each player a chance to say how she shall be pulled, asking:

'Do you want egg, bacon, or chips?' ('*Egg is a slow twizzle, bacon a fast one, chips a very fast one.*')

Or in Bristol: 'Do you want bread, or honey, or wedding cake?' ('*Fast or slow or medium.*')

Or in Edinburgh: 'Salt, pepper, mustard, or vinegar?' ('*Vinegar is an extra hard pull with a swing to it.*')

Or in Market Rasen: 'Do you want blood, water, pop, or curtsy?' ('*Blood you pull the person out fast, water gently, pop you go round in a circle with them, and curtsy you take them out gently and leave them in a curtsy position.*')

When the player has been pulled she must 'stand how she has been thrown', remaining motionless as if a statue. The puller then tells the statues what she wants them to be, for instance, monster, fairy, mouse, clown, or dancer, either giving all the players the same part or each a different one, and commands 'Clockwork begin'. The statues have to come to life, and do the things they think monsters or fairies, or whatever they are supposed to be, would do, until the puller commands 'Clockwork stop', when—as one boy put it—'you've got to take a shape': the players must adopt the posture they think most appropriate to their set character. The puller then commands 'Lights out' or 'Lanterns shut' and they have to close their eyes. It is now the puller's job to judge which statue has the ugliest or funniest or most beautiful or most monstrous face and posture according to the pose that has been ordered. She goes round tapping those she does not think are so good until only one player is left standing motionless with her eyes shut. The rest silently make a circle round this last statue, and then suddenly begin dancing round, and shouting 'Wake up sleeping beauty', 'You're the pretty lady' (in Wigan, 'You're the big soft jelly baby'), and this person is the winner and becomes the puller in the next game.

Names: 'Statues' (the usual name); 'Bread and Honey' (Bristol); 'Clockwork' (Scarborough); 'Clockwork Statues' (Basingstoke); 'Egg and Bacon' (Helston, Norwich, Wolstanton); 'Egg, Bacon, and Chips' (Alton); 'Hot, Cold, or Boiling' (Wandsworth); 'Pepper and Salt' (Isle of Arran); 'Penny in the Slot' (Peterborough); 'Salt, Mustard, Vinegar, Pepper' (Swansea); 'Stookies' (Edinburgh, Forfar, and New Cumnock).

'Statues' is also played in simpler forms in which the test is no more than to try who can keep still the longest (in St. Andrews this is known as 'Dead Lions'), or who can keep a straight face the longest despite being laughed at or tickled.

In Glasgow a similar game, usually played on a slope, is called 'What Do You Want?' A player on the top of the slope is asked 'What do you want?' and may reply 'Gun' or 'Arrow' or 'Atom bomb' or whatever he fancies. He is then shot at 'and has to make a good fall, and when he has finished falling he lies very still as if he was dead'. The player whose act of dying was the most spectacular is then declared the winner. In Edmonton, Canada, the game is called 'The Deadest', and after each of the players has been killed the competition is to find which player is the deadest.

** The present elaborate form of 'Statues' seems to be a recent development and was not known to our correspondents whose childhoods covered the period 1890–1930. In *London Street Games*, 1916, p. 41, the players were merely pulled, and took up a posture 'sometimes pretty but mostly ugly', which was then judged. In Austria, today, players are simply twizzled, freeze where they land, and then perform appropriate actions at command ('Figurenwerfen' in *Oberösterreichische Kinderspiele*, 1965, p. 162). In Italy the children are not even pulled, they merely adopt poses as instructed ('Le Belle Statuine' in *Giochi*, 1967, pp. 193-7).

LEAPFROG

They say leapfrog can be played with any number, large or small (even by only one if there are posts or milk churns about), and it is a good game with just two: one bending down and the other vaulting over, running forward a few steps and making a back for the first boy to vault over. In this way leapfrog is 'quite useful', or so an 11-year-old asserts, 'because if anyone has to go a long distance walk it helps one to get along'. But it is more fun if there are more players. They begin by lining up; one person bends down, and the first in the line jumps over him and runs forward a few steps, and makes a back himself; the second in the line jumps over each of these two players and runs forward a few more steps and makes a third back. Each player who follows has one more back to leap over, and makes another back himself, until everyone is stooping. Then the player who first bent down stands up, and himself jumps along the line of backs, and makes a further back at the other end of the line, and the second person who bent down gets up as soon as he has been jumped over, and starts jumping, and so on. In this way the game is continuous. Whoever is last in the line gets up and jumps over those in front of him, and whoever has cleared all the backs makes a new back at the far end of the line. The sport becomes more exciting if the last in the line stands up as soon as he has been jumped over, and straightway follows after the boy who has jumped over him: then there may be almost as many boys jumping as there are boys bending down, and when a boy comes to the end of the line of backs he has to be quick about bending down himself, for the person he first jumped over may be close on his heels ready, in his turn, to complete the line of jumps. Those who can are gasping 'Keep the kettle boiling' to encourage the flow of jumpers which—in theory—may ripple onwards around the playground without stopping until the school bell rings. But the sport is not as simple as it looks. If a player does not vault properly, placing his hands evenly on the back of the bent figure and leaping lightly at the same time, the result, says an 11-year-old, will be 'an unartistic heap of humans' on the ground. And as another boy remarks, 'There is only one special rule and that is "Do not push people over when you are jumping over them".' In fact, holding oneself steady while being jumped over requires almost as much effort as does the jumping. When a person bends down he presents either his backside to the jumper, placing one leg forward with knee bent, resting his hands firmly on his knee, and tucking his head well in;[1] or, perhaps more often, he bends down sideways to the oncomers, and grips an ankle. The higher the back he makes the more difficult it will be to jump over him, and the more likely he is to be knocked over. Likewise the nearer he is to the previous person who bent down the more difficult it will be to jump over him; and sometimes they insist that a player runs five or six yards before he bends down, while sometimes, to make the sport more difficult, the rule is that he must take only three or four steps after he has jumped. Sometimes they play 'Higher and Higher', when those bending down make their backs a little higher after each turn, and a person who can no longer get over has to drop out. Sometimes they play in a circle and one person goes on jumping until he falls. Occasionally the rule is that a jumper may use only one hand when he vaults, a feat which needs practice. And sometimes those making the backs are allowed to sink down as the leaper places his hands on their back which, far from making it easier for the leaper 'may mean that he goes flat on his face'.[2]

***** The only evidence that leapfrog was played in the ancient world, according to de Fouquières, is a decoration on a cyclix showing a child crouching down while another child jumps over him (*Cabinet Durand*, de Witte, no. 706). In the Middle Ages, too, the sport seems to have been uncommon, but it had become popular by the sixteenth century. Gargantua played 'a crocqueteste', a name happily glossed in the caption to a woodcut depicting leapfrog in *Les Trente-six figures*, 1587:

> Ils sautent tous, en criant couppe teste,
> L'vn par sus l'autre, est-ce pas ieu honneste.

Bruegel shows six boys playing leapfrog in his picture of children's games painted in 1560. Borcht shows anthropomorphized apes playing it in his picture 'Spelende Apen' about 1580. And Shakespeare's Henry

---

[1] 'Tuck yer napper in, I'm a-coming' was the cry in London streets sixty years ago, and before that (*vide* 'The Gondoliers') it was 'Tuck in yer Tuppenny'.

[2] The 'baneful trick' of stooping without warning, just as a player is making his leap, used to be known as 'fudging', and, declared *The Boy's Handy Book*, 1863, deserved 'whatever punishment playground justice may award . . . We have heard of a collar-bone broken through *fudging*'.

the Fifth seems to have regarded excellence at 'leape-frogge' almost as a token of manliness (v. ii, in the Folio).[1]

Other references occur in Samuel Rowlands, *The Letting of Hvmours Blood*, 1600, D8b; Thomas Dekker, *The Seuen deadlie Sinns*, 1606, iv; Ben Jonson, *Bartholomew Fayre*, 1614, I. i (1631); William Hawkins, *Apollo Shrouing*, 1626, p. 5; *The Independent Whig*, 1720, no. 32, ¶ 13, 'Hop Frog'; Abel Boyer, *Dictionnaire*, 1727, s.v. *La Poste*, 'Skip Frog'.

The game is depicted in the first children's book of games *A Little Pretty Pocket-Book*, 1744 (1760, p. 44); it is nicely described in Henry Brooke's *The Fool of Quality*, vol. i, 1766, p. 79; and a character in *The Book of Games*, 1805, p. 110, says that although there are many types of jumping 'leap-frog beats all those hollow'. In *The Boy's Country Book*, 1839, p. 219, William Howitt, who went to Ackworth School in Yorkshire around 1805, recalled that he had 'often seen the whole number—180— making one long line at leap-frog'.

Dialect names: 'Hog over Hie' (Suffolk, 1823); 'Hop-Frog-over-the-Dog' (Lincolnshire, 1894); 'Lankister-lowp' (Lancashire, 1882); 'Lantie Lawp' (Cumberland, 1915); 'Frog Jump' (County Cork, 1938); 'Frog-Lope' (Yorkshire, 1892).

In France the game is generally 'le saute-mouton' or 'le coupe-tête', in Italy 'saltamontone' or 'salta cavalletta', in Denmark 'springe buk', and in Holland 'haasje-over'. In Germany, according to Böhme (*Deutsches Kinderspiel*, 1897, p. 591) it is 'Hammelsprung' when a boy stands sideways to the leapers, and 'Bocksprung' when he is turned away from them, a distinction which is borne out by Kampmüller's description of 'Bockspringen' in present-day Austria (*Oberösterreichische Kinderspiele*, 1965, p. 110).

The leapfrog games which follow were clearly more popular before the First World War than they are today, even if they were not as numerous as appears from Norman Douglas's catalogue in *London Street Games*, 1916 (pp. 24–34).

GENTLE JACK

This is leapfrog with additional hazards. In north London they start by dipping and one boy is picked to bend down, and one to be leader. They jump one at a time over the boy who has made a back, and run round

[1] Today leapfrog is chiefly the sport of the younger boys and, judging by the number of accounts received from girls, is becoming a feminine pastime. Probably its image has not been enhanced by its inclusion amongst gymnasium games, along with the vaulting-horse. As a Spennymoor girl remarked, 'The boys play a rougher game on the same idea as leapfrog called "Hum-a-dum-dum" ' (see hereafter).

ready to jump over him again, the leader coming last. After the leader has jumped he gives a command such as 'kick him', 'pull his hair', 'give him a cauliflower' (twist his ear), and each boy as he jumps, has to do as instructed. Should one of them while vaulting fail to offer his respects to the stooping figure, he himself must make a back and become the object of attention. Thus the game proceeds, needing to stop only when 'we've all had enough of it'.

 ⁂ It is apparent from Edmund Routledge's *Every Boy's Book*, 1868, p. 9, that responsibility for devising this amusement is not to be credited to a modern young bully. In this enumeration of the feats that can be performed while jumping over a boy's back, Routledge writes:

'The next trick is "knuckling",—that is to say, overing with the hands clenched; the next, "slapping", which is performed by placing one hand on the boy's back, and hitting him with the other, while overing; the last, "spurring", or touching him up with the heel.'

We find that Victorian writers for the young were, on occasion, more considerable realists than those who have succeeded them. Routledge prefaced his description with the remark: 'This game is capable of being varied to any extent by an ingenious boy.'

SPANISH LEAPFROG

'Spanish leapfrog is similar to ordinary leapfrog but requires more skill', says an 11-year-old Dulwich girl. 'You play this with as many players as possible, and while leaping over the person's back the leader places a cap (or any object) on her back. The other players have to leap over the object as well as the person's back, trying not to knock it off. If they do knock it off they change with the person bending down.'

 ⁂ This game was formerly known as 'Spanish Fly' (and is so termed in *Games and Sports for Young Boys*, 1859, p. 16; *The Boy's Handy Book*, 1863, p. 6; *The Boy's Own Paper*, 12 November 1887, p. 103). There was a second way of playing the game in which every player as he jumped deposited his cap or rolled-up handkerchief on the person's back, so that the difficulty of surmounting the pile increased with each turn. Sometimes further complications were introduced, particularly regarding the orderly removal of the caps, which might, for instance, be picked up by the teeth as the jump was made.

Other names: 'Accroshay' (Cornwall, *Folk-Lore Journal*, vol. v, 1887, p. 60); 'French Flies' (*Antrim and Down Glossary*, 1880); 'Leap Frog with Bonnets' (Golspie, 1892, and still played there 1953); 'Cappy' (*Northum-*

berland Words, 1892); 'Cap-it' (*Warwickshire Word-Book*, 1896); 'Hot Pies' (*Games of Argyleshire*, 1901), and 'Chimney Pots' (Vauxhall Bridge Road, *c.* 1905).

FOOT-AN-A-HALF

Of the eighty games described by pupils of Spennymoor Grammar Technical School, 'Foot-an-a-Half' was fourth in popularity, after 'Hum a Dum Dum', 'Split the Kipper', and 'Skipping'. This game, an exacting one, needs about eight well-matched players, one of whom is chosen to be 'down' (often 'the last person who arrives at the place where we are going to play'), and he takes up the leapfrog stance, bending over with his back to the other players, while a line is drawn or a stick placed on the ground to mark his position. He chooses one player to be 'foot-an-a-half', and sometimes a second player, designated 'leader', who he thinks is a good jumper; for this second player may be able to relieve him of his position in the game, which is merely that of vaulting-horse. The foot-an-a-half leaps over him, and the player who is down has to move forward to wherever foot-an-a-half's back foot landed. Foot-an-a-half then looks at the distance between the stick or line and the doubled-up figure, and calls the type of step, or steps, which the others may take beyond the line to make their leap.

'He may shout a number like "a oner", this means that the rest of the team have to jump over the person's back from the line, taking only one step. He may say "a standing oner", which means that the team has to stand on the line and jump. If however he says "a running oner" then it is usually a difficult jump, and the persons jumping are entitled to a run.

The players jump in turn, the leader jumping last, and if they all do it successfully, the foot-an-a-half jumps again, and, as before, the boy bending down moves to wherever the foot-an-a-half lands. Since there is now more space between the mark and the boy to be jumped over, the foot-an-a-half probably allows a further step or hop to be taken beyond the line, and this goes on until one or more players fail to do the leap: either funking it, or not getting over, or taking too many steps beyond the mark, or knocking over the bending boy. When this happens the last person in the line to fail makes the new back.

Properly played, the point of the game is that the foot-an-a-half has to call a jump which, he hopes, one of the players will be unable to manage, although it must be a jump he himself is able to do; while should the leader be able to do it with less steps than the foot-an-a-half has decreed, the foot-an-a-half has to become the vaulting horse. Whenever there is a change in this role the game starts at the beginning again.

At Ponders End, Enfield, where the game is known as 'Long Man', the steps which the 'captain' names are customarily fancy ones, similar to those in 'May I?' (q.v.). Thus he may say that the jump should be 'Broad and straight over' (one standing jump from the line, and then leap), or 'a scissor' (a jump landing with feet astride, and another bringing feet together again), or a certain number of 'pigeon toes' (heel-to-toe steps), or 'an umbrella' (a jump landing face-about, always given in pairs so that the player ends facing the right way), or, if the distance between the line and the bending boy (known as 'it') warrants it, he gives instructions for a sequence of movements such as 'One broad, six pigeon toes, and straight over'.

In Forfar, where 'Foot-an-a-Half' is played in much the same way, a 14-year-old says it is a rule that if the person jumping should knock over the one who is bending down, 'It is a "bull" and he is out, but if the person that is "down" collapses under the person's weight, he can cry "Touch Bill" and the other person can cry "No Touch Bill" before the other person can say his part.'[1]

⁎ Several Spennymoor boys remarked that they could not understand why the game was called 'Foot-an-a-Half'. This was formerly the name of a similar game in which, after each player had had his turn and jumped successfully, the boy bending down moved forward a standard 'foot and a half' (the length of his boot plus its width) and automatically increased the difficulty of the jump for the next round. 'Foot and a Half' was played at Sedgley Park School, about 1805 (Husenbeth, *History*, 1856, p. 106); while in the south of England the game was known as 'Fly the Garter' (referred to by Keats in a letter to John Hamilton Reynolds, 3 May 1818; and by Bob Sawyer in *Pickwick Papers*, 1837, ch. xxxvii).[2] The fore-runner of the present game seems to have been 'Foot and an O'er' described in *The Youth's Own Book of Healthful Amusements*, 1845, pp. 35–8:

---

[1] This terminology is traditional. The Revd. Walter Gregor, born 1826, recalled that in his boyhood at Keith, Banffshire, 'the boy that stooped his back was called "the bull" pronounced *bill*' (Gomme, vol. ii, 1898, p. 440). In Wigan the game is called 'Cutter', which is the cry of the leader when determining where the boy making the back shall move to next. This term, too, is old, being recorded in *The Boy's Own Book*, 1855, p. 25 (see hereafter). At King's School, Canterbury, *c*. 1875, the game was called 'Cut Throat'; in Canning Town, *c*. 1910, it was 'Cut Lump'; in *London Street Games*, 1916, it was 'Cut-a-Lump' or 'Cutter'; and in Dundee today it is 'Cut Foot and Guide'.

[2] In France 'Le Saute-Mouton à la Semelle', the sheep similarly moving forward 'la longueur de son pied droit, posé en équerre au milieu du gauche' (*200 Jeux d'enfants*, *c*. 1892, p. 140).

'This is a famous play in Lancashire; for go where you will on the cold, wintry day, or the clear summer's morning, there are sure to be some employed at "Foot and an O'er". As many may play as please. In the first place, the lads draw lots as to who shall stoop first; and he, you know, who is unfortunate enough to lose, is obliged to set his back. The first youth who jumps over is called the leader, after whom all the rest must follow: the last that leaps is styled the Footer, who when he has jumped, marks with the toe of his shoe the distance he leaped beyond the lad stooping, and the latter has to remove, and put his heel exactly to the mark of the footer . . . But great as is the distance, if the leader does it not, and the footer does, the former must put down his back, and take the place of the lad that is now down. If the footer cannot perform what the leader has failed to do, the latter will *not* have to stoop, but may take the leap at one and an o'er, as it is termed; which means that the leader jump once from the mark as near to the boy as he pleases, and then takes a stand jump over: that is called one and an o'er.'[1]

The rule that the first jumper shall decree the steps which the others may take between the line and the boy who is down, appears in the version of the game called 'Last Man's Jump' in *The Boy's Own Book*, 1855, p. 32:

'The one who is down . . . goes to the spot where the last player, who is called the "cutter", alighted at. The best player is generally chosen by him for the cutter, so that he may, by manoeuvring, or jumping a good distance, cause another to be down, and so relieve him. The one who goes first has the privilege of taking a jump, a hop, and jump,—or a hop, stride, and jump between the garter [the starting mark] and the back, before going over; but whichever he does, the others must follow, except the cutter, who is allowed to run to the back, so as to enable him to take a better spring.'

Brewster, *American Nonsinging Games*, 1953, p. 106, gives the name of the game in Illinois as 'One and Over'; and Culin, 'Street Games of Boys in Brooklyn', *Journal of American Folklore*, vol. iv, 1891, pp. 227–8, gives the name 'Head and Footer'.

JUMPING GAMES

There are several competitive jumping games, on the principle of 'Foot-an-a-Half', for which sudden crazes will arise, and then, when it has become clear who the best jumpers are in the school, everybody becomes sick of the craze, and the games seem to be forgotten, only to reappear afresh two or three years later.

One a Foot. This might be described as the girls' version of 'Foot-an-a-Half'. 'Out of the people playing a leader is chosen, usually a volunteer.

[1] The writer concluded with a warning, necessary for boys in the middle of the nineteenth century: 'Whenever you play at "Foot and an O'er", always loosen the straps at the bottom of your trousers, if any you wear; because if you don't, you may very easily do them a serious injury, by tearing them.'

The girl puts two sticks together, then she runs and jumps from the front of the sticks, and where her heel lands the second stick is placed. Then she says whether it is a "noner" or "oner" or "twoer". The people following must jump as she did. If it is a "oner" the people jumping are only allowed to put one foot once in between the two sticks. The game goes on until one person is left in. This person is the leader for the next game' (Girl, 13, Spennymoor).

Fly. ' "Fly" is the name of a game where you use six sticks and place them about a foot apart, and have to jump over each stick without touching it and without jumping over a space between two sticks. The last girl to jump over takes a long jump and calls the number of a stick. They put the stick of this number where she jumped to, and the game goes on until the spaces between the sticks get so big that people can't jump them. The person who can jump furthest is the last girl next time' (Girl, *c.* 11, Portsmouth). 'Fly' appears to be well known in Australia, but there only the last stick (or stone) is moved forward.

Leapers. 'You get a piece of chalk and draw an oblong square, draw a line across two inches past the middle, get a dice, and throw on to the line. You stand about two feet back from it and try and leap over the dice. If you can leap over the dice you are a leaper and have one point. You keep moving the chalk line back and see how many points you get in half an hour' (Girl, 9, Wolstanton).

Baby Leapfrog. 'A lot of people lie down with their backs up in the air. The very last person who is at the end jumps over the others' backs. When she has jumped over the last one she lies down. The first person she jumped over, gets up and starts to jump over the others and so on till they have all had a shot and then they begin again' (Girl, 10, Kingarth, Isle of Bute).

** This trivial amusement might be passed over as the artless invention of a group of little girls seeking an afternoon's diversion, were it not that the following report, from nearby Argyllshire, appeared in *Folk-Lore*, vol. xvi, 1905, p. 78:

'*Leum Maighiche.* (Hare's Leap.) Several take part in this. One lies down on the ground on his back; another jumps over him and lies down where he has landed parallel to the one already down. Another leaps over both and likewise lies down till all are down, or the distance that can be leapt is covered. The first to go down then rises and leaps over the party, followed in rapid succession by the remainder, the fun largely consisting in the rapidity in which they follow each other. If one is slow in clearing the way for his successor, he is said to be "run down", and must retire from the game. This was also played in Orkney.

This, in turn, may be compared with 'Loup the Bullocks' in Mactaggart's *Gallovidian Encyclopedia*, 1824, p. 320, in which the players plant themselves in a row on 'all fours' about two yards from each other, and are jumped over in similar fashion.

HI JIMMY KNACKER

Of all street games this is the one grown men recall most readily, and with the greatest complacency, possibly because it is the toughest of the games, the one in which players are most frequently hurt, and which requires the greatest amount of stamina, *esprit de corps*, and indeed fortitude.[1] Two sides are chosen with four to six boys, or sometimes girls, on each side ('It's best to get the big hefty ones on your side'), and they toss to see which side shall jump first. The side which loses has to be the 'down' side and 'make a back'. One boy, variously known as the 'pillar', 'pillow', 'cushion', or 'buffer', stands with his back against something firm, as a tree, wall, or lamp-post, and sometimes interlocks his fingers in front of him to form a cup. The next in the team, usually one of the smaller boys, stoops down placing his forehead in the cup, with the crown of his head against the cushion's stomach, and holds on to the cushion's waist or legs. The next boy in size comes behind him, placing his head under the second boy's legs and gripping his thighs; the rest of the side follow behind, interlocking in similar fashion to present one long back, the biggest boy taking last place, since he generally has to support most weight, and makes a higher and more awkward back for the other side to jump over. The other side have now to jump on to the backs of the 'down' side, an obligation for which they show no reluctance, rushing forward one at a time, placing their hands on the back of the end boy, and leaping with as much force and weight as they have in them, for if they crush the backs of those they jump on, and bring them to the ground, they can command another turn. Usually the best jumper in the side runs first, for he must jump over three or four backs, landing far enough forward to leave room for the other members of his side to mount behind him. This they attempt to do, among 'many angry shouts and groans' from those underneath, and often try to concentrate their weight on one of the weaker boys who they think can be made to collapse. Thus a boy who has not tucked his head in properly may find himself with

---

[1] H. E. Bates has said it was a game of which 'I can hardly speak without emotion'. Howard Spring imagined it the splendid invention of his own boyhood gang in Cardiff. Readers of *The Times*, January–February 1951, wrote more letters about it to the editor than could be printed.

another lad far heavier than himself sitting on his neck. Arguments start, and retribution may be taken if a jumping boy lands with his knees on someone's back, or if anyone starts edging forward (known as 'creeping') to make room for those who are to follow, or when anyone takes a long time before he jumps. In some places those about to jump give notice of their approach with cries such as 'Olley olley' or 'More weight' or 'Warnie, I'm a-comin' (London), or 'Two stone heave-ho!' (Wigan), or 'Weights coming on' (Accrington), or 'Rum-stick-a-bum, here I come' (Nottingham). But it is not wholly a joy-ride for the mounting side. Should one of them fall off, or put a foot on the ground to steady himself, the whole side pays forfeit, and has to dismount and make a back for the others to jump on.[1] Frequently there is little room for the last player to jump on and he clings precariously to the backs of the riders in front of him, while those underneath shake and bounce to intensify his discomfort. Usually the team has to stay mounted for a given length of time. A Croydon boy claimed that they must remain seated for five minutes while those underneath strove to dislodge them. In general they have to remain seated (or at least keep their feet from the ground) for twenty seconds after the last boy has mounted, or while the player at the back counts up to ten, or while the mounted team give vent to a chant of local prescription, for instance in Stepney and Poplar:

> Hi Jimmy Knacker, one, two, three,
> Hi Jimmy Knacker, one, two, three,
> Hi Jimmy Knacker, one, two, three.

In Enfield:

> Onk, onk, horney, one, two, three,
> Onk, onk, horney, one, two, three,
> Onk, onk, horney, one, two, three.
> —All over!

In Grimsby, 'Bung the barrel, bung the barrel, one, two, three'. In

[1] The question whether or not part of a boy's anatomy has touched the ground can be the subject of urgent debate. A writer in the London *Evening News*, 21 December 1931, p. 11, reported the following dialogue when a member of the mounted side had lost his seat, and was clinging on desperately underneath his mount:

'Steve's touched!'
'Garn, I 'aven't!'
'Just look at yer!'
'All right—look!'
'Yer back's on the ground!'
'Ye're blind!'
'Wait till ye're up, an' I'll give yer a claht on the jaw!'

As a Walworth boy confessed to us (1961): 'The reason why I like Jump Jimmy Knacker is because it is not always a fair game and it gets very exciting.'

Newcastle, 'Mountikitty, mountikitty, one, two, three'. In Warwick, 'Mollie, Mollie Mopstick, all off! all off!' In Nuneaton:

> Mopstick, mopstick, bear our weight,
> Two, four, six, eight, ten.

At Meir, in the Potteries:

> Badger, badger, badger, one, two, three,
> All off and have another go.

At Tunstall, where girls are notable enthusiasts, 'One, two, three, four, five, six, seven, eight, nine, ten—Cock Robin!' And in Forfar:

> Huckie duck, huckie duck,
> Three times aff an on again.

Alternatively, in Manchester and Birmingham, and in places generally where the game has an equine name, victory is claimed by the under side, who declare themselves 'Strong horses' if they manage to support the 'up' side for a prescribed length of time; while if they collapse the mounted side cry 'Weak horses' or 'Weak donkeys' or, in Scotland, 'The cuddy's broke'. In Scotland, too, they are particular about the conduct of the jumping side. Sometimes no talking is allowed, or no talking or laughing, or, in Kirkcaldy, 'no tickling, talking, laughing, or kicking'. In Wigan, according to a young informant, he who stands with his back to the wall acts as a kind of umpire, and should he see any rider show as much as his teeth, he shouts 'Lall-i-ho!' and the riders have to dismount and become the under team.

The game is in fact remarkable not only for being played throughout Britain,[1] but for the almost royal burden of names it bears, sometimes several known in one district. For instance, Croydon boys call it not only 'Hi Jimmy Knacker', but 'Bung the Barrel', 'Hi Cockalorum', 'Jump Teddy Wagtail', and 'Trust Weight'. In Kirkcaldy it has no less than seven names: 'Cuddy's Wecht', 'Cuddy gie Wecht', 'Loup the Cuddy', 'Leap the Horse', 'Hunch, Cuddy, Hunch', 'Camel's Back', and 'All Aboard'. Some names are clearly regional: 'Mountikitty' on Tyneside, 'Bumberino' in south Wales, 'Pomperino' in Cornwall. Other names, such as 'Hi Cockalorum' and 'Bung the Barrel', can be described as standard names; but even setting these aside, the nomenclature is so mixed, regionally, it has not been found possible to produce a distribution map. Thus the north-country 'Muntikitty' is found in East Anglia

---

[1] Although not in Camborne and Helston when we were there in 1961. The boys said they used indeed to play it, but someone at school had broken an arm (or leg), the game had been banned, and this had, for the time being, dissipated enthusiasm.

where it is meaningless ('Muntikitty' becomes 'Mad-a-kiddy', 'Mud-a-giddy', and even 'Mother Giddy'), while in Ayrshire 'Bung the Barrel' has taken the place of 'Hunch, Cuddy, Hunch' which was the name there half a century ago. The fact is that the game is much played by Boy Scouts, and enthusiastic Scoutmasters have probably assisted in disrupting the traditional terminology.

The following are the names known to be current: 'All Aboard', 'All on the Horses', 'Apple Bombers' (Weymouth), 'Badger' (general around Stoke-on-Trent), 'Bucking Bronco', 'Buckle Up' (Petersfield, for past fifty years), 'Bull Rag' (Manchester), 'Bumberino' or 'Bomberino' (south Wales, especially Cardiff for past sixty years), 'Bump a Cuddy' (Gateshead), 'Bung Billy Barrel' (Camberwell), 'Bung the Barrel' (London since 1900, and elsewhere especially Surrey, Hampshire, and Oxfordshire), 'Bungle Barrel' (Hampshire, west of Winchester), 'Carabuncle' (Stirling), 'Cock Robin' (Tunstall), 'Cuddie Backs' (Portpatrick), 'Cuddie Hunkers', 'Cuddie's Loup', 'Cuddie's Weight' or 'Cuddie gie Way' (Edinburgh and around Firth of Forth), 'Donkey' (Morpeth), 'Funking Cuddie' (Ballingry), 'Hi Cockalorum' (Croydon and Sittingbourne—see hereafter), 'Hi Jimmy Knacker' (East London since 1890), 'Hi Johnny All On' (Nuneaton), 'Hacky-Duck' (Teignmouth), 'Huckie-Duck' (Dundee, Forfar, Helensburgh), 'Huckey Buck' (around Ipswich), 'Hunch, Cuddy, Hunch' (Glasgow, Stirling, and Lanarkshire), 'Jackerback' (Peterborough), 'Jimmy Knacker' or 'Jimmy, Jimmy Knacker' (predominant in north London since nineteenth century), 'Jockeys' (Brixton), 'Johnny Knacker' (Greenwich), 'Jump Jimmy Knacker' (predominant in south London, e.g. Camberwell and Kennington), 'Jump the Cuddie' (Aberdeen), 'Jump the Long Horse' or 'Jump the Long Mare' (Lydney), 'Lall-i-ho' (Wigan), 'Leapfrog' (Wimborne, boys emphatic this was the name), 'Loup the Cuddie' (Kirkcaldy), 'Lumps' (Wareham), 'Marching Army' (Pontefract), 'Mollie Mollie Mopstick' (Warwick), 'Mont-a-kiddy', 'Mudikiddy', 'Mountikitty', or 'Munt-a-cuddy' (general in Cumberland, Northumberland, Westmorland, Durham), 'Onk, Onk, Horney' (Enfield), 'Piecrust' (Bootle and Pontefract), 'Pig's Whistle' (Norwich, for past seventy years), 'Piggy 'gainst the Wall' (Nuneaton), 'Polly on the Mopstick' (Birmingham and Nuneaton), 'Pomperino' (Cornwall), 'Rum-stick-a-bum' (Nottingham), 'Strong Horses' (various places), 'Strong Horses, Weak Donkeys' (Glamorgan and Monmouthshire), 'Trust' (Luton), 'Trust Weight' (usual name, Sale, Manchester), 'Weak Horses' (Camberwell, Leicester, Manchester, Swansea), 'Weights' (Accrington), and 'Wooden Horse' (Manchester).

One boy in Camberwell called it, graphically, 'Jump Jimmy Knocker-bone'; a boy in Liss swore the name was 'Squashed Guts'. In Guernsey it is known as 'Lamp-Post' or 'English Jumpbacks', formerly it was 'Saute Mouton'—in Jersey rendered 'Saltey Mayou'. In New York it is 'Johnny on the Mopstick' or 'Johnny on the Pony'.

*** The early history of the game is bound up with 'Buck, Buck, How Many Fingers do I Hold Up?' (q.v.). Urquhart renders Rabelais's *Cheveau fondu* as 'Trusse', a name seemingly common in the seventeenth century, being known to William Hawkins in 1627 ('The Waues in the sea play at trusse and at leapfrogge on one anothers backe'), and to John Cleveland, about 1658 ('Or do the Iuncto leap at truss-a-fail?'). In *The Daily Advertiser*, 7 November 1741, a writing-sheet was offered to schools having pictures of youthful diversions including 'Truss-Fail'. The present-day 'Trust' and 'Trust-Weight', known to have been current for three generations, presumably stem from this, a truss being a measure or weight of hay or straw, and on the Cheshire–Lancashire border the game was still known as 'Truss and Weight' in the 1920s. Another name in the seventeenth century seems to have been 'Leaping the Nag', since this is given in a contemporary manuscript annotation to 'Le Cheval fondv' in a copy of Stella's *Jeux de l'enfance*, 1657, pl. 23.

The earliest account of the game, which is in *School Boys' Diversions*, 1820, under 'Leap-Frog', is not satisfactory. But it is adequately des-cribed in the fourth edition of *The Boy's Own Book*, 1829, p. 23, under 'Saddle My Nag'; and very fully in the 1855 edition under 'Jump, Little Nag-Tail' together with the formula for the mounted side:

> Jump, little Nag-tail, one, two, three,
> Jump, little Nag-tail, one, two, three,
> Jump, little Nag-tail, one, two, three,
> —Off, off, off.

The details are given that boys about to jump should shout 'Warning', and that those who cannot bear their burden are dubbed 'Weak horses'. However, the best-known name for the game in the nineteenth century seems to have been 'Hi Cockalorum': indeed this name became almost proverbial, and the Great Macdermott had a song about it.[1]

---

[1] Written by John Stanford and published 18 May 1878. Edward Moor, 1823, listed 'Hie Cocolorum Jig' amongst the sports of his Suffolk childhood. 'High-cock-a-lorum' was the name known to Tom Brown before he started his *School Days* (1857, p. 63). The children down 'Deadman's yard' devoted their energies to 'Hi Cockolorum' (*Illustrated London News*, 7 January 1860, p. 24). So did Michael Ernest Sadler at his Winchester prep school in 1871; also the boys of St. Paul's School, c. 1885; of Repton College, c. 1905; of Cheltenham College, c. 1935; and the choir boys of Westminster Abbey, c. 1930.

The following names are also known to have been current: 'Agony Oss' (Midlands, *c.* 1910), 'Backs' (Staffordshire, 1920s), 'Bom Bom the Barrow' (Stratton-on-the-Fosse, Somerset, 1922), 'Blind Rabbit' (Windermere, 1937), 'Bring the Basket' (Christ's Hospital, *c.* 1850), 'Broken Down Horses' (London, 1908), 'Bull Loup' (Crosby Ravensworth, Westmorland, 1937), 'Bum Bum Barrel' (Newark, *c.* 1900), 'Bumeroo' (Weymouth, 1920s), 'Bumsy Barrels' (Oxford, *c.* 1910), 'Bung the Bucket' (Barnes, 1894), 'Bungs and Barrels' (Oxford, *c.* 1912), 'Challey Wag' (East Anglia, *c.* 1890), 'Charley Ecko' (Paddington, 1910), 'Charley Nacker' (London, 1903, Hampstead, *c.* 1910), 'Char char Wagtail' (south Devon, *c.* 1915), 'Churchy' (Belfast, *c.* 1900), 'Cockit' (Wellingborough, 1920s), 'Donkey Jump' (Askam-in-Furness, 1937), 'Dumb Weight' (Bradford, *c.* 1910), 'Gipsy Bunker' (Leicester, 1930), 'Heavy on Ton Weight' (Liverpool, *c.* 1912 and *c.* 1925), 'Here Come I, Ship Full Sail' (Southampton, *c.* 1915), 'Here Comes My Ship Full Sail' (*London Street Games*, 1916), 'Hi Bobberee' (East London, *c.* 1900), 'Hi Diddy Jacko' (Beckenham, 1914–16), 'Hipperay-Ho' (Swansea, 1930), 'Hopopop' (Bembridge, Isle of Wight, *c.* 1905), 'Horney-Dorney' (Surrey, 1927), 'Horny Winkle's Horses' (*Living London*, vol. iii, 1903, p. 267), 'Hotchie Pig' (Scottish Border country, *c.* 1925), 'Huckaback' (North Walsham, Norfolk, 1880s), 'Hunching Cuddy' (Coalburn, Lanarkshire, *c.* 1915), 'Hunk, Cuddy, Hunk' (Dalkeith, *c.* 1900), 'Iron Donkeys' (Chardstock, Devon, 1922), 'Jack upon the Mopstick' (Warwickshire, 1892), 'Jacky, Jacky, Nine Tails' (Alton, *c.* 1906, *c.* 1920, Medstead, *c.* 1935), 'Jimmy Wagtail' (*London Street Games*, 1916), 'Jump Jimmy Wagtail' (Gloucestershire, 1902), 'Johnny on the Mopstick' (Worcester, *c.* 1930), 'Kicking Donkey' (Crosthwaite, Westmorland, *c.* 1910, 1935), 'Lanky Lowp' or 'Lancaster Lowp' (Kirby Lonsdale and neighbourhood, *c.* 1900, 1937), 'Lip-Toss Coming In' (Redruth, *c.* 1900), 'Long-Back' (East Yorkshire, 1890), 'Long-Tailed Nag' (Devonshire, *c.* 1920), 'Mopstick' (Kettering, *c.* 1915), 'Mount Nag' (*Every Boy's Book*, 1856), 'Piggy-Wiggy-Wagtail' (Framlingham College, *c.* 1915), 'Ride or Kench' (Lancashire, 1920s), 'Sally on the Mopstick' (Broadway, Worcestershire, *c.* 1901, Birmingham, *c.* 1930), 'Ships' or 'Ships a-sailing' (Huddersfield, 1883, Henfield, Sussex, *c.* 1930), 'Three Ships' (Huddersfield, *c.* 1910), 'Thrust' (Pendlebury, *c.* 1910), 'Tick-Tock-Tovey' (south Devon, *c.* 1915), 'Ton Weight Coming' (Newton-le-Willows, 1930s), 'Trusty' (Kidsgrove, 1920s), 'Warnie I'm a Coming' (*London Street Games*, 1916), 'Weak Horses' (Bradshaw, Birkenhead, and Wandsworth, *c.* 1910), 'Weights On' (Worcester Cathedral Choir School, *c.* 1925).

Our information about the game in other countries is less extensive than about 'Buck Buck' but seems to show that its hardihood (or fool-hardiness) appeals to the youth of all nations. Certainly in France 'Le cheval fondu' has long been played in this exacting form. It was not only one of Gargantua's games, but was mentioned by Calvin's teacher, Mathurin Cordier, in his *De corrupti sermonis emendatione*, 1531, ch. 38, and many times depicted by French artists in the seventeenth and eighteenth centuries. A detailed description appears in Belèze's *Jeux des adolescents*, 1856, pp. 15–18. In Belgium, Pinon records 'Plus fort cheval', and in Holland 'Bok-sta-vast' (*La nouvelle Lyre Malmédienne*, vol. ii (pt. 3), 1954, pp. 98 and 103). In Germany, Böhme gives 'Das lange Ross' and 'Baumhopsen' (*Deutsches Kinderspiel*, 1897, p. 591). In Moscow, or anyway in the Vnukovo district of Moscow, the game is known as 'Sloná' (Elephants). In Italy it is played with splendid enthusiasm, by youths as well as boys, and is known as 'Il cavallo lungo', long horse, or in Capresi dialect 'U' cavall luong'. Here each player before he jumps shouts the warning 'Mo' vir che me ne vengo', and when the team are mounted they endeavour to remain seated while calling '1 – 2 – 3 – 4 – 5 – 6 – 7 – 8 – scarica a' botta!' The game seems to have been played in Naples for centuries. According to Penzer 'cavallo luongo' is mentioned in Del Tufo's MS. *Ritratto di Napoli nel 1588*; while Basile refers to it as 'travo luongo' in *Il Pentamerone*, 1634 (1932 edition, p. 133, note 23). At Massafra, in the heel of Italy, the game is known as 'Pes u chiumm', weigh the lead (Lumbroso, *Giochi*, 1967, pp. 13–14). Thomas Hyde described the Turkish version, 'Uzun Eshek', long donkey, in *De ludis Orientalibus*, ii, 1694, p. 242; and mention of this name, as we have found, instantly produces a gleam in the eyes of Turkish youth today. A correspondent informs us he saw the game played in India. And a Japanese boy told Howard Spring that a similar game 'Uma-Nori', restive horse riding, is 'one of the most popular winter games of Japanese boys' (*Country Life Annual*, 1958, p. 128).

SKIN THE CUDDY

This variation of 'Hi Jimmy Knacker' is traditional in the north-east of Scotland, and is also played in Wandsworth as 'Jumping Jupiter'. All the boys except one, sometimes two, bend down and make a cuddy or long-back against a wall, in the manner of 'Hi Jimmy Knacker', although the back they make is longer, as more boys are bending down. The player who is out, who is sometimes called the 'bronco buster', has to jump on to the line of backs and attempt to work his way along the line and touch the head or take the cap of the boy who stands at the far end as 'post' or 'pillar'. Since the chain of boys is constantly heaving about, and each player in turn feels it an indignity to be sat upon, the rider is not always successful in his object. In fact edging along the line of backs is a perilous undertaking; and the determination of the players to make it so can be judged by the displeasure expressed when a rider manages to reach the other end and wins a second turn. 'This game is very rough,' confided an 11-year-old, 'but I enjoy it.'

** The game has been known as 'Skin the Cuddy' in Aberdeenshire for three generations. In Argyllshire it was 'Bull the Cuddy' (*Folk-Lore*, vol. xvi, 1905, p. 346). At Dyke in Morayshire it was 'Saddle the Nag', the players in one team trying in turn to wriggle along the backs of the other team; and at Barchory in Kincardineshire it was 'Skin the Goatie', a single boy on the back of another trying to 'crown' the player who stood upright supporting the boy giving the back (Gomme, *Traditional Games*, vol. ii, 1898, pp. 147 and 199–200).

9

Daring Games

'All the games that are played are all harmless (in a way) and very good fun.'

Boy, 14, St. Peter Port

CHILDREN seem to be instinctively aware that there is more to living than doing what is prudent and permitted. In a boy's world trees are for climbing, streams are for jumping over, loose stones are for throwing, and a high wall is a standing challenge to be mounted and walked along. To a child it is more interesting to hang out of a window than to look out of it, to jump down a flight of steps than to walk down it, to defy a park-keeper than to hide from him. Exuberance and curiosity lead him into many of his scrapes; yet a boy's love of daring must arise from something still deeper. The glory he sees in danger is that it seems to be linked somehow with his becoming mature. If he did not do what was forbidden how could he be sure he was a person with freedom of choice? So it is that the juvenile tribe holds ritual games, almost as part of the process of growing up, in which the faint-hearted are goaded into being courageous, and the fool-hardy stimulated to further foolhardiness. On the face of it little can be said for these entertainments. Nevertheless, if juvenile folly sometimes takes the unhappy forms it does owing to children's inability to appreciate the consequences of an action until they occur, it may be that these sports serve a purpose. In these games fixed procedures are generally adopted: the dares are proposed and accepted, the feats are closely watched by the rest and appraised; and it may be that these games, by their very formality and emphasis on daring, help the majority of children to understand the nature of risk-taking.

TRUTH, DARE, PROMISE, OR OPINION

No game is more revealing of the childhood community than this sport, apparently known everywhere (descriptions from forty places), which they impose on themselves 'when we want to play something lively'. Any number may take part, although normally not more than five or six of them are in it together; and the severity of the ordeal varies with the occasion and the players. Sometimes a group of girls go off to the park, find a dry place, and sit down in a circle, as if nothing more was intended than a picnic; at other times the participants are a mixed group of boys and girls ('It is much better if you play with girls,' declares a 14-year-old boy, 'then you have more scope'), and they meet after dark in a disused air-raid shelter, or in somebody's backyard or home.

One person is named 'leader', 'master', or 'questioner', and turning to the player nearest him asks, 'Will you have truth, dare, promise, or opinion?' If the person's choice is 'Truth', he has to answer truthfully any question (usually personal) that the leader may ask: 'Is it true that you love Betty Matthews?' If the choice is 'Promise' he is safe for the present, but has to bind himself to something in the future: 'Will you promise to give me six sweets next time we meet?' If the choice is 'Opinion' he is likely to be assigned the delicate task of giving his opinion of somebody present. And if the person's choice is 'Dare' (the usual choice with boys and hoydens), he is liable to be instructed to knock at a front door and run away, or to punch a passer-by on the back, or to ring up the telephone operator and tell him to 'get off the line as a train is coming', or to engage in some other provocation of the adult world, for there is a curious feeling in this game of having to respond to the challenge 'Are you one of *us*, or one of *them*?' Thus a 13-year-old Scots girl, telling how they meet in the street after dark and play 'Truths and Dares', says:

'Sometimes you [are dared to] go to a lady's house and ask for a "jellie piece".

'Sometimes the question master tells you to go to a wifie's window and knock on it and cry "Are you in Nellie?" and then run along and knock at the wifie's door and act as you're walking up the street, and say that a boy did it and went away round the corner.

'Then the question master says you have to throw water on the wifie's doorstep and she threatens you "I'm going awa for the Bobby," and she shammies she's going away for the police. Then you play your game on another auld wifie until the other wifie come back.

'Then sometimes you're dared to go and tie the wifies' doors together (of course its the wifies you do not like) and knock on them and run awa, and the auld wifies come out both together. What a laugh you get when they come out and discover the string tied to the door.'[1]

Each player in turn is asked his choice, and having made it, has to comply with the demand which follows. If he refuses he is jeered at

[1] For a summary of door-knocking pranks and their names see *The Lore and Language of Schoolchildren*, pp. 377–92.

('Called a coward until we next play the game'), or expelled from the gang, or in some places, what may seem to him providential, he is made to pay a forfeit, usually a piece of clothing 'like a belt or a shoe', which he has subsequently to redeem by accepting a new challenge. Sometimes the temper of the game hardens, almost accidentally, and without realizing what is happening the players have become hell-bent on extending themselves, encouraging dares to remove more clothes or to jump from yet higher places. Then on some desperate occasion, the invisible line may be crossed beyond which there is no stepping back, and a dare ends sadly as a court case or newspaper story. In Swansea a 13-year-old schoolboy who had been dared was found to have swallowed eight two-inch nails (*Yorkshire Post*, 31 October 1960, p. 3). At Ossett an 8-year-old schoolboy admitted setting fire to a stack of unthreshed oats for a dare (*Yorkshire Post*, 10 November 1960, p. 14). At Frimley a 7-year-old boy woke up in the middle of the night screaming that he had seen a body. He had been dared to peep through the window of the public mortuary adjoining a recreation ground (*The Times*, 13 December 1963, p. 9). And in Portsmouth a particularly unfortunate game of 'Dare, Truth, or Promise' ended with two boys of twelve and thirteen being prosecuted for improperly assaulting two young girls (*News of the World*, 14 February 1954, p. 7).

It seems too, from the varying names of the game, that in different places, or on different occasions, different types of challenge are predominant. On the Scottish Border the game is sometimes called simply 'Finding Out' and the demands are purely (or impurely) for factual information.[1] In other districts the emphasis is on their relationships with each other. In Kidderminster the game is occasionally called 'Truth, Dare, Promise, or Kiss'; in Peterborough and Swansea, 'Do, Dare, Kiss, or Promise'; in Gloucester, 'Truth, Dare, Warning, Love, Kiss, or Marriage' ('If you choose kiss you must kiss whoever is named'), and in West Ham it is 'Dare, Truth, Love, Kiss, Promise, or Demand' ('If he picks "Love" you must tell him to do something with love'—Girl, 10). In some places there are degrees of enforcement, thus in Inverness, 'Truth, Dare, or Got to'; in Watford and Orpington, 'Truth, Dare, Promise, or Must'; in Aberdeen, 'Truth, Dare, Force, or Promise' ('If it was "Dare" I would dare her to climb a lamp-post, ring a bell, or kiss a boy, but she can refuse. At "Force" it goes the same as "Dare" but one is forced to do it'—Girl, 12). At Spennymoor, where the game is called 'Truth, Dare, Will, Force,

---

[1] This may be equated with the 'game' (so styled) known as 'Confessions', where a person is tortured until he reveals something considered sufficiently shameful to justify his release; and for it to be someone else's turn to be made to confess.

and Command', three of the choices seem indistinguishable, for with each of them the dare is enforceable. In Edinburgh, where the game is 'Truth, Dare, Promise, or Repeat', the player who chooses 'Repeat' has to repeat his task a certain number of times, the task usually being to make a statement such as 'I'm my ma's big bubblie bairn', which a 14-year-old boy confessed he found exceedingly embarrassing. And in some forms of the game there are no alternatives to accepting the dare, as in 'Got to' in north London, 'Dare, Double Dare' in Plymouth ('After the person has done one dare he has to do another'), and 'King's Command', 'Get the Coward', and 'Chicken' (see hereafter).

** The forerunner to this game seems to be 'Questions and Commands', otherwise known as 'King I am', a popular social sport in the seventeenth century, mentioned by Burton, Wycherley, Herrick, and Randle Holme. It seems to have been as audacious a game as 'Truth, Dare, Promise, or Opinion', and, since it was played by adults, decidedly more witty. Thus in *Gratiae Ludentes*, 1638, p. 65, 'a question proposed to a gentlewoman at the play of *questions and commands*' was:

'Suppose you and I were in a roome together, you being naked, pray which part would you first cover?' *An.* 'Your eyes, sir.'

In the eighteenth century 'King I am' was looked upon as a childish amusement. It was included in the juvenile *Little Pretty Pocket-Book*, 1744 (1760, p. 30), and was played by the children in Brooke's *The Fool of Quality*, vol. i, 1766, p. 72. Probably its formula was that quoted in *The Craftsman*, 4 February 1738, p. 1:

'King I am, says one Boy; another answers, I am your Man; then his Majesty demands, what Service He will do Him; to which the obsequious Courtier replies, the Best and Worst, and All I can.'

The author of *Round about our Coal Fire* (1731, p. 10), who considered the game suitable for Christmas, suggested that the forfeit should be fixed at a certain sum, or that the face should be smutted, so that he who did not wish to comply with a demand could be 'easy at discretion'. Strutt, in 1801, also termed 'Questions and Commands' a childish pastime, and noted, as earlier writers had done (e.g. William King and John Arbuthnot), that the game appeared to be related to the 'Basilinda' of the Greeks, in which a king, elected by lot, commanded his comrades what they should perform (Pollux, ix. 110). The name 'King's Command' seems to have survived to the present time among certain teenagers, being referred to in a British Medical Association report in 1964.[1] Mention should also be

[1] *Venereal Disease and Young People*, B.M.A., 1964. The rules of this unseemly game appeared in greater detail in the *Daily Express*, 6 March 1964, p. 8.

made of 'Roi qui ne ment' played by Froissart about 1345; 'Questions'
mentioned by Drayton in 1598; 'Wadds and Wears' in *The Gallovidian
Encyclopedia*, 1824; and the game 'Truth' played by the party at Camp
Laurence in *Little Women*, 1868, ch. xii. However, in the United States
a boy is liable to be 'stumped' instead of dared, as in W. M. Thayer's
From Log-Cabin to White House, 1881, p. 73, where Edwin holds up
a pullet's egg and challenges James Garfield, the future President, 'I stump
you to swaller it'. In Edmonton, Alberta, the game is called 'Hot Box', the
penalty for failing to comply with a demand being to crawl between the
commander's legs and be hit while doing so.

FOLLOW MY LEADER

'Follow my Leader' can be a mere nursery game with a line of little players
walking wherever the leader walks, jumping when he jumps, hopping when
he hops, dancing, turning somersaults, making faces, crowing, and very
often—in accord with the first child's lead—acting like silly-billies.
Amongst older children the sport is more testing.

In some versions of 'Truth, Dare, Promise, or Opinion', the challenger
has himself to be prepared to carry out the dare if the other refuses, and
fundamentally 'Follow my Leader' is the same; but the leader always per-
forms the feat first, and then by implication, and sometimes in so many
words, challenges the rest to do likewise. If any follower is unable to perform an action he either goes to the back of the line, or is sent out of the
game and has to stay in the den. Sometimes the actual point of the game is
that the leader should go on doing difficult things 'until all the followers
are in the den. Then the game starts again'. Quite often the game is played
on bicycles, the leader choosing the most slippery paths he can between
trees and up banks, and the others having to follow: 'If you fall off you're
out.' And sometimes the game becomes a test of nerves. The leader,
chosen because he has a stout heart, leads them along the top of the para-
pet of a railway bridge, through the narrow gap between two lime kilns
when they are burning, across a weir, or over the roofs of buildings; and
sometimes the expedition does not end happily. 'An 11-year-old boy . . .
was leading four other boys who had walked along the side of a stream and
climbed on to the roof, playing "follow my leader" ', reported *The Yorkshire Post*, 12 April 1960, p. 12. The leader 'stepped on to a glass panel on
the roof and disappeared into the works . . . He landed on the only space
on the floor which is clear of machinery. At the hospital late yesterday the
boy was stated to be "comfortable" '.

Other names: 'Jack follow my Leader' (Leeds), 'Pied Piper' (Swansea), 'Cappers' (Penrith).[1]

A variation is 'Leading the Blind' (Slough) or 'Guide me with Guilt' (Wilmslow), in which those who follow are blindfolded or must keep their eyes shut, and hold on in a line each to the one in front. Their guide looks for as many obstacles as he can find to lead them over or under, and put them in disarray. 'This game is best to play in a park or in a wood,' explained a 10-year-old, 'where you can fall into bushes or trip over trees.'

** 'Follow the Leader' is often referred to in children's literature, usually in an admonitory manner, e.g. in *Youthful Recreations*, 1789, p. 122, 'If Tommy Heedless had any fault, it was that of his being too fond of a game we call Follow the Leader'; *Dangerous Sports*, 1801 (1807, p. 85), 'Follow the leader is a game which, as generally played, is full of danger'; *School Boys' Diversions*, 1820, pp. 26–7, 'Follow the Leader . . . certainly affords occasions to display courage, and even temerity; but such excesses will ever be shunned by prudent lads.' However, the editor of *The Boy's Own Book*, 1829, p. 24, remarked that 'Without a bold and active leader this sport is dull and monotonous'.

In Brockett's *North Country Words*, 1829, the game is styled 'Jock and Jock's-man'.

GET THE COWARD

This game ('the game I like best', says a Wigan 10-year-old) is a form of 'Truth, Dare, Promise, or Opinion' but without the options and without a leader. 'It is a game for five or six players. You start off by putting the boys' names in a cap. When you have done this you pick a boy to pick a name out of the cap. When you have done this you hold a council of boys who decide what to do with the boy who's been picked out of the cap. Sometimes the council might say that he has to ring a doorbell or knock off a policeman's helmet. If he does not do it you shout:

> Iggly piggly poo,
> Put the coward in your shoe,
> When you're through let him go,
> Inny tinny let him know.

[1] The name 'Cappers' apparently comes from *cap*, meaning to emulate, to surpass, in the way schoolboys used to *cap verses*. 'In puritanical times of old,' remarks Forby in his *Vocabulary of East Anglia*, 1830, '*capping of texts* was a favourite, and doubtless very edifying, sort of pious pastime. With us at large, the word is used on a great variety of profane occasions. An idle boy leaps a ditch, or climbs a tree, and if his play-fellow cannot equal or out-do him, it is a *cap*; he has *cap'd* him.' In Westmorland two generations ago the game was known as 'King Cappers' (*EDD*).

Then you run at him and start fighting him, but you don't hurt him. I like it', adds the 10-year-old, 'because it is a rough and very daring game.'

LAST ACROSS

'Last Across' is the game in which, to the consternation of motorists, children line up on a pavement, wait until the leader has selected a particular car or lorry, and then 'when it gets rather close you all run across the road and the one who gets nearest to the front bumper wins'. Another name for the game is 'First to the Cemetery'.

The question whether there are deeper motives for this game than mere devilment is touched upon in the discussion on play in the street in the Introductory chapter (see pp. 11–12). Certainly when children take part in street games it is not they who are afraid of the traffic, it is the traffic that is terrified of them; and the children are aware of this, and willing to take advantage of it.

The game of 'Last Across the Railway Line', however, the pleasure of which is even less comprehensible to the adult mind, might seem to dispose of the theory that children are in part giving vent to a frustration, or making a form of protest, when they run in front of cars, for children are not ordinarily thought to be antipathetic to trains. Nevertheless railway property with its untrodden railroad, its quiet green slopes, trees, and bushes, its tantalizingly exposed machinery, must often appear to a child, and in fact be, the most attractive playground in the district; and any investigation into juvenile trespass on the railway might well start with a study of the geography of the places where trespass most frequently occurs. Certainly in some districts the game of 'Last Across' in front of trains is almost normal child's-play. In 1957 the sport was reported to be so common on the stretch of line between Devons Road and Poplar Station that engine drivers were threatening to refuse to drive trains along it for fear of being involved in an accident (*The Times*, 12 March 1957, p. 4). In 1961 the secretary of the Derby branch of the Associated Society of Locomotive Engineers and Firemen stated that playing 'Last Across' the line was 'prevalent throughout the whole of the Midlands', and he added that a new variation had been observed, which intensified juvenile rashness: 'A group of children put pennies into a hat and put it down on the side of the track. They dash across at the last possible moment in the path of the train and the last one to move picks up the cap and claims the pennies' (*Yorkshire Post*, 28 April 1961, p. 9). In 1964 a British Railways spokesman stated that during that year, in Essex alone, three children had been killed and seven injured while playing 'Last Across the Line' (*Daily Express*, 11 December 1964, p. 11). In 1968 at Fryston, near Castleford, where 'the trains go very fast . . . and some of the children play a game in which they dare each other to get out of the way at the last minute', eight children saw two of their friends killed by a train (*The Times*, 5 April 1968, p. 1). And in Plymouth, the Western Region became so concerned by the number of incidents, that in July 1968 they launched a 'Rail Safety Campaign', emphasizing the danger of children playing 'Last Across' and 'Chicken', although well aware that such publicity can be self-defeating.

* Both Norman Douglas in *London Street Games*, 1916, p. 142, and Newell in *Games of American Children*, 1883, p. 122, refer to playing last-across-the-road, but under 'Follow the Leader'. Norman Douglas called it the 'only really dangerous game we have', dangerous because the bravest boy was chosen leader 'who crosses the road just in front of some heavy van'; and it is worth recalling that in the days of horse-drawn traffic crossing the road was scarcely less hazardous than it is today. (In 1865, for example, no fewer than 232 people were killed by carriages in the streets of London alone.) In America, Newell reported, the game was played in a peculiarly reckless fashion in the southern states, 'the leader will sometimes go under a horse's legs or between the wheels of a wagon', whereupon the driver knowing what to expect, would find himself having to stop for the rest of the players to follow.

CHICKEN

'If you want something done and you are too scared to do it yourself you go up and "Chicken" somebody. You tell them what to do, and if they don't do it you call them "Chicken" and you dare them again, and if they don't do it you call them "Double chicken". And they go off and everybody calls them a chicken, and you call them all names, "Quack quack" and names like that. The sort of thing you tell them to do is to go and hit somebody, or to go and tell a boy you like him.'

Thus two girls of 13 and 14 in Petersfield. Unhappily, by undefined degrees, the practice or threat of 'Chicken' can lead to less innocuous impositions.

A 13-year-old boy in Grimsby says that if two boys are fighting one of them may challenge the other to a 'Chicken Run'. They take their bicycles and go to some traffic lights, and 'when the lights change to red they both go straight through the lights, and the first one to stop is "Chicken", but if one of them keeps going he is the winner of the game'.

Sometimes the duel in which two boys charge each other on bicycles is called 'Chicken'; whoever swerves first as they come at each other is tagged with the odious name, and is even likely to feel he deserves it for his faint-heartedness.

Sometimes 'Last Across' is called 'Chicken', hence the schoolchild joke (current 1963): 'Why didn't the *cock* cross the road?—Because he was chicken.'

And sometimes the game takes yet more senseless forms. In Halifax a boy was caught playing 'Chicken', swerving his bicycle in front of a car and then jumping off and lying on the ground. When the car stopped he jumped on his bicycle and rode off, and he and his friend had a laugh about it. Unfortunately for him (or perhaps fortunately) the fourth and last time he did this was in front of a police car (*Daily Express*, 2 August 1961, p. 4).

In Liphook, Hampshire, a youth playing 'Chicken' with a mixed group of younger children, lay down in the middle of the road at night, deliberately staying there until a car came, but another car which was following overtook it, ran over his head, and killed him. 'The idea of the game,' a 14-year-old boy told the Coroner, 'is to frighten the motorist' (*Hampshire Telegraph*, 12 November 1964, p. 9).

When a group of 11 to 14-year-old boys at Harold Hill in Essex were playing a game to see who dared stay longest on the railway line before an oncoming train, two stayed too long and were killed, and a third was only saved by being knocked off the line by the bodies of the others. The fourth boy, who said he was alive because he had lost his nerve, added, 'This has been going on most nights since a gang of us started going over to a railway bridge on Harold Court road a few months ago. Another test of courage was to lie at the side of the track and let the overlays of the train pass over your head' (*Daily Express*, 21 and 25 November 1960, pp. 5 and 16).

In 1957 an official of the London Midland Region of British Railways reported that they had instances 'where boys lie on the track and put their heads on the line. When the train stops they jump up and run away' (*The Times*, 12 March, p. 4).

In 1966 when thirteen boys and two girls were arrested for trespassing on railway property at Glen Parva in Leicestershire, one boy, aged 13, was caught while actually lying between the tracks, with his feet over one track and his head over the other (*The Times*, 27 May 1966, p. 13).

It is also not unknown for children to lie flat in the middle of the tracks and allow trains to go over them. An express driver on the London to Leicester line was reported in *The Times*, 4 May 1966, p. 10, as saying that at Newton Harcourt, 'I came round a bend at 80 m.p.h. and saw a boy lying face down between the lines with his hands behind his head. I braked, but it was too late to pull up. The train passed over the boy and afterwards I saw him get up and disappear on his bicycle.' A railway spokesman, stating that this was not the only case that had been reported, added that fortunately the train was a diesel, 'any other type would have cut his head off'. Indeed children sometimes seem to be as unaware of danger as the first-day fledglings in one's garden. One boy of 6, who showed his head-mistress how to lie under a train, commented: 'All you have to do is lie still with your arms stretched out and keep your eyes closed as the train goes over. It doesn't hurt a bit.'

* The term *chicken* for the faint-hearted, although old, was not found among children in our survey in the 1950s (when 'yellow' and 'windy' were the popular expressions), but was suddenly known to everyone in 1960, even as the name for the game (see, e.g., *Yorkshire Post*, 24 September 1960, p. 4). Apparently the term came from the United States, where it had been current anyway since 1935; yet it was a return export. In 1707 in *The Beaux' Stratagem* (IV. iii) Gibbet says 'You assure me that Scrub is a Coward', and Boniface replies, 'A Chicken, as the saying is.' And in 'The Prisoners' Van', in *Sketches by Boz*, 1836–7, a young girl criminal up-braids her still younger sister, who is showing distress, with the words, 'Hold up your head, you chicken. Hold up your head, and show 'em your face.'

MISPLACED AUDACITY

In children's games as in other departments of life, the player who takes no risks can expect little admiration; and few children have the ability to judge whether a particular game is worth playing, let alone whether it is worth playing courageously. In a diversion such as the 'Knife Game', which is commonplace among boys in the north country, the sport has little interest if not played with a certain *élan*. *Sine periculo friget lusus*.

'You need a knife, preferably a large one with a sharp edge and a good point,' reports a 13-year-old. 'First of all you spreadeagle your left hand. Then you take the knife in your other hand and as quick as you can stick the knife in the spaces between the fingers without cutting your finger. When you are experienced enough you make it harder, for instance you thrust the knife between the third and fourth fingers, then between the first and second, then between the second and third fingers, and then between the thumb and first finger.'

Likewise on the swings in the park, children do not merely sit on the seats and see who can swing highest, but try who can climb furthest up the chains while swinging, and who can best twizzle the swing while swinging, and who will jump off his seat from the greatest height, and who, by swinging hard, can leap the furthest off the swing, a sport sometimes called 'Parachutes'. ('It is very dangerous for I have fallen off and cut my arm and had to have it in plaster for two weeks'—Boy, 12.) In County Durham, a person is pushed as high as he dare go, and then the pusher stops pushing and cries:

> The cat dies once,
> The cat dies twice,
> The cat dies three times over.
> If you don't get off I'll give you a knock,
> It's time your swing was over.

The person has then to jump. Jumping off the swing is also part of the game called 'Countries' (Liverpool) or 'All Over the World' (Glasgow), where several children go on swings beside each other, and a player on the ground calls out 'England', 'Scotland', 'Ireland', or 'Wales', the names given to the several concrete slabs on which the pushers ordinarily stand, and the swingers must attempt to land on the particular slab that has been named. If there are protecting bars to prevent the swings going too high they also play 'bumps', swinging higher and higher until the chain or rod hits the bar; and one 13-year-old Stoke Newington girl was killed while attempting actually to jump from the swing and catch hold of the crossbar (*The Times*, 14 June 1960, p. 15).

Yet roller-towels, when children make free with them, can be as dangerous as swings. They put their head in the loop of the towel and wind themselves up until they can lift their feet off the ground. They play 'Dangling Man', a game of extraordinary attraction, twisting the towel tight and letting their neck take the weight of their body until they go blue in the face. At Southwell, an 11-year-old boy had a go in the towel after his companions had left, and took twelve hours to regain consciousness (*Yorkshire Post*, 13 October 1961, p. 1). In New Romney a 9-year-old died while playing the game, which his classmates described as one of their favourites (*Daily Express*, 10 December 1963, p. 11). In Islington a 9-year-old died at his school, apparently also while playing this game with a roller-towel, which was there referred to as a 'Spinning game' (*Daily Express*, 7 October 1966, p. 18).

Frequently, as we have observed, it is not bravado that occasions foolishness so much as wonder and curiosity, those twin attributes of inexperience which, for instance, prompt a child to turn round and round until he is giddy and can 'see the world going round'. So it is, when a rumour sweeps through the school that a person who stands on wet blotting-paper, or who puts wet blotting-paper in his shoes, is likely to faint, repeated trials are made; and although the experimenters may be encouraged by the thought that anyone who faints will miss the next lesson, this is undoubtedly not the mainspring of their research.

In 1961, for instance, when a craze for making people faint broke out in a Lancashire primary school, the boys were eager to demonstrate even to their headmaster. 'We can make a boy faint for a minute, sir,' they declared.

'Oh yes?' said Sir, tolerantly.

They sat on their hocks, knees bent, arms outstretched, and took ten deep breaths, then stood up holding the tenth breath, and someone from behind squeezed them round their waist. One boy was flat out for a minute. ('Every thing went all white, sir, and the next I knew I was on the ground.') As the headmaster remarked later, this was another craze that had to be killed as soon as it was born. 'There was a lump the size of an egg where the boy's head had hit the ground.' But five years later the 'Faint Game' was reported in *The Times*, 27 June 1966, p. 11, as newly current amongst schoolboys in north Kent. The procedure was precisely the same as in Lancashire, and apparently equally effective. One boy, *The Times* stated, had already been taken to hospital with a suspected fractured skull.

IO

Guessing Games

CHILDREN's preference for exercising their bodies rather than their minds is obvious enough, and their participation in intellectual guessing games, even of the humble order of 'Coffee Pots', 'Animal, Vegetable, and Mineral', and 'I Spy With My Little Eye', is apt to be limited to occasions when they are restricted, and unable to play anything else. Out in the street or playground their guessing games are little more than preludes to action: the means of starting a race or initiating a scrimmage. Even in the simplest of their out-door guessing games, such as 'Squeak, Piggy, Squeak' (or 'Puss, Puss, Miaow'), in which a player has to identify a disguised voice; or 'Guess Who', in which a person is obliged to name whose hands have unexpectedly been placed over his eyes; or 'Leading the Blind' or 'Dumping', in which a blindfold child must guess where he has been led or dumped, the appeal is to the senses rather than the intellect. The fact is that in their guessing games children do not greatly care whether a person guesses correctly or not; no scores are kept, as in gambling games; and, quite often, the games have a dramatic or ritualistic form in which the guesser is happily guessing in the guise of somebody else.

FILM STARS

'Film Stars' is the most popular guessing game in Britain. It is played anywhere out of doors, often across a road, and sometimes at night. One girl goes to the far side of the road (the game is played mostly by girls), and calls out the initials of a film star or singer, for instance 'C.R.' for Cliff Richard. As soon as a player thinks she knows who the initials stand for she races across the road and back again, and then shouts out the name. If her guess is right she changes places with the person in front; but if her guess is wrong—this is the fun of the game—she has run for nothing. In some places the guessers have to run across the road and back again, and then across the road a third time, and whisper their suggestion to the person in front so that the others cannot hear it. Sometimes several children will be racing as fast as they can against each other, and all of them find that they have thought of the wrong name. ' "Film Stars" is marvellous,' said an 11-year-old, 'I play it as often as I can.' And she is typical. Almost everywhere, it seems, the game becomes an obsession with one group after another of little girls who have recently become starry-eyed, but are still athletic.

Other names: 'Initials', 'Pop Stars', 'TV Stars', and, in Liverpool, 'Filmy', a typical scouse apocope. In Edmonton, Canada, and doubtless elsewhere in the New World, it is called 'Movie Stars'.

VARIANTS OF 'FILM STARS'

A game such as 'Film Stars' which is in the ascendant may not only out-shine and eventually replace its progenitor, it can cast its glamourie over similar games, and five of them are now played under the name 'Film Stars'; or, to put it the other way round, 'Film Stars' is now played in five variant forms.

(i) The girl who chooses the initials stands only just in front of the others, and as soon as someone guesses correctly she starts to run, but must stop immediately the person who guessed right shouts 'One, two, three, stop!' The guesser then attempts to reach the caller with a certain number of strides, usually three, but sometimes with as many 'pigeon toes', or other specified steps, as there are letters in the film star's name. The guesser does not change places with the caller unless she manages to reach her. This game is surprisingly popular, particularly in the London area.

(ii) In some places, for instance Knighton and Forfar, the chase is more energetic. All the players line up against the wall except the caller, who stands on the kerb. When a player thinks she knows the film star's name she chases the caller. If she catches her before the caller reaches the wall on the other side of the road, she gives her the name of the film star, and if right, becomes the new caller. In Forfar, if none of the guessers can think of the name of the film star, the caller says 'Steppie for the first time', and is allowed to take a step into the road before she gives the star's Christian name.

(iii) This version, which is little more than a way of starting a race, is particularly common in Scotland and the north country, and also in Guernsey, and is played by boys as well as girls. The caller stands on the kerb on the opposite side of the road, and as soon as someone names the film or TV star correctly, he runs across the road and back, while the guesser does the same in the opposite direction. Whichever of them arrives back first is the next caller. (Cf. 'Time' in Race Games.)

(iv) Sometimes, as reported from Aberdeen, Accrington, Netley, and Liverpool, there is no running, only guessing. The initials called are usually for difficult names, and the caller has to give hints when requested, such as the sex of the star, the names of films in which the star has appeared, and sometimes must even adopt the characteristic stance of the hero or heroine. 'If' says a 9-year-old, 'the people want to know the first or last name, the person who has ask them has to tell them otherwise the person who has ask them is not allowed to play any more that day.'

(v) In Guernsey they have a hybrid variety, called 'Step for a Hint', which seems to be descended from both the foregoing game and the race game 'Aunts and Uncles'. In this game it is the player in front who guesses the name. The others decide who the film star shall be, line up, and call out the initials. The one in front asks each player in turn a question about the film star: 'Is he a man?' 'Is he tall?' 'Is he fat?' 'Is he dark-haired?' and each time if the answer is 'Yes' the player addressed takes a step forward. The player in front tries to guess the name of the star before somebody reaches her.

I SENT MY SON JOHN

This is the principal game from which 'Film Stars' seems to have evolved, and it is perhaps remarkable that it continues to retain a separate identity.

One person stands out, usually some distance away, and thinks of an article that is sold in a particular kind of shop, for instance, as a 10-year-old West Ham girl suggested, 'If she was thinking of Bird's Eye frozen peas she would say "I sent my son John to the grocer to buy some B.E.F.P." ' When somebody thinks she knows what B.E.F.P. stands for, she runs to an agreed place and back to the person in front, and tells her what she thinks it is. 'You go on like this till someone gets it.'

Known as 'I Sent My Son John' in West Ham, Brighton, Orpington, Plympton St. Mary, and Forfar (where it has been played under this name for fifty years), it is also called 'My Son John' (Oxford and London, S.E. 7), 'My Mother Works' (Old Kilpatrick, Glasgow), and 'Shops' (Banbury).

⁎ In the nineteenth century this game was generally known as 'I Apprentice My Son', e.g., in *The Girl's Book of Diversions*, 1835, pp. 51–3, and *Cassell's Book of Sports*, 1888, p. 775. It stems, apparently, from a game called 'Two Poor Tradesmen', played in Regency days, and described in *School Boys' Diversions*, 1820, pp. 15–16. In this, two boys addressed the rest, 'We are two poor men, come out of the country, and want to set up a trade.' 'What's your trade?' they were asked. They named a trade and gave the first and last letter of the article they required. Guesses were

then made in rotation, and the person who guessed the article, or anyone who named an article not belonging to the trade, was 'buffeted by the rest to an agreed distance' for his trouble.

Cf. 'Three Jolly Workmen' (pp. 280–4).

SHOP WINDOWS

This game is hardly to be distinguished from 'I Sent My Son John' except that it is played in front of a shop window, and the object which has to be guessed is chosen from the window. Usually one person studies the window while the others line up on the kerb. He chooses an object such as a tin of salmon, and says 'I see T.O.S.' If someone guesses right he may have to run to the wall the other side of the road without being caught (Abertillery), or race the person who gave the initials; or, when a person thinks he knows, he may have to run to the other side and back before he makes his guess (Kirkcaldy). In Birmingham, where the game is called 'In the Window', all the players choose the object, except one, who has to guess the item they have chosen before he can give chase. In Colne, Lancashire, where the game is known as 'Hot Peas', the children standing on the kerb chorus 'Try again' each time a wrong guess is made, but when the guess is correct they shout 'Hot Peas', and dash to the other side of the road, hoping to get there without being caught.

ANYTHING UNDER THE SUN

This, too, is basically 'I Sent My Son John', but the caller may choose any subject she likes. 'You can have towns, rivers, counties, countrys, seas, capitals, birds, ships, children's names, plants, and trees.' The caller may say ' "It's a girl's name beginning with G," and if any of the players think they know it,' says a 10-year-old Ipswich girl, 'they run up to the person and back to their place and up to the person again, and if they guess right they are "in". If two people guess right at the same time they choose colours, flowers, or vegetables, for instance, Kingfisher Blue and Lavender Green, not saying who has chosen which. The other players pick one of the colours, and the person whose colour has the most votes is "in".'

In Oxford and Banbury the game opens with the set formula, 'John Peel's favourite animal's name begins with A', and is known as 'John Peel', or sometimes 'John Bull'. In Aberdeen the game is called 'Odds and Ends'. And in a number of places children are insistent that the game is known only as 'Flowers' or 'Animals' or 'Cars' or 'Footballers' or 'Soap Powders' because these alone are the permitted subjects of the initials, as if TV and film stars had never existed.

In Edinburgh, and some other places, the game is played under the name 'Advertisements', and the clues are not initials but slogans such as 'The Sunshine Breakfast Food', and the players have to guess the product sold with the slogan.

CAPITAL LETTER, FULL STOP

'Capital Letter, Full Stop' is the distinctive name in Ipswich of the form of playing 'Film Stars' in which the caller runs away as soon as a correct guess is made, and the guesser shouts 'Stop' (variation i). The person who guessed correctly then spells out the name she guessed, taking a 'pigeon step' for each letter. She then does 'capital letter', 'full stop', and jumps. 'Then she tries to touch the person's toes.' The game is played in precisely the same manner in Plympton St. Mary, but is there known as 'Colours'; and it is played in a very similar manner in Market Rasen, where it is called 'Hop, Skip, and Jump'.

BIRDS, BEASTS, FISHES, OR FLOWERS

In this game every player is physically involved whether or not he himself makes a guess. Two sides are chosen of equal size, and line up on opposite sides of the road, facing each other. One side thinks of a bird, beast, fish, or flower (or, sometimes, the name of a film star, famous person, or 'child in the school'), and advances line abreast to the other side and gives the initial (or initials). If the guessing side makes a wrong guess the callers stand fast. But if the guess is correct they sprint back to the safety of their own side of the road, for any who are caught on the way back have to join the opposing side, whose turn it becomes to give the initial of a subject.

The game is particularly popular in Scotland and the Isles under the name 'My Father' or 'My Father's Occupation', when the subject of the initial is not an object but a trade, or a worker in some trade.

⁎⁎ In the nineteenth century the game was known as 'Flowers' or 'Bird Apprentice' (Gomme, *Traditional Games*, vol. i, 1894, pp. 33 and 129).

AMERICAN TIMES

'American Times' is here described by a 12-year-old girl in Fraserburgh:

'You play on the pavement with this one because it is mostly played outside. You have to get about six children and they all stand against a fence or a wall except two children. The two children that are out, they have to whisper to themselves the time that they choose and when they have thought of a time they go and ask the children "What time is it?" and they go right through them all and if they still can't answer it, well they go right back to the beginning until a person gets it right. When the person gets the time right she has got to choose names of things. She gets to pick her own choice between girls' names, boys' names, colours, fruit, car names, schools, and cigarettes. When she's picked what she wants, well she tells the two girls. Then they go away and whisper again. If she had picked colours then the two girls have to pick two colours. One girl takes red and the other takes blue, and they go up to the girl who got the right answer to the time and say "What colour do you want, red or blue?" and if she says "red" well the girl who picked blue has got to go in that girl's place and the girl has to go with the other girl. When they have changed places they start all over again and they play until everybody gets a chance to be the "man".'

Appallingly tedious as this game may seem, it is nevertheless popular in Scotland, reports coming particularly from Ayrshire and Lanarkshire, and it is not unknown in England.

** It is possible that the game is American in more than name, for it is described by Charles F. Smith, of the Teachers College, Columbia University, as a 'favourite after-supper game', in *Games and Game Leadership*, 1932, p. 267. However, Matizia Maroni Lumbroso, in her excellent collection of *Giochi descritti dai bambini*, 1967, pp. 230–1, although she has no records of the game in Italy, reports it being played in Sardinia (under the name 'Corochè') exactly as it is played in Scotland.[1]

THREE JOLLY WORKMEN

Even if the pedigree of this pavement 'Dumb Crambo' was unknown, its age would be apparent from the old-fashioned language children continue to use while playing it. In Walworth two or three in a group of youngsters, and sometimes all but one, go to the other side of the road and decide what trade they shall mime. As they come skipping back across the road they

[1] Since no game, worthy of the name, could be more innocuous than this one, Mr. Smith provides an interesting example of the dedicated adult's inability to leave well alone. Having been shown how to play 'Time' by his young daughter, he not only felt it necessary to alter the game to make it suitable, as he considered, for a schoolroom, but recommended that the teacher should take part, should indeed take the principal role, and should retain the principal role throughout the course of the game. For example: 'When one of the players finally guesses the time decided upon, the teacher says to the fortunate guesser, "which do you like better, pears or apples?"' Such small amusement as the game possesses is thus effectively monopolized by the teacher.

call, 'We're three jolly workmen come to do some work.' 'What work can you do?' demand the other side. 'Anything to please you', reply the workmen. 'Pray let us see you do it', say the employers. The workmen then mime the trade, and if one of the employers guesses correctly what it is they turn and run, with the employers after them. If one of the employers manages to catch them before they reach the wall on the other side of the road they change roles and the guessers become the mimers.

This game seems to be played almost everywhere. In Swansea and Welshpool, and other parts of Wales, it is usual for one player to stand alone, while the others, perhaps three players, go away 'to think'. The mimers then return chanting: 'Three jolly Welshmen come to look for work.'

Foreman: 'What can you do?'

Welshmen: 'All sorts of things that you can't do.'

Foreman: 'I'll show you one.'

Welshmen: 'We'll show you two, that's more than you can do.'

Foreman: 'Do it then.'

In West Ham, where the game is known as 'Here we come to London Town', the workmen are equally cocky. One person is chosen 'It', and all the rest are the workmen, and retire out of hearing to choose what action they shall mime, 'for instance' (says a 10-year-old) 'robbing a bank or dress-making'. When they have chosen they walk towards 'It' and announce themselves: 'Here we come to London Town'.

'What to do?'

'Eat plum pudding as fast as you.'

'Set to work and do it', says 'It', and they start their mime. If 'It' guesses right he chases them, and whoever he catches takes his place; but if he cannot guess they think of another mime, and 'It' has to guess again.

In Kirkcaldy the mimers announce themselves: 'We're the men from Buckley Bane, guess what we are doing today.' In Forfar they say, 'Here we come to knock at the old man's door.' In Caistor: 'We're three wise witches come to work.' In Alton: 'We're three men in the workhouse, don't know what to do.' And the dialogue continues: 'What's your trade?'

'Monkey Moses.'

'How do you work it?'

In Accrington: 'We're three jolly sailor boys out of work.'

'What is your cargo?'

'Anything but workship.'

In Glasgow: 'Here are two broken matchsticks.'

'Where from?'

'Hong Kong.'
'What's your trade?'
'Lemonade.'
'How do you work it?'

And each time, when a correct guess is made, the actors are chased back across the road.[1]

Names: 'Three Jolly Workmen' (Walworth, Bermondsey, Manchester, Runcorn), 'Three Jolly Welshmen' (Swansea, Knighton, Welshpool, Bristol, Lydney), 'Three Jolly Watchmen' (Guernsey), 'Three Wise Men' (Garderhouse, Shetland), 'Three Wise Witches' (Caistor), 'Three Jolly Sailor Boys' (Accrington), 'Two Broken Matchsticks' (Glasgow, Cumnock, Stornoway), 'Matchstick Man' (Inverness), 'Buckley Bane' (Kirkcaldy), 'Here we come to London Town' (West Ham), 'London Town' (Ipswich and Poplar), 'Here comes an Old Woman to do some Work' (Brighton), 'Old Woman in the Wood' (Wells), 'Come to Knock at the Old Man's Door' (Forfar), 'Come to Learn a Trade' (Birmingham), 'Learn a Trade' (Enfield), 'Jack of All Trades' (Market Rasen), 'Chinky, Chinky Chinaman' (Perth), and 'Cobbler, Cobbler' (alternative name, Accrington).

Also, with activities rather than trades, and a king or queen guessing: 'Queen' (Birmingham); 'Queenie, Queenie' (Aberdeen); 'Good Morning, O King' (Bristol), and 'King and His Subjects' (Inverness).

In Whalsay, Shetland, the game is known as 'Aggie Waggie', which is the actors' retort when a wrong guess is made.

** In *The Folk-Lore Journal*, vol. vii, 1889, p. 230, under the name 'Dumb Motions' or 'An Old Woman from the Wood' (the same name we learnt from a girl in Wells, Somerset, in 1964) the game is said to have been played for several generations in Dorset: 'Here comes an old 'oman from the wood.' 'What cans't thee do?' 'Do anythin'.' 'Work away.' In Sussex the game was known as 'A Man Across the Common'. In *Traditional Games*, vol. ii, 1898, p. 305, three versions are given under the name 'Trades', including one from Ogbourne, Wiltshire, beginning:

> Here are three men from Botany Bay,
> Got any work to give us today;

and one from Fraserburgh in which the performers proclaimed themselves 'Three poor tradesmen wanting a trade'. It will be noticed that the work-

---

[1] In the United States this last dialogue is so common it gives the game its usual name, 'Lemonade'. Indeed the ridiculous words have been repeated times without number, for as long ago as 1903 Newell set down the following from the recitation of a 12-year-old girl in St. Paul, Minnesota: 'Here we come!' 'Where from?' 'Jamestown, Virginia.' 'What's your trade?' 'Lemonade.' 'Give us some.' (*Games of American Children*, 1903, pp. 249–50.)

men in the nineteenth century described themselves as 'poor' or 'broken', while those today are 'jolly' and generally pleased with themselves. In *London Street Games*, 1916, p. 40, the game appears under the names 'Dumb Motions', 'Guessing Words', and 'Please we've come to learn a trade'; and the performers not only mime the trade but give its initials, as 'P.H.' for picking hops. In the 1855 edition of *The Boy's Own Book*, pp. 51–2, where the game is styled 'Trades and Professions' ('sometimes called "Dumb Motions" in London'), the guessers are similarly assisted with initials and mime. But the 1872 edition says: 'It may also be played without giving the initials of the trade represented.'

Clearly the game is another offshoot of 'Two Poor Tradesmen', played in George III's time (see p. 277), and described in Halliwell's *Nursery Rhymes*, 1844, p. 108, where the performers claim to be:

> Two broken tradesmen, newly come over,
> The one from France and Scotland, the other from Dover.

In this game 'played exclusively by boys' the tradesmen told their trade, and what had to be guessed was the implement they required. 'The fun is, that the unfortunate wight who guesses the tool is beaten with the caps of his fellows till he reaches a fixed goal.' (Compare 'Wadds and the Wears' in Mactaggart's *Gallovidian Encyclopedia*, 1824, pp. 460–2.)

The game is also well-known on the Continent, and has apparently long been so. Otto Kampmüller gives four versions played in present-day Austria. In one, 'Es kommen zwei Damen aus Wien', a trade is imitated and its initials given. The player who guesses correctly then chooses either 'apple' or 'pear' (as in 'American Times'), and takes the place of whichever of the principals the choice denotes (*Oberösterreichische Kinderspiele*, 1965, pp. 67–8). In another version, 'Meister und Geselle', there is no miming, and the players have to guess what tool is needed for a given trade, which is precisely as Halliwell described the game in Britain more than a century ago. The game is also popular in Berlin, where the children say or sing, 'Meister, Meister, gib uns Arbeit', or 'Wir kommen aus dem Morgenland, die Sonne hat uns braun gebrannt'. Here, if the master guesses the activity being mimed, the performers run, and the master tries to catch one of them, who becomes master for the next turn (Reinhard Peesch, *Berliner Kinderspiel der Gegenwart*, 1957, pp. 35–6). Further versions are given by F. M. Böhme, *Deutsches Kinderlied und Kinderspiel*, 1897, pp. 667–9, and it seems clear from one of them called 'Botschimber, Botschamber', recorded in 1891, that this is the game Goethe's mother was referring to when she wrote to her grandchildren in Weimar in 1786:

'Wenn ich bei euch wäre, lernte ich euch allerlei Spiele: Vögelverkaufen, Tuchdiebes, Potzschimper, Potzschemper und noch viele andere.'

Further A. de Cock and I. Teirlinck, *Kinderspel in Zuid-Nederland*, vol. iv, pp. 7–10, give a version, 'Stom-en-ambacht', in which two children perform a trade while the rest guess its nature. It thus seems possible that this was one of the games Bruegel depicted in 1560, where he shows a row of little girls sitting against a wall watching two others who are standing in front of them. Certainly the game was current about this time for not only did Fischart list 'Das Handwerck ausschrenen' and 'Handwerksmann, was gibst dazu', in 1590 (*Geschichtklitterung von Gargantua*), but the game is well documented in France. 'Les métiers' is described in Rolland's *Jeux de l'enfance*, 1883, in *Les Jeux de la jeunesse*, c. 1814, p. 71, and in *Les Récréations galantes*, 1672. In Cotgrave's *Dictionarie*, 1611, appears the entry:

Mestiers, a certain Game wherein all trades are counterfeited by signes.'

And Rabelais listed 'mestiers' among the games played by Gargantua (1534). Indeed there is a pleasant story that Louis XIII, as a child, showed 'une imagination singulière' when playing the game, by suggesting that his side should imitate the trade of pickpockets. (Compare the West Ham 10-year-old today.)

In Italy, where the game is also widespread, there are versions called 'Nonnina' and 'Buon giorno re', although usually it is 'Mestieri muti' (M. M. Lumbroso, *Giochi*, 1967, pp. 198–9, 329–30). Here two children act a trade and, as in the Austrian game of the two ladies from Vienna, when a girl guesses correctly she becomes instrumental in determining which of the two actors she shall replace. She is asked whether she will have 'flowers, fruits, cats, actresses, singers, cars, name-plates, or sweets'. If her choice is 'flowers' the pair decide privately the kind of flower each shall be, and then ask her, for example, 'Will you have a rose or a violet?' and she takes the player's place according to the one she has named. In Sardinia, however, the game is 'La regina delle api'. The players perform before a single child, the Queen, who, when she guesses their occupation correctly (for example 'ironing'), chases after them and tries to catch one of them. It is curious that this is how the game is played in Sweden in the present day, where a single player, the King, must guess the occupation of his sons (not daughters); and, as our own survey has shown, it is also one of the ways the game is played today in Britain, particularly Scotland, where a single player, usually designated 'King' or 'Queen' has to guess the *activity* rather than trade which the other players are miming.

FOOL, FOOL, COME TO SCHOOL

In this game the guesser, who is styled the 'fool' or sometimes the 'little dog', moves out of earshot while each of the others assumes a fancy name. The guesser has then to decide which person has acquired a given name, and, if, as is most likely, his guess is incorrect, they laugh at him for his trouble. Thus a 10-year-old girl in Hanley writes:

'A popular game in our street is "Fool, Fool, come out to play". First of all somebody goes to the other side of the street. Then someone names the others such things as the golden ring and the purple peacock. Then the girl that told them what they were shouts "Fool, Fool, come out to play, and find the golden ring." And if the Fool picks the wrong one they all shout "Go back and learn your lessons." But if the person picks the right one she takes her and puts her in her den.'

Similarly, a 12-year-old girl in Cumnock, where the game is called 'Little Black Doggie', states:

'To begin with there are about half-a-dozen players. Two of the players are picked to be what we call the "man" and the "little black doggie". The "man" gives each player a romantic name, for example "golden moon", "tiger lily", etc. When everyone has received a flattering name they chant "Little black doggie come a hop, hop, hop, and pick out the golden moon", or whatever name they happen to chant. The "doggie" then hops up in the hope of picking out the right one, which is very difficult to do. If she doesn't pick the right one, the players shout at the "doggie", "Away you go you horrible man", and the poor little "doggie" hops back in disgrace to let somebody else take her place. If on the other hand the "doggie" does pick the right one, the "man" tells her to take the player by the nose or the ear, etc., thus upholding her position of "Little Black Doggie".'

⁂ Both these forms of the game are traditional. *The Folk-Lore Journal*, vol. v, 1887, pp. 49–50, gives 'Fool, Fool, come to School' from Cornwall. Gomme, *Traditional Games*, vol. i, 1894, pp. 330–1, has 'Little Dog I call you' from Sheffield. The Macmillan MSS., 1922, contain a description of 'Dunce, Dunce, Double Dee', played at Kilmington in Devon. Further, J. M. McBain in *Arbroath, Past and Present*, 1887, p. 344, describing the game as it was known in his youth about 1837, shows it to have been played in essentially the manner it is in Cumnock today. (The players even adopted golden names such as 'Golden Rose' and 'Golden Butter-plate'.) He adds that the player who guessed did so in the guise of 'a weird wife hirpling on a broomstick', to whom the children called, 'Witchie, witchie, warlock, come happin' to your dael'. Thus it appears that the absurd hopping of the 'little black doggie' at Cumnock today is not the product of

a modern schoolchild's imagination, but a deep-rooted memory of a limping night-hag; and this is apparently confirmed by reference to the game
'Jams' which follows. See also 'Goldens' in *Games of Argyleshire*, 1901,
pp. 236–7 (players named 'Golden Slipper', 'Golden Ball', etc.); 'Golden
Names' in *Folk-Lore*, vol. xvii, 1906, pp. 224–5, where the guesser is
'Ruggy Dug'; and—a possible relative—'William-a-Trimbletoe' in *American Nonsinging Games*, 1953, pp. 177–8.

JAMS

The distinction between this game and 'Fool, Fool, come to school' is that
the guesser, who is here openly an old woman or witch, knows the kind of
names the children have adopted (usually varieties of jam), and has to
guess precisely what ones they are. The game, which is apparently exclusive to Scotland and the Isles, ends in one of three ways. In Stornoway,
it takes the form of a playlet akin to 'Mother, the Cake is Burning':

> 'The witch comes and knocks at the door and asks if there is any jams today
> The maid says yes, and the witch tries to guess the names of them. If she does
> not guess the names of them she does not take them.'

On the other hand, should the witch, who 'has to live in a creepy corner',
succeed in guessing the names, she gains possession of the 'jams', takes the
children away, and hides them in her house.

> 'The mother comes to the witch's house and asks her if she has seen her
> children. The witch says she saw them at a house or somewhere and when the
> mother goes to see, the witch sends the children home. The mother comes
> back and the witch says she saw them running into her house, so the mother
> goes home and asks the children where they have been. They tell their mother
> something like they have been eating all the sweets in the house' (Girl, 11).

At Westerkirk in Dumfriesshire, where the game is known as 'Limpety
Lil', the guesser is summoned with the words 'Limpety Lil come over the
hill and see what you can find'. At Langholm, where the game is called
'Limping Jenny', a 12-year-old reports:

> 'Old Jenny comes into the shop and asks for something. If she asks for lard
> and somebody is lard she takes that person away with her. When she gets home
> if she likes the product she puts it in the pantry, and if she doesn't in the dust-
> bin. She then goes for more. At the end of the game all the products that were
> put in the bin and Jenny all chase the shopkeeper and "bash" him.'

In Scalloway, however, the game becomes a form of 'Honey Pots'
(q.v.). When the witch has guessed someone's identity, 'that person is then

swung ten times by the other two, being held by the arms (the "jam" clasps her hands below her knees). If the "jam" falls before the end of the swinging then she is rotten and is out' (Girl, *c.* 12).

*_** Such details as the division of the products between the pantry and the bin, and the declaration that some of the jams are rotten, indicate that the players were formerly divided into two sides. In fact this game is significant not only for its relationship to the latter part of 'Mother, the Cake is Burning' (p. 317), but is central to the appreciation of child-lore in Europe, with its ancient dramatization of the conflicting forces of good and evil. In Gomme's *Traditional Games*, vol. i, 1894, p. 8, appears a version called 'Angel and Devil' from Deptford, Kent, in which the angel comes to the door seeking ribbons. The angel is asked what colour she wants; and if there is a child in the house of that colour, the angel leads her away. This is akin to the game 'Colors' in the United States, where the chief characters are an Angel, Devil, and Mother. The angel and devil each seek to acquire children (under the guise of colours) for their respective sides which, at the end of the game, have a pulling-match, members of each side lining up behind each other, and holding on to the waist of the player in front. In Newell's *Games of American Children*, 1883, pp. 213–14, the Good Angel or 'Angel with the Golden Star' opposes the Bad Angel or 'Angel with the Pitchfork'. Newell suggested that the game had come from Europe only recently, since German children in New York used the same imagery, 'Der Engel mit dem goldenen Strauß' and 'Der Engel mit dem Feuerhaken'. Indeed the game 'Farbe angeben' was well known in Germany at this time (Böhme, *Deutsches Kinderspiel*, 1897, p. 636). Newell also pointed out that the game was played in Austria, where the practice was for the children to be taken to Heaven or Hell according to who guessed their colour, while the division between these places was marked by a piece of wood, called 'Fire', over which the subsequent tug-of-war took place. According to Kampmüller versions of this game, with Angel and Devil alternately choosing colours, continue to be played in Austria, and are called 'Engel und Teufel' or 'Das Farbenaufgeben' (*Oberöster-reichische Kinderspiele*, 1965, p. 166). The game is also much played today in Italy, where it is known simply as 'Colori', although the Madonna or an Angel likewise take turns with the Devil to ask for colours, and the side wins whose leader guesses the most colours correctly. However Signora Lumbroso in her *Giochi*, 1967, pp. 222–9, adds the interesting information that in Umbria the children are not just colours, but 'coloured pots'; and that at Pesaro, near San Marino, each child whose colour has been guessed correctly is tested (as she is in the Shetlands), by being made to

put her hands round her knees while the Angel and the Devil swing her a certain number of times according to the number of years old she is. If the player passes this test she goes with the Angel to 'Paradiso', but should her feet touch the ground while she is being swung she goes with the Devil to 'L'inferno'. The game has not only long been played in Italy (described by G. Bernoni, *Giuochi popolari veneziani*, 1874, p. 51), it is traditional in Spain, where the protagonists are likewise an Angel and the Devil (F. Maspons y Labrós, *Jochs de la Infancia*, 1874, p. 91); and it is said that 'Los Colores' was mentioned by Alonso de Ledesma Buitrago in 1606. It is remarkable that when Eugène Rolland reported the game in France the principal players were 'la Sainte-Vierge' and 'le Diable', and the third player was 'une marchande de rubans' (cf. Deptford, 1894). In Germany, as long ago as 1827, a version was recorded called 'Blumen verkaufen' in which the principal players were a Gardener and a Buyer, and the players adopted the names of flowers (H. Dittmar, *Der Kinder Lustfeld*, 1827, p. 270). From this it appears that even the minor variations of this game can be traditional, for in Piedmont, in the present day, Signora Lumbroso reports that children play a game called 'Fiori', in which the principal players are, similarly, a gardener and a buyer, and the children adopt the names of flowers, only here, if the buyer guesses incorrectly three times in succession the gardener becomes the buyer (*Giochi*, 1967, p. 224).

Compare 'Coloured Birds', which follows, 'Honey Pots', and 'Eggy Peggy'.

COLOURED BIRDS

The strangeness of the two preceding games is not lessened by the coexistence of the following drama, known as 'Coloured Birds' or 'Coloured Eggs', which is particularly popular in Bristol. One child becomes a mother, one a wolf or fox, and the rest are birds or eggs and choose which colours they will be. The wolf, who has not been a party to their variegation, comes to the door and knocks. Sometimes in Bristol the mother also acts the role of the door, which opens and shuts in keeping with the story. Thus the wolf taps on her head, and the door turns round and says 'Go away'. The wolf taps again, and the door says 'Come in'. In Wolstanton, Staffordshire, instead of knocking at the door, the wolf seeks admittance apparently by pulling a door bell, for he tugs the person's hair. He is asked, 'What do you want?'

'Have you any coloured birds?'

'What colour?'

The wolf names a colour. If there are no birds (or eggs) of that colour everybody stands firm, and the wolf has to go away and try again; but if one of the players has assumed this colour she runs for her life, either across the road to safety, or in a circle round the playground, or runs until caught. When she is caught she has to go 'to a place' until the catcher has caught the rest of the colours.

** This game (which several children describe as 'my favourite game') might be cited as an example of the shallowness of modern juvenile fantasy, were it not that in *The American Anthropologist*, vol. i, 1888, p. 281, W. H. Babcock noted it being played by children in Washington D.C. The visitor was there an angel, and the game called 'Birds'. The angel came up to the other principal player (the 'namer') and touched her on the back.

Namer: 'Who is that?'

Angel: 'It's me.'

Namer: 'What do you want?'

Angel: 'I want some birds.'

Namer: 'What color?'

Angel: 'Blue' (for example).

The angel, having guessed one of the colours, chased the person, exactly as in England today. In France, Belèze reported, the Devil sought animals in the same fashion, his dialogue with the seller commencing, 'Pan, pan.'

'Qui est-ce qui est là?'

'C'est le diable avec sa fourche.'

'Que veut-il?'

'Un animal.'

'Entrez.'

In this version, if the Devil guessed an animal's name correctly, he had first to buy it, with so many taps on the dealer's palm, before he could set off in pursuit; and if he caught the animal he made it captive with three blows on the head and tail (*Jeux des adolescents*, 1856, pp. 43–4). At Ottensheim in Austria, according to Kampmüller, the game is 'Vogel verkaufen', and the buyer has first to count a certain number before he chases the bird (*Oberösterreichische Kinderspiele*, 1965, p. 165). 'Vogelverkauf' or 'Vögelausjagen' has often been recorded in Germany (Böhme, *Deutsches Kinderspiel*, 1897, pp. 587–8); and it seems to be another of the games Goethe's mother spoke of in 1786 (see previous, under 'Three Jolly Workmen'). It was fully described by Gutsmuths in his deservedly renowned *Spiele für die Jugend*, 1796 (1802, pp. 271–2):

'The leader of the company is the Seller. All the little ones are birds, except one who plays the role of Buyer. The Seller stands in front of his birds and gives each a bird-name: you are a citril-finch, you a starling, you a finch, etc. The Buyer must not hear this. Then he comes and asks the Seller: "Have you birds for sale?"—"Yes!"—"Have you a raven?"—"No!"—"A sparrow?"— "No!" Always "No!" until the buyer asks, for instance, for a starling which is amongst the number. Then they all cry "Yes!" and the bird so-named flies off as quickly as possible; that is, he runs hard to get away from his new master. To allow him a start, the Buyer must first pay the Seller a certain number of gold pieces, by tapping his hand. Then he starts off. If he catches the bird, it is his: otherwise the bird goes back to the Seller: so he belongs to the Seller again and he gives him a meal, apples, pears, plums. The party which is the strongest at the end has won.'

It is apparent that this game has also a long history in Britain. In 1825 Jamieson reported it being played at Abernethy under the name 'All the Wild Birds in the Air'. Here, after the players had been given the names of birds, 'the person who opposes tries to guess the name of each individual. When he errs, he is subjected to a stroke on the back. When his conjecture is right, he carries away on his back that bird, which is subjected to a blow from each of the rest. When he has discovered and carried off the whole, he has gained the game' (*Scottish Dictionary*, *Supplement*, vol. ii, p. 681). Furthermore, the game 'All the Birds in the Air', so named, is poetically described, and clearly depicted in *A Little Pretty Pocket-Book*, 1744, the first book of games for children, and one of the first books to be published for juvenile entertainment. The description, on p. 47 of the 1760 edition, reads:

> Here various Boys stand round the Room,
> Each does some favourite Bird assume;
> And if the *Slave* once hits his Name,
> He's then made free, and crowns the Game.

The accompanying woodcut, despite the rhyme, shows the game being played out of doors. Five boys, clad in skirted coats, breeches, and tricorn hats, are to be seen standing against a wall, with one lad out at the side conducting the game, and there is one in front, 'the slave', who is apparently guessing.

QUEENIE

'Queenie' is the perpetual delight of little girls aged eight and nine, for it has the great recommendation of combining the mysterious pleasure of guessing with the ball-play to which they are so much addicted. One girl is selected 'Queenie', given the ball, and stands with her back turned to the rest. Without looking behind her, she throws the ball back over her head, and the other girls scramble for it. They then form up in a row, with their

hands behind their backs so that the girl in front will not know which of
them has the ball; and they let 'Queenie' know they are ready, sometimes
addressing her in doggerel verse:

Queenie, Queenie,
Who has the ball?
I haven't got it,
It isn't in my pocket,
Queenie, Queenie,
Who has the ball?
> *Scarborough, Lincoln, Plympton St.
> Mary, Ruthin, Penrith. In Scotland the
> verse begins 'Alabala, alabala, who has
> the ball?'*

Queenie, Queenie,
Who's got the ball?
Is she big or is she small,
Is she fat or is she thin,
Or is she like a rolling pin?
> *Chiswick, West Ham, Alton. In Nor-
> wich they ask 'Or has she got a double
> chin?'*

Queenie, Queenie, on the wall,
Who has got the golden ball?
> *Welwyn and St. Peter Port*

Queenie, Queenie,
Who has got the ball?
Cashee loo, throw it away,
Boney's behind the wall.
> *Stockton, Co. Durham*

'Queenie' then turns round, and picks out the player she thinks is concealing the ball. If her guess is correct she usually remains Queenie; if she is
wrong, the girl who has been successful in concealing the ball takes her
place. If a girl manages to catch the ball when Queenie throws it, that girl
takes her place in front, provided that she says 'Caught ball' or whatever is
the term locally prescribed. (In Liverpool it is 'Copper', in Lincoln
'Kings', in Welshpool 'White Horse', in Swansea 'Cabbage', in Stoke-on-
Trent 'Fish'.) Should the ball, by chance, roll back to the person who
threw it, they pick it up and return it to Queenie to throw again, although
in Norwich they must first cry 'See ball', in Lincoln 'Carrots'. When
Queenie turns round to guess who has the ball she need not make up her
mind immediately; she can ask one girl, and then another, to stretch out
an arm, or to open her legs, or even to do 'a twist' (a quick half turn). And
this is considered part of the sport. The girls pretend they have the ball
when they have not got it; and the girl who has it feels safe, as long as she
does not giggle, because 'we usually hide it under our jumper'.

Names: From Pendeen to Penrith the usual name is 'Queenie', although
in Bristol it is 'Queenie-ball', in Liverpool 'Queenie-i', in Preston
'Queenio', and in some places 'Queenie-o-co-co' especially in north Wales
and the Potteries. In Scotland and north-east England it is 'Alabala' or,
in Helensburgh and Inverness, 'Ali Baba'. In Lincoln it is 'Ellabella
Cinderella'.

✲ In the nineteenth century the game seems to have been less active and more poetic. The girls divided into two sides. One side had the ball or a similar object, the other side sat with the Queen on the grass. The first side advanced line abreast singing:

> Lady Queen Anne, she sits in the sun,
> As fair as a lily, as white as a swan;
> We bring you three letters, and pray you read one.

The Queen answered:

> I cannot read one unless I read all
> So pray, Miss . . . (guessing who has the ball) deliver the ball.

If the Queen guessed correctly the two sides changed roles; if incorrectly, the one who had the ball replied:

> The ball is mine, it is not thine,
> So you, proud Queen, may sit on your throne
> While we poor gipsies go and come.

The side with the ball then retired, chose a new player to hold the ball, and advanced again, while the others kept their seats for what must have seemed eternity.

There are accounts of the game being played in this way in *The Girl's Book of Diversions*, 1835, pp. 2–3; Halliwell's *Popular Rhymes*, 1849, pp. 133–4; Chambers's *Popular Rhymes of Scotland*, 1869, p. 136; Newell's *Games of American Children*, 1883, p. 151, 'My Lady Queen Anne'; Walter Crane's *Little Queen Anne and Her Majesty's Letters*, 1886; Emmeline Plunket's *Merrie Games in Rhyme*, 1886, pp. 14–15; Gomme's *Traditional Games*, vol. ii, 1898, pp. 90–102 and 453; and Norman Douglas's *London Street Games*, 1916, p. 43, 'Queen Mab', and p. 65, 'Queen Anne, Queen Anne, she sits in the sun'. Today the old verses appear only in nursery rhyme books. Our correspondents state that the game was already being played in its modern form in London during the First World War.

STROKE THE BABY

'One boy turns his face to a wall. Then one child strokes his back while the other children say:

> Stroke the baby, stroke the baby,
> I will tip it.

Then the boy turns round, and points to the one he thinks stroked him. If he points to the right boy, the boy he has pointed to turns his face to the wall and he is the baby. But if the boy points to the wrong boy he is still the baby.'

Boy, 12, Welshpool

Guessing who touched one is common today in the opening phase of 'I Draw a Snake upon Your Back' (q.v.), but is sometimes, as above, a game in itself. A student recalls that in Darlington three or four children together would often play the game in the school yard 'in the short time between the whistle and lining up to go in'. One child would turn his back while the rest chanted:

> Strokey back, strokey back,
> Which hand will you tak':
> Be you right, be you wrong,
> Which hand stroked you?

'Then you had to guess who poked you, which hand was used, and what finger.' And she adds that the first person to be 'on' was usually the least popular person in the group: the term 'strokey back' apparently being somewhat euphemistic, for the players would take the opportunity to give the person a hard poke, and 'occasionally there would be vicious punches with clenched fists'.

Names: 'Hot Cockrel' (Ipswich), 'I Draw a Snake upon Your Back' (Windermere), 'North, South, East, West' (Bradford), 'Stroke the Baby' (Welshpool), 'Strokey Back' (Darlington).

In the United States, where the game is not uncommon, two of its names are 'Bore a Hole' and 'Punchboard'.

** It is almost an axiom that the more insignificant a game appears, the more remarkable is its history. The sport of giving a person a clout and having him guess who did it, has been popular in England for anyway the past three or four centuries, usually under the name 'Hot Cockles'. It appears to be referred to by Coverdale in 1549 (*OED*); and it is named in Sidney's *Arcadia*, 1590; Rowlands's *Letting of Hvmovrs Blood*, 1600; Marvell's *Mr. Smirke*, 1676; and Arbuthnot's *Memoirs of Martinus Scriblerus* written 1714. Traditionally the game was played at Christmas, the guesser being blindfolded and kneeling on the floor, the other players in turn slapping him with some force on his head or back, and hoping that their blow would not be identified. Steele in *The Spectator*, 28 December 1711, and Gay in *The Shepherd's Week*, 1714, are characteristically witty about the game's possibilities when played by simple folk. Brooke in *The Fool of Quality*, vol. i, 1766, p. 72, describes the guesser holding out his hand behind him to be 'well warmed' by the other players until he makes a correct guess. Indeed in France the game is generally known as 'La Main chaude'. In *Les Trente-six figures*, 1587, the caption calls it the 'jeu de frappe main, Où deviner il fault celuy qui frappe'. In Froissart's boyhood,

c. 1345, the game was simply 'Je me plaing qui me feri' ('L'Espinette amoureuse', l. 223). In Italy today it is 'Mano calda'. In modern Greece the sport is so popular among sailors that even tourists notice it: one player puts his right hand round to the left side of his waist, and shields it from his view by extending his left hand against the side of his face; the others gather round in a group and repeatedly strike or pretend to strike his hand, until he guesses correctly who last hit him. In ancient Greece the game was 'Kollabismos'. Pollux says that one player covered his eyes with the palms of his hands while another hit him and asked him (as do children today in Darlington) to identify which hand it was that dealt the blow (*Onomasticon*, ix. 129). It seems more than likely that this sport was familiar to the men guarding Jesus, when they blindfolded him, and 'smote him with the palms of their hands, saying, Prophesy unto us, thou Christ, Who is he that smote thee?' (Matthew xxvi: 67–8; Mark xiv: 65; Luke xxii: 64). Indeed the game was possibly already old. One of the pictures on the wall of a tomb at Beni Hassan, *c.* 2000 B.C., shows a player on his knees while two others, unseen by him, thump or pretend to thump his back with their fists (J. G. Wilkinson, *Ancient Egyptians*, vol. ii, 1878, p. 61). It is difficult to think what kind of game they are playing if it is not one like 'Kollabismos' or 'Stroke the Baby'.

Other names in the British Isles have been 'Slap-of-the-Ear' (north country, *EDD*, vol. v, 1904), 'Buaileadh Am Bas' (= Striking the palm of the hand, *Games of Argyleshire*, 1901, pp. 130–1), 'Bosuigheacht' (= Hitting on the palm, Co. Mayo, *Béaloideas*, vol. viii, 1938, p. 134), 'Soola Winka' (*Irish Wake Amusements*, 1967, pp. 51–2, where several related games are given), 'Handy-Croopen' (*Shetland and Orkney Glossary*, 1866), and 'Pick the Craw' (Miss Jean Rodger, Forfar, *c.* 1910). This last name may be compared with the Austrian 'Krähenrupfen', the name of a similar game in which the players gather round the one who cannot see, who must guess which of them is plucking his hair (*Oberösterreichische Kinderspiele*, 1965, p. 66). It is possible that this game was depicted by Bruegel in 1560 where, in his picture of children at play, he shows a group with their hands on one boy's head, but with only one of them, it seems, actually pulling his hair.

HUSKY-BUM, FINGER OR THUMB?

This curious game survives chiefly in the north and midlands. At Wednesbury in Staffordshire, where it is known as 'Bugs', one child leans forward putting his hands up against a wall, and the child playing with him leaps on to his back, holds up a certain number of fingers, and shouts 'How many bugs?' The one underneath makes a guess. If his guess is correct the two change places; if incorrect, he stays underneath, and his rider holds up a different number of fingers. In Wigan, and also in Peterborough, Ponte-fract, and Bishop Auckland, three or four boys may play at once. One boy makes a back, and the boy who jumps on calls 'Finger or thumb?' If the one underneath guesses correctly he stands up, and the rider bends down in his place. If he guesses incorrectly, another boy jumps on his back and holds up a finger or thumb. 'So we carry on with the game as long as we like.' In Bishop Auckland the one on top calls 'Hum, dum, dum, finger or thumb?' At Retford in Nottinghamshire the cry is 'Finger, thumb, or dum', and either two boys play together, or two teams. When two teams are to play they toss a coin to see which side shall have first jump, and the boys who have to bend down shout 'Baggy front man', because the one in front gets no weight on him, but stands with his back against the wall supporting the second in the team, who bends down and hangs on to him.[1] The others in the team line up, each putting his head between the legs of the boy in front, and his arms around his thighs, so that they present one long close-knit back. The members of the other team jump on to the 'down' side, one at a time, until all are astride. If the bottom side collapses, the players have to form a new back, and the jumpers have another turn; if they stay steady, the leader of the mounted team calls 'Finger, thumb, or dum?', and holds up a finger, a thumb, or a clenched fist. If the down side guesses correctly they become the jumpers, if incorrectly, they are jumped upon again. But if one of the riders' feet touches the ground while he is mounted, there is a cry of 'Touch!' and the ones underneath become the jumpers, as in 'Hi Jimmy Knacker' (q.v.). At Holmfirth in the West Riding, when all are mounted, the cry is 'Finger, thumb, or rusty?' —'rusty' being a clenched fist; at Elsecar, near Barnsley, it is 'Finger, thumb, or rusty bum?'; in Nottingham City it is 'Husky-bum, finger or thumb?'; at Brinsley, north-west of Nottingham, it is 'Husky fusky, finger or thumb?'; at Nelson in Lancashire it is 'Thumb, stick, or roger?' —'stick' being the index finger, and 'roger' the little finger; while at Barrow-in-Furness it is 'Stick, roger, or dodger?'—'stick' being the thumbs pointed upwards, 'roger' pointed sideways, 'dodger' pointing downwards. At Spennymoor, where this 'rough and exciting' game is immensely popular (fifty-six accounts received), they play with as many as

[1] The 'front man' is elsewhere known as the 'standard', 'post', 'pillar', 'pillow', or 'cushion'. These names are in themselves long established, e.g. the term 'pillow' appeared in a description of the game, under the name 'Ships', in the *Almondbury and Huddersfield Glossary*, 1883.

ten boys on each side. Since they allow 'no creeping', that is to say, no creeping forward once a boy has jumped on to the line of backs, the first boy to mount has to make a terrific jump along the line of backs to leave

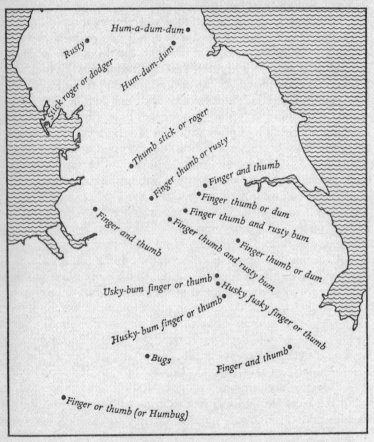

Fig. VIII. Names for 'Husky-bum, Finger or Thumb?' in northern England.

room for the others to mount behind him, and 'it is the job of the "pillar" or "cushion" to stop the players from hitting their heads on the wall'. Moreover the jumping side, says a 14-year-old, will try to make the 'down' side collapse. 'This is done either by jumping high into the air and landing with force on their backs or, when everybody except one is mounted, the last boy will run hard and try to push his team-mates forward, this sometimes results in the boy at the front being nearly bent double, and he has to give way to prevent injury.' It is only then—if the side which is down does not collapse—that the point of the game is reached. 'The last man on the jumping side shouts,

> "Hum a dum dum,
> A finger or a thumb,"

and he holds either a finger or a thumb up in the air.'

As we have said the game is now played mostly in the north and midlands (Fig. VIII). In other parts of the country it has been progressively overlaid by 'Hi Jimmy Knacker' (q.v.), a game in which victory comes only to the strong, not to successful guessers; but it continues to be played in Penzance (1961) where the leader of the mounted team holds up a certain number of fingers and cries:

> Willy, Willy Whiteman,
> How many fingers do I hold up?

** Local journals, and our older correspondents, give general confirmation of the dialect terms used by boys today. In the *Leeds Mercury Weekly Supplement*, 5 June 1897, the leader of the mounted side is said to hold up a digit and exclaim 'Dick, Prick, Anawny, Cherry, Berry!' (Thumb, first finger, second, third, fourth). In York, around 1915, he cried 'Dick, Prick, Polony'. In Forfar, *c.* 1890, the cry was 'Big Dick, Little Dick, and the Bozer' (a circle with the thumb and forefinger). In Barrow-in-Furness, in the 1930s, it was 'Stick an' Roger'. In the Scottish Border country, *c.* 1925, the game 'which involved a wee bit of second sight' was called 'Pinkie, Finger, Thoomb, or Knuckle' (*Southern Annual*, 1957, p. 28). In parts of Derbyshire, *c.* 1900, the cry was 'Husky, fusky, finger or thumb'. In general in Derbyshire, Nottinghamshire, Lindsey, and Sheffield, the game was called 'Rusty' or 'Rustibum, Finger or Thumb'. In Morpeth, 1912, it was 'Kittyback, Kittyback, finger or thumb'. In the neighbourhood of Halifax, up to the 1920s, the leader of the mounted team challenged the boys underneath with the words:

> Inkum, jinkum, jerry mi buck,
> How mony horns do I hold up?

And if their guess was incorrect he responded:

> *Two*, tha ses, an' *three* it is;
> I'll leean thi how to play
> At Inkum, jinkum, jerry mi buck.

At the Calder Ironworks, near Glasgow, about 1855 (*Rymour Club*, vol. i,

1906, p. 5), one boy used to 'make a back' and the others in succession jumped astride of him singing:

> Bairdy, Bairdy, buckety-buck,
> Hoo mony fingers stan's up?

A remarkable feature of this game is that its oral formula, including the meaningless word 'buck', appears to have survived from classical antiquity. Although the name 'Buck, Buck' has not been found amongst English children in the 1960s, both the name and the challenge 'Buck, Buck, how many fingers do I hold up?' are recollected by some of our adult informants, e.g. in Leeds in 1920, and in Launceston, Cornwall, in 1915 ('Bucksebuck, how many fingers do I hold up?'). It is also reflected in the names 'Buckle Up' and 'Huckey Buck' current today amongst the boys of, respectively, Petersfield and Ipswich, for the related game 'Hi Jimmy Knacker'.[1] And it survives in the sidewalk rhyme picked up by Dr. Howard in the United States:

> Buck, buck, you lousey muck,
> How many fists have I got up,
> One, two, or none?[2]

Indeed, 'Buck, Buck, how many fingers do I hold up?' seems to have been the standard formula in London up to the First World War (*Living London*, vol. iii, 1903, p. 269; *London Street Games*, 1916, p. 31). In Randle Holme's list of sports 'used by our countrey Boys and Girls' (*Academie of Armory*, Book III, 1688, ch. xvi, § 91), appears the entry:

'Runing or Leaping, Hoping, Skipping, Buk or Runing Leaps.'

This is presumably a reference to the game, whose popularity in eighteenth- and nineteenth-century England is attested by notices such as the following:

c. 1781 *Nancy Cock's Pretty Song Book*, p. 37 (almost certainly first printed 1744):
　　　　　Buck, buck, How many horns do I hold up?
　　　　　Three.
　　　　　Three you say and two there are:
　　　　　Buck, buck, How many horns do I hold up?
　　　　　One.
　　　　　One you say and one there is:
　　　　　Buck, buck, rise up.

---

[1] 'Huckey Buck' was in fact also a name for the present game. Children in Suffolk used to cry:
> Huck-a-buck, huck-a-buck,
> How many fingers do I hold up?

G. F. Northall, *Traditional Rhymes*, 1892, p. 401.

[2] *The New Yorker*, 13 November 1937, pp. 32–42.

1796 Introduced into a caricature by Gillray on 17 June.

1819 Keats, in a letter describing the production of Charles Bucke's tragedy *The Italians* at Drury Lane on 3 April: 'It was damn'd—The people in the Pit had a favourite call on the night of "Buck Buck rise up" and "Buck Buck how many horns do I hold up".'

1820 *School Boys' Diversions*, pp. 48–9, 'Any number of boys may play this, if half of them are bucks, and half riders; or, if when either buck guesses right, another rider will take his place.'

1829 *Boy's Own Book*, pp. 38–9. ' "Buck" . . . a sport for two boys only . . . It is usual, but, we think, quite unnecessary, for the player who gives the back to be blindfolded.'

1856 *Every Boy's Book*, p. 9. 'The boy who plays the Buck gives a back with his head down, and rests his hands on some wall or paling in front of him.' The boy who mounts him then cries 'Buck! Buck! How many horns do I hold up?'

In *The Satyricon* of Petronius Arbiter, written about A.D. 65, in the time of Nero, there is an incident at Trimalchio's feast involving his favourite serving boy:

'Trimalchio, not to seem moved by the loss, kissed the boy and bade him get on his back. Without delay the boy climbed on horseback on him, and slapped him on the shoulders with his hand, laughing and calling out "Bucca, bucca, quot sunt hic?" '

The coincidence that the same amusement should be played with the same word-sound in the first century A.D. and in the twentieth, can, it seems, only be explained by admitting the stamina of oral tradition. Thus investigations show that the game is international. It is played, not only throughout the English-speaking world, but in, for instance, France, Germany, Switzerland, the Netherlands, Scandinavia, Spain, Portugal, Italy, Yugoslavia, Albania, Greece, Turkey, Russia, India, and Japan.[1] Furthermore the word *buck* or *bock* (Old English *buc* or *bucca*) is recurrent. In Germany a common name and formula is 'Bock, Bock, wieviel Hörner hab ich?' In Sweden (according to Eric Gamby, *Kråken satt på tallekvist*, 1952, p. 71) it is:

> Bulleri bulleri bock
> Hur många horn står upp?

[1] See F. M. Böhme, *Deutsches Kinderspiel*, 1897, pp. 633–4; J. W. P. Drost, *Het Nederlandsch kinderspel*, 1914, pp. 41–6; Stavre Th. Frashëri, *Folk-Lore*, vol. xl, 1929, pp. 370–1 (Albanian account, the game being called 'Raqe, Raqe, Hypa, Zdrypa'); B. L. Ullman in *Classical Philology*, vol. xxxviii, 1943, pp. 94–102; Paul G. Brewster in *Béaloideas*, vol. xiii, 1943, pp. 40–79, and in *North Carolina Folklore*, vol. i, 1952, pp. 58–9; Roger Pinon, *La nouvelle Lyre Malmédienne*, vol. ii, pt. 3, 1954, pp. 17–25; Brian Sutton-Smith in *Folk-Lore*, vol. xlii, 1951, pp. 329–33 (New Zealand variants); Jeanette Hills, *Das Kinderspielbild von Pieter Bruegel*, 1957, pp. 18–19; R. T. Allen in *Maclean's Magazine*, 22 October 1960, p. 32 (Canadian account); M. M. Lumbroso, *Giochi*, 1967, p. 55 ('Quanti corni' played in Lombardy). In Istanbul it is 'Çatti Patti Kaç Atti'. In the Moscow district of Vnukovo, a boy told our correspondent he called the game 'Kozyol' (Goat).

(the response being similar to that in the West Riding in 1920:

> *Fyra* du sa, *tre* det var).

And in South Africa, where 'Bok, Bok' is almost a national sport, it is

> Bok, Bok, staan styf,
> Hoeveel vingers op jou lyf?[1]

Moreover the game did not go unnoticed on the Continent even in the sixteenth and seventeenth centuries. In *Il Pentamerone*, Day II, 1634, one of the games the Prince played to pass the time till dinner was 'Anca Nicola'. In Bruegel's picture of children at play, dated 1560, five boys are shown playing the game, two making a back, two mounted (one holding up his hand), and the fifth seated, acting 'pillow'. It is also probably the game 'Peertgen wel bereyt' mentioned by Adrianus Junius (*Nomenclator Octilinguis*, 1567); the game 'Pferdlin wol bereit', and perhaps 'Eselin beschlagen', named by Fischart (*Geschichtklitterung*, 1590); and Roger Pinon makes out a good case for it being Froissart's 'Charette Michaut' alluded to in 'L'Espinette amoureuse', written *c.* 1380.

There is also a form of the game, played indoors among women and small children, in which the child who is to guess, kneels down and buries his or her face in the lap of a second player who is seated. This gentle form of the game, consisting purely of guessing the number of fingers held up, is described in the American *Girl's Book of Diversions*, 1835, p. 8, under the name 'How Many Fingers?'

'This is a very simple play, and can be understood by children of three years old ... One lays her head in the lap of the other, in such a manner that she can see nothing. Her companion claps her several times on the back,[2] holding up one or more fingers, saying,

> "Mingledy, mingledy, clap, clap,
> How many fingers do I hold up?" '

This nursery sport, familiar in England, too, in the nineteenth century, continues to have life on the Isle of Barra in the Outer Hebrides where it is known in Gaelic as 'Imricean Beag'. The adult sits the child on her lap, and keeping her hand on his head, raises some fingers asking 'Cia meud

---

[1] In 1942 English members of the Eighth Army in the Western Desert were surprised to see South African troops playing an apparently juvenile leap-frog game which, when they joined in, was found to be 'not a game for weaklings'. The game has, in fact, been banned in South African schools owing to the number of girls as well as boys who have been injured while playing it.

[2] Cf. Petronius: 'manuque plena scapulas eius subinde verberavit'. Also 'Old Johnny Hairy, Crap in!' in *Traditional Games*, vol. ii, 1898, pp. 449–50; and 'Hurly-burly, Thump on the Back' listed as a standard name in *A Handbook of Irish Folklore*, 1942, p. 673.

adhairc air a' bhoc?' (How many horns are upon the buck?) If the child guesses right he is tapped gently ('Is firinneach am boc'), if wrong he gets a harder tap ('Is breugach am boc')—*An Gaidheal*, December 1952, p. 4. This form of the game is, in fact, the usual one in Sweden, and is also known in other parts of Europe. Compare, for instance, the American formula with that in Silesia:

> Mingeldi, mingeldi, hopp, hurräh,
> Wieviel Finger sind in der Höh?

At Maastricht, on the Dutch border with Belgium, it is said to have been customary for three players to take part, one sitting, one squatting with his face in the first one's lap, and the third striking the back of the guesser and challenging him to tell the number of fingers raised (A. de Cock and I. Teirlinck, *Kinderspel in Zuid-Nederland*, vol. i, 1902, p. 299). This has every appearance of being the game depicted on fol. 52 of 'The Romance of Alexander' (MS. Bodley 264) illuminated at Bruges about 1340.[1] Further, the game as played today in the Hebrides, seems analogous, to say the least, to one of the games depicted in the rock-cut tombs at Beni Hassan in Middle Egypt, which belong to the XII Dynasty. Here, about 2000 B.C., a player hides the fingers he is holding up by shielding them with his other hand, which is placed against the guesser's brow, and the inscription reads 'Putting the *Atep* on the forehead' (Edward Falkener, *Games Ancient and Oriental*, 1892, pp. 104–5).

HOW FAR TO LONDON?

In this game, which is played mostly by girls, one child who is blindfolded tells the others where they must go, and then has to guess the identity of the person with whom she happens to come into contact. As played in Liverpool the blindfolded girl stands facing the wall on one side of the street, while the others go to the wall on the other side and shout 'How far is it to London?' The blindfolded girl tells them perhaps 'Six miles forward', and everyone has to take six paces forward. She then might say 'Five miles to the left' and 'Two miles back', and 'Four miles to the left again', and the players carry out these instructions. Then the girl who is blindfolded is led across the road to the spot where the others started, and she must herself carry out the directions she gave the others, and, when she thinks she has reached London, try to touch somebody. The players are not allowed to move their feet while she tries to touch someone, but are

---

[1] Sometimes thought to be 'Hot Cockles', a game more obviously depicted at fol. 97.

allowed to sway about or bend down to avoid her touch. When she finds someone, she 'feels all over', and makes a guess who the person is. If her guess is correct that player becomes 'man'; but if her guess is incorrect she has to be 'man' again herself.

The exact procedure in this game varies in different places. Sometimes the blind player calls out the number of miles the children are to walk but they may go in whichever direction they like; sometimes she gives each player individual instructions; and sometimes she gives no instructions, but turns round fourteen times while the players take up positions wherever they like (Wolstanton), or she has to wait until the players call 'Coo-ee', while they go off and actually hide themselves (Durham). Just occasionally (a single account) there is no guessing of identity when she finds someone, and whoever she touches automatically becomes the new blind man. In Oxford and Headington, although there is guessing, it does not seem to matter whether she guesses correctly or not, for when she has found someone, and thinks she knows who the person is, she calls out her name, opens her eyes, and dashes to the starting-place. The person she has caught must run too, and whichever of them arrives first makes the other player the next blind man.

Names: 'Cuckoo' (Durham), 'Lion's Den' or 'How Many Miles to the Lion's Den' (Oxford), 'London' or 'How Far to London?' (Liverpool), 'Number Fourteen' (Wolstanton). In Indiana, 'Fourteen, Stand Still' (Brewster). In Edmonton, Alberta, 'Bee Find Honey'.

** This game has been recollected by only one of our older correspondents (at Midgley, near Halifax, about 1895, where it was called 'How Many Miles to London?'), yet the form of the game apparently belongs to classical times.

Pollux, in his *Onomasticon* (ix. 113), has a rather obscure passage under 'Muinda' (Blindman), in which he seems to describe three games, or three versions of a game, that feature a player who has his eyes shut. In the first: 'One of the players shuts his eyes, cries "Look out!" and begins to chase the others. The one he catches takes his place and shuts his eyes in his turn.' Such a game, to be successful, would have to be played in a confined space, and would seem to be a game like 'Blind Man's Buff' (q.v., especially for the reference to Pollux, and the description he gives of the old form of 'Blind Man's Buff', which bears no relation to 'How Far to London'). In the second: 'One of the players shuts his eyes and the others hide themselves. The blind man searches for them until he finds one of them.' Commentators on this passage have not been able to decide whether, as appears from the Greek, this means that the child kept his eyes closed even when searching for the hidden players; or whether it means, as seemed practical, that he opened his eyes when he began searching. On the evidence, just given, of our report of 'Cuckoo' played today in Durham, the first hypothesis seems the more likely. In the third of Pollux's descriptions: 'The blind man, whether he touches one of his comrades, or whether he points at him, must guess what his name is.'

If these three accounts, written about A.D. 180–90, are taken in conjunction, particularly the last two, it may be felt that they are a not unreasonable description of the versions children play today of 'How Far to London?'

II

Acting Games

THE difference between pretending games and acting games is that in pretending games children make-believe they are other people, and extemporise; in acting games they are allotted parts, they enact a particular story, and for the most part they repeat the words that were spoken when the game was last played. Thus the content of the acting games, as well as their style, is traditional; and children may be preserving for us, in however vestigial form, some of the most genuine folk-plays performed in Britain. Certainly their playlets are no new compositions. Similar passages of dialogue are to be heard repeated in different corners of Britain, and in the different languages of Europe. And at no time, it appears, have the plots been modified to suit the changing susceptibilities of the adult world. Indeed the scenes which oral tradition and the juvenile mind have thought fit to perpetuate over the years are strange to observe. In these playlets children are stolen to be eaten, mutilation is accepted almost as commonplace, and the supernatural is ever-present and constantly interfering with workaday activities. Time and again it is apparent that children either have a need to act out their fears, or actually enjoy the pretence of being frightened. The younger ones, in particular, can be seen relishing these games (they do not think of them as plays), and lingering over the more beastly details. Yet it will be noticed that, as in adult folk-drama, the suspense or horror never piles up without relief. There are repeated interludes for slapstick, when parental or ghostly authority is flouted, when children get spanked (there is much glee in this), or when pure nonsense is spoken. And it seems that these interludes are very necessary. Children of a certain age enter into their parts so intensely that they will, as it were, 'double-live' them. They will know from the story that, for instance, a witch is about to knock at their door to beg matches; and they will act beautifully the role that they are unconcerned and do not know. Yet a close observer will perceive that for all the bold face they are putting on, they are shivering with fright just as if a real witch were approaching.

OLD MAN IN THE WELL

The best version of this acting game was collected in Swansea. The characters are a Mother, her Children, and an Old Man. The Old Man goes off and secretes himself in some suitably dark and mysterious place which is designated 'the well'. Thereupon a set dialogue takes place.

Children to Mother: 'Please, mother, can we have a piece of bread and butter?'

Mother: 'Let me see your hands.'

The children hold out their hands for inspection.

Mother: 'Your hands are very dirty. Go to the well and wash them.'

The children go to the well, where they spy the Old Man crouching down. They rush back to the Mother screaming: 'Mother! Mother! There's an Old Man in the well.'

Mother: 'Don't be silly, children. There isn't an Old Man in the well.'

Children: 'But we saw him.'

Mother: 'It's only your father's under-pants. I hung them out to dry. Go again.'

The children go again, see the Old Man, and again come back screaming: 'Mother! Mother! There's an Old Man in the well.'

The Mother again reassures them (repetition is a feature of these playlets), and they go to the well perhaps twice or three times more, until they persuade the Mother to come herself. She sends one of the children to fetch a candle (a twig), lights it, and goes to the well with the children. When they reach the well, and she is about to look in, the Old Man blows the candle out.

Mother to child nearest her: 'What did you want to blow out my candle for?' She cuffs the child who sets up a howl.

The Mother relights the candle, holds it over the well, and the Old Man blows it out again. Mother to another child: 'What did you want to blow out my candle for?' A second child is cuffed and sets up a howl. This happens three or four times, so that three or four children, or perhaps all of them, are crying at once, which gives a fine opportunity for dramatics.

Eventually the Mother manages to look in the well. The Old Man jumps up with a horrible shriek and gives chase. Whoever he catches is the next Old Man.

The game is also known as 'Frog in the Well' and 'Ghost in the Well'.

** Previous recordings include: *Folk-Lore Journal*, vol. v, 1887, p. 55, 'Ghost at the Well' described by 'a little girl in west Cornwall in 1882'; *Traditional Games*, 1894–8, 'Ghost in the Copper' (London), 'Mouse and the Cobbler' (Deptford), and 'Deil amo' the Dishes' (Aberdeen); *Games of Argyleshire*, 1901, p. 215, 'Ghost in the Garden'; and *London Street Games*, 1916, pp. 45-7, 'White Shirt', 'Old Devil in Fire', and 'Light Mother's Copper Fire'.

The following additional dialogue, lacking in the Swansea version or in the above recordings, has long been traditional in Devon (Macmillan MSS.):

Mother, on seeing the Old Man in the well: 'What are you doing here?'

Old Man: 'Picking up sand.'

Mother: 'What do you want sand for?'

Old Man: 'To sharpen my needles.'

Mother: 'What do you want needles for?'

Old Man: 'To make a bag.'

Mother: 'What do you want a bag for?'

Old Man: 'To keep my knives in.'

Mother: 'What do you want knives for?'

Old Man: 'To cut off your heads.'

Mother: 'Then catch us if you can.'

This additional dialogue was also part of 'Ghost in the Garden' played in Dunedin from *c.* 1900 until anyway 1949 (*Games of New Zealand Children*, 1959, pp. 32-3). It was known in Adelaide, South Australia, *c.* 1900 (Howard MSS.). And it was a common component of the game 'Fox and Chickens', described hereafter, e.g. in the Lincolnshire version called 'Pins and Needles', the Galloway 'Auld Grannie', and the Belfast 'Grannie's Needle' (*Traditional Games*, vol. i, 1894, pp. 201-2; vol. ii, pp. 404-5).

The dialogue also occurs in the game 'Die Hexe im Keller' which is said to be much played today in Upper Austria (Kampmüller, *Oberösterreichische Kinderspiele*, 1965, pp. 166-9). Here one child says she wants some bread and butter. The mother says the butter is in the cellar. The child goes down the steps into the cellar and is shoved by the witch who says 'Huuuu'. The child runs back: 'Mother, there is somebody down the steps.' 'Who then?' 'An old witch.' The mother sends a second child with her, who is similarly treated, and then a third and a fourth. Eventually she goes down into the cellar herself. She shakes the witch by the head, saying to the children, 'Isn't this the butter churn?' Then she asks the witch: 'Was tust du denn da?'

The witch replies: 'Steine klauben!'

Mother: 'Was tust du mit den Steinen?'

Witch: 'Messer wetzen!'

Mother: 'Was tust du mit dem Messer?'

Witch: 'Leut abstechen!'
Mother: 'Stichst mich auch ab und meine Kinder auch?'
Witch: 'Ja.'

A correspondent, recalling the English game from her childhood, commented: 'The queer thing was that we were all *really* terrified by the Old Man in the well. We knew he was a playmate, we had even been him ourselves, and we played many jumping-out games and were not frightened by them. But the Old Man was terrifying. He was the personification of evil. In fact we seldom played the game, it scared us too much.'

GHOSTIES IN THE GARRET

In this game, as played in Forfar, and also in Whalsay, Shetland, under the name 'Ghost', one player goes away to be the ghost, as in 'Old Man in the Well', but the mother sends only one of her children, the eldest, to fetch her shirt from the garret. When the eldest child does not return she sends the next eldest, and so on, one child after another, until she sends the last and youngest child. When the last and youngest does not return she goes to the garret herself, and she, too, is captured by the ghost. The ghost then says he intends to eat them, but first he will fetch a friend to share the feast. While he is away the children start escaping; suddenly the ghost reappears, and the chase begins.

In a game called 'Old Man' played at Matching Green in Essex the chaser has similar carnivorous inclinations. After the Old Man has succeeded in catching all the children, he sets about eating them, but finds, to their unconcealed delight, that they 'taste of horrible things like frogs, snakes, and slugs'.

OLD MOTHER GREY

This game, which was highly popular in the nineteenth century, remains fairly general today among younger children (accounts from seventeen schools) and it is the only playlet in which the dialogue is humorous throughout. Usually one child is appointed 'Grandmother Grey' or 'Granny, Granny Grey' or 'Old Mother Grey', and the other children gather round her chorusing:

'Old Mother Grey, may we go out to play?'
Mother Grey: 'No, it's raining.'
Children: 'No it isn't, the sun's shining.'
Mother Grey: 'All right, you may go out to play.'

The children rush off, and play around, perhaps moving out of sight.

Old Mother Grey calls: 'Children, it's dinner time!'

Children: 'Coming, mother.'

Mother: 'Where have you been?'

Children: 'We've been to London to see the Queen.'

Mother: 'What did she give you?'

Children: 'A loaf of bread as big as our head, a piece of cheese as big as our knees, a lump of jelly as big as our belly, and a teeny weeny sixpence.'

Mother: 'Where's my share?'

Children: 'Up in the air.'

Mother: 'How shall I get it?'

Children: 'Stand on a chair.'

Mother: 'What if I fall?'

Children all laugh, and shout out: 'We don't care.'

Old Mother Grey runs after them, and whoever she catches becomes the next Mother.

In another version, also widespread (current alike in Bristol and Aberdeen), the children sing:

> 'Granny, Granny Grey, may we go out to play?
> We won't go near the water and shoo the ducks away.'

Granny Grey: 'No, it is my washing day.'
The children plead a second time:

> 'Please, Granny Grey, may we go out to play?
> We won't go near the water and shoo the ducks away.'

Granny Grey: 'No, it is my ironing day.'
The children plead a third time, with even greater earnestness:

> '*Please*, Granny Grey, may we go out to play?
> We *won't* go near the water and shoo the ducks away.'

Granny Grey relents:

> 'Yes, you may go out to play.
> But don't go near the water and shoo the ducks away.'

The children run off, generally to the other side of the road, and immediately start shooing the ducks away.

Granny Grey, very angry, shouts at them to come back at once.

The children shout back impudently: 'We can't come across the water.'

Granny: 'Swim across.'

Children: 'We can't swim.'

Granny: 'Come by boat.'

Children: 'We haven't got a boat.'

Granny: 'Paddle across.'

Children: 'Yes, we can paddle across.'

One child is then usually caught and spanked for his disobedience. 'It seems senseless,' one teacher remarked, 'but they play it over and over again.'

* Previous recordings: *Folk-Lore Journal*, vol. v, 1887, pp. 55–6, 'Mother, Mother, may I go out to play?' 'No child! no, child! not for the day', etc.—'I thought this game was a thing of the past, but I came on some children playing it in the streets of Penzance in 1883.' *Yorkshire Folk-Lore Record*, vol. i, 1888, pp. 214–15, 'I saw today three little girls, aged 3, 5, and 7, play a new game, or, at least, a new one to me.'—'Please, mother, may I go out to play?' 'No, my loves, it is a very wet day', etc.; *Folk-Lore Journal*, vol. vii, 1889, pp. 219–22, two Dorsetshire versions; *Traditional Games*, 1894, pp. 390–6. *London Street Games*, 1916, p. 101; *Games of American Children*, 1883, p. 172.

The game has been described with affection by a number of our older correspondents, the following being a Hampshire text of about 1910:

'Please, Mother, can we go out to play?'

'No, it's wet and cold.'

'Please, Mother, can we go out to play?'

'No, it's wet and cold.'

'Please, Mother, can we go out to play?'

'Go then!' (The children go off, and after a while return.)

'Where have you been?'

'Grandmother's wedding.'

'What did she give you?'

'Cake and wine.'

'Where's mine?'

'On the shelf.'

' 'Tisn't there.'

'Cat's got it.'

'Where's the cat?'

'Behind the stack.'

'Where's the stack?'

'On fire.'

'Where's the fire?'

'Water's quenched it.'

'Where's the water?'

'Fox has drunk it.'

'Where's the fox?'

'Behind the churchyard cracking nuts and you can eat the shells.' (On this the children ran away and were chased by the Mother with a stick.)

This is another game that is not confined to English-speaking children. In Bremen in the nineteenth century girls had an amusement which they called 'Der Hühnerhof'. One girl was Grandmother, and the others asked: 'Grotmooder, wat makst du dar?'

Grandmother: 'Ich spinne.'

Children: 'Wo is denn din Mann?'

Grandmother: 'Open Hönerhof.'

Children: 'Wat makt he dar?'

Grandmother: 'He futtert de Höner.'

Children: 'Könt wi denn nig ok en beten hen gaan?'

Grandmother: 'Ja, awert jagt se mi nig weg!'

The children scampered into the poultry-yard, and immediately began chasing the hens, shooing them with their aprons. The Grandmother ran after them, and, as in England, the child she caught became the next Grandmother (F. M. Böhme, *Deutsches Kinderspiel*, 1897, p. 592).

FOX AND CHICKENS

In this playlet, unlike the others, the catching is an integral part of the game and, except in places where the words have been passed on by an adult, the drama is becoming a speechless shadow of its former self. In 1922, when the dialogue was still a feature of the game, A. S. Macmillan could write: 'It is no exaggeration to say that scores of young people have sent me descriptions . . . all very much alike but called by many different names, such as "Old Woman, What's the Time?", "Polly (or Betsy), What's o'Clock?", "The Wolf and the Sheep", "The Fox and the Hen", "Hen and Chickens" &c.'

In the standard game one of the bigger children is the fox, another the hen, and the rest make up the hen's brood. The chickens form up in single file behind the hen, each holding on to the waist or garment of the one in front. They march up to the fox, who is crouching on the ground, and chant:

> Chickany, chickany, crany crow,
> I went to the well to wash my toe,
> When I came back a chicken was dead.

They stop in front of the fox, and the hen asks: 'What are you doing, old fox?'

Fox, in a gruff voice: 'Picking up sticks.'

Hen: 'What for?'

Fox: 'To make a fire.'

Hen: 'What do you want a fire for?'

Fox: 'To cook a chicken.'

Hen: 'Where will you get it?'

Fox: 'Out of your flock.'

The fox springs up, and tries to seize the last chicken in the line. The hen, with the chickens hanging on behind her like a tail, does her best to guard them, swinging round to face the fox whichever way he goes, and holding out her wings to prevent him slipping past. This active part of the game is now popular in gymnasiums, where the sport is usually called 'Fox and Geese', and the turnover in foxes is frequent, for 'when the end person is touched he becomes the fox, and the game continues until everyone has had a turn at being fox'. In the traditional game, however, the fox has to catch each of the chickens in turn, and take them away to his den. Thus, as the line of chickens diminishes, his predatory activities become more difficult, and it needs a determined fox to succeed in catching the last ones.

** There is good reason to think that this game was current in Queen Anne's time. Up to a generation or so ago it was played not only under the names 'Fox and Chickens', 'Fox and Goose', 'Wolf and Sheep', and 'Hen and Chickens', but under such curious denominations as 'Chicken come Clock' in Kiltubbrid, Co. Leitrim (Gomme, vol. ii, 1898, pp. 410–11), 'Chickamy, Chickamy, Chany Trow' in Surrey (George Bourne, *William Smith*, 1920, pp. 33–4), 'Wigamy, Wigamy, Waterhen' (*Old Surrey Singing Games*, 1909, p. 11), and 'I'm going to the Pippen to water my chickens' in Wales, *c.* 1915. It is thus likely that 'Chicken a Train Trow' listed amongst children's games in William King's *Useful Transactions in Philosophy*, pt. i, 1708/9, p. 43, is this game, as also 'Come water my chickens, come clock' in 'The Nurse's Song' in *The Tea-Table Miscellany*, vol. iv, 1740; and 'Water my chickens come clock' in Brooke's *The Fool of Quality*, 1764–70 (1859, vol. i, p. 272). It was certainly the 'singular game played at country schools' in Galloway under the name 'Gled Wylie' for it is well enough described in Mactaggart's *Gallovidian Encyclopedia*, 1824; as also in Jamieson's *Scottish Dictionary, Supplement*, 1825, under 'Shue-Gled-Wylie' (Fife), and 'Shoo-Gled's Wylie' (Teviotdale). (A *gled* is a kite, compare continental 'hawk' versions.) In addition J. O. Halliwell, *Popular Rhymes*, 1849, p. 132, gives an 'Old Dame' version in which the children start by singing:

> To Beccles! to Beccles!
> To buy a bunch of nettles!
> Pray, Old Dame, what's o'clock?

The Dame replies 'One, going for two', and the game does not proceed until the old dame replies 'Eleven going for twelve'—another detail that has continental and American parallels.

Indeed the game is a favourite in the United States (referred to at least ten times in the *Journal of American Folklore*), the usual name being 'Chickamy Chickamy Craney Crow'. It was described as long ago as 1835 in the American *Girl's Book of Diversions*, pp. 27–8; and Newell, *Games of American Children*, 1883, pp. 155–8, gives interesting accounts in which the predators are variously a fox, an 'Old Buzzard', and a witch. He reports that in Georgia the game began with the rhyme:

> Chickamy, chickamy, crany, crow,
> I went to the well to wash my toe,
> And when I came back my chicken was gone;
> What o'clock, old witch?

The witch named an hour, and the children repeated their question until the witch replied 'Twelve o'clock'.

'What are you doing, old witch?'

'I'm making a fire to cook a chicken.'

'Where are you going to get it?'

'Out of your coop.'

'I've got the lock.'

'I've got the key.'

'Well, we'll see who will have it.'

The witch tried to get past the hen, and seize the last of the line; the mother, spreading out her arms, barred the passage. The witch cried, 'I must have a chick.' The hen retorted, 'You shan't have a chick.' Each child caught dropped out, and, as Newell remarks, 'as the line grows shorter the struggle becomes desperate'.

The similarity of continental and British versions can scarcely be coincidental. In Denmark, where the predator is a hawk, the game (recorded almost a century ago), began as in England with the bird crouching down and scratching in the earth.

Hen: 'What are you scratching for?'

Hawk: 'For an old rusty needle.'

Hen: 'What are you going to do with it?'

Hawk: 'Mend my pots and pans.'

Hen: 'What are you going to use them for?'

Hawk: 'Boiling chickens.'

Hen: 'Where will you get the chickens?'

Hawk: 'From you!' And the hawk tried to seize the chickens who were hanging on behind the hen (*Børneses Musik og Sange, samlede af en Moder*, 1879). In Yugoslavia, Czechoslovakia, Romania, and Hungary, the pattern, according to Brewster, is similar. Thus in Hungary:

'Hawk, hawk, what are you doing?'

'Making a fire.'

'What's the fire for?'

'To warm water.'

'What's the water for?'

'To scald a chicken.'

'From whose?'

'From yours.'

'I'm their mother; I won't let them.'

'I'm a hawk; I'll get them!'

'Then the hawk dashes for the chickens, and the hen spreads out her arms to protect them' (*American Nonsinging Games*, 1953, pp. 71–6). The game is also traditional amongst German-speaking peoples. Kampmüller says it continues to be played in Austria, where it is called either 'Henne und Geir' or 'Fuchserl und Hahnl', the game starting with the predator crouching on the ground, and stirring with a stick (*Oberösterreichische Kinderspiele*, 1965, pp. 139–40). In Germany, too, the game generally starts with the predator squatting down and scratching in the earth. 'Was grabst?' asks the hen. 'Ein Löchle,' replies the fearsome bird. 'A hole.'

'What are you looking for in the hole?'

'A stone.'

'What will you do with the stone?'

'Sharpen a knife.'

'What will you do with the knife?'

'Kill all your chickens.'

Zingerle, who speaks of the vulture game ('Geierspiel') as being beloved throughout Germany, remarks that the number of times it was named in the sixteenth and seventeenth centuries proves it to have been well known in the past. One description, cited by Böhme, occurs in Ammann's *Kinderspiele*, 1657 (a rendering of Cats's *Kinder-spel*), where such familiar details are noted as that the vulture's part is taken by the boldest player, the clucking hen's by the most intelligent (*Deutsches Kinderspiel*, 1897, pp. 569–71). In France, unlike in Germany, the predator is usually a wolf. The game is best known as 'La Queue du loup', and the version recorded by Rolland (*Rimes et jeux de l'enfance*, 1883, p. 156) follows the familiar pattern:

La mère: 'Que fais-tu là?'
Le loup: 'Je ramasse des bûchettes.'
La mère: 'Pour quoi faire ces bûchettes?'
Le loup: 'Pour allumer mon feu.'
La mère: 'Pour quoi faire ce feu?'
Le loup: 'Pour chauffer mon eau.'
La mère: 'Pour quoi faire cette eau?'
Le loup: 'Pour affiler mon petit coutiau.'
La mère: 'Pour quoi faire ce petit coutiau?'
Le loup: 'Pour couper la langue à tes petits agneaux.'

However, there is also a version called 'L'Émouchet', in which the players are 'un émouchet, une poule, et des poussins'; and it seems that both forms of the game have coexisted in France for some centuries. In *Les Trente-six figures contenant tous les jeux*, published in 1587, two groups of youths are separately depicted playing versions of the game, the one called 'poussinets', the other *vulgairement* 'la queuë leu leu'. Rabelais listed 'la queue au loup' in 1534, as one of the games played by Gargantua; and it appears that the game was known in France in the Middle Ages, for around 1380 Froissart, recalling the pleasures of his childhood in 'L'Espinette amoureuse', wrote:

> Puis juiens à un aultre jeu
> Qu'on dist, *à la Kevve leu leu.*

In fact, the distribution of the game is probably world-wide. Further countries in which it is known to have been played, either with or without the preliminary dialogue, include Italy, Sicily, Sardinia, Spain, Sweden, Russia, Poland, Cuba, Haiti, Ceylon, and China. See, e.g., M. M. Lumbroso, *Giochi*, 1967, p. 73, 'Il lupo, il pastore e gli agnelli'; K. L. Bates, *Spanish Highways and Byways*, 1900, pp. 5–7 (chief players kite, mother, and smallest girl); Caroline Crawford, *Folk Games*, 1908, p. 38 (chief players in Sweden are a fox and geese); *Folklore*, vol. lxv, 1954, p. 69, 'Les Petits Oiseaux' in Haiti; Shufang Yui, *Chinese Children at Play*, 1939, 'Eagle Catch the Chicks'.

JOHNNY LINGO

The players stand in a ring with one child within the ring and one child outside. Those forming the ring are said to be sheep, chickens, or pigs (each of these pretences is traditional), while the child outside the ring has a name such as 'Johnny Lingo', 'Bobby Bingo', or 'Jacky Jingle'. The child within the ring is a farmer, and begins the game by chanting: 'Who's that walking round my stoney wall?'

Child outside the ring: 'Only little Johnny Lingo.'

Farmer: 'Don't you steal any of my fat sheep or I shall make you tingle.'[1]

Child outside: 'I stole one last night, and I'll steal another tonight', or, 'Neither shall I do so, except I take them one by one', and he touches one of the children on the back, saying 'Come on', or 'Come on little chick, chick, chick' (Matching Green), or 'Come whip' (Roe, Shetland), and the child joins on behind him, holding his waist.

'Who's that walking round my stoney wall?' demands the farmer again, and both children reply, 'Only little Johnny Lingo'.

'Don't you steal any of my fat sheep', cries the farmer.

'I stole one last night, and I'll steal another tonight', reply the pair, padding round the outside, and touching another member of the circle on the back.

So the game proceeds until everybody in the ring has joined behind Johnny Lingo, the challenges and warnings being repeated over and over again, which presumably accounts for the preservation of their archaic wording. There follows a chase in which, usually, the farmer tries to regain his sheep one by one from the end of the line, as in 'Fox and Chickens', while Johnny Lingo does his best to prevent him.

The game is known either by the name of the principal character, or as 'Who Goes Round my Stoney Wall?'

⁂ This game was clearly popular in the nineteenth century. A writer in *Blackwood's Magazine*, August (pt. ii), 1821, p. 36, reported that in Edinburgh 'the little actors' spoke the following by way of question and answer:

> Who goes round my house this night?
> None but bloody Tom;
> Who stole all my chickens away?
> None but this poor one.

Washington Irving in *Bracebridge Hall*, vol. ii, 1822, p. 37, recorded almost the same words:

> Who goes round the house at night?
> None but bloody Tom!
> Who steals all the sheep at night?
> None, but one by one.

Robert Chambers said that the game 'Bloody Tom' was common among

---

[1] Or, as a small stolid lad in Somerset was heard warning, 'Don't yew come stealing my gurt hog'.

boys 'all over Scotland' (described in *Popular Rhymes of Scotland*, 1842, pp. 62–3); and J. O. Halliwell, the same year, said the children called it, in the slang of the time, 'a regular tearing game' (*Nursery Rhymes*, 1842, pp. 122–3). The thief, and hence the game, has also been known as Johnny Jingo, Johnny Ringo, Johnny Nero, Johnny Able, Johnny Winkle, Jack and Jingle, Jack and Jingo, Jackie Lingo, Daddy Dingo, Jack the Lentern (*sic*), Jenny Langal, Tommy Jingle, Limping Tom, Poor Old Tom, King Sailor, and, in Tasmania, Old Black Joe.

Several of our older correspondents have described the game as played fifty or sixty years ago, sometimes with additional dialogue. At Bridgwater in Somerset a shepherd took part, as well as a farmer and wolf. The wolf approached the shepherd and said, 'Sheep, sheep, sheen-o'.

The shepherd replied, 'What's want here-o?'

The wolf: 'A good fat sheep.'

Shepherd: 'Take the best and leave the worst and come again tomorrow.'

The farmer, who had not seen the wolf, inquired: 'What's that? What's that?'

Shepherd: 'Only calling the dog, sir.'

The farmer turned away, and the wolf dragged off one of the sheep, although not without a struggle, for the Bridgwater sheep resisted being taken. When the last sheep had gone the farmer looked in the fold, found it empty, and 'with much abuse' sacked the shepherd.

In the west country, too, the farmer or master of the flock sometimes had words with the thief ('Bobby Bingo') after the sheep had been stolen, and while they were clinging on behind his back.

Master: 'What's that behind your back?'

Bobby Bingo: 'Only a bundle of straw.'

Master: 'What do you want a bundle of straw for?'

Bobby: 'To light a fire.'

Master: 'What do you want to light a fire for?'

Bobby: 'To boil my kettle.'

Master: 'What do you want to boil a kettle for?'

Bobby: 'To make myself a cup of tea.'

Master: 'Let me look through your key-hole.'

Bobby placed one hand on his hip, and the master looked through the space between his arm and body.

Master: 'I see my sheep.'

The sheep bleated 'Baa, baa', and the master tried to regain them by catching the hindermost in the line.

This action and dialogue links the game, not only with 'Fox and

Chickens' and 'Old Man in the Well', but with 'Mother, the Cake is Burning' described hereafter.[1]

Versions of the game are also well known in Austria where it is called 'Lamperlstehln' or 'Wer geht? Wer geht?' (Kampmüller, *Oberösterreichische Kinderspiele*, 1965, pp. 162–4); and in Italy where it is 'Mamma Polleone' (Lumbroso, *Giochi*, 1967, pp. 223–4). The French game 'Le Boucher et les moutons' is similar, but the butcher first weighs the sheep, agrees with the shepherd to buy them, and then finds he has a business to take them, for they are in a line behind the ram (*200 Jeux d'enfants*, c. 1892, pp. 125–6).

MOTHER, THE CAKE IS BURNING

This pantomime of a game contains enough myth and madness to gratify both folklorists and 9-year-olds. Briefly the plot is as follows: A mother goes to market, leaving her seven children in the care of a maid or eldest daughter. While she is away an evil visitor comes to the door, enters the house on some pretext, and snatches one of the children ('the youngest child' or 'the most precious child'). The mother returns, beats the maid or eldest daughter for allowing the child to be stolen, and goes off again. While she is away the second time, a second child is stolen. The mother, who seems to be simple-minded, returns and again beats the girl in charge, and again leaves home. This occurs seven times until all the children have been stolen. The mother then seeks out the kidnapper who, it is becoming clear, is a magical person. For a while the kidnapper obstructs the mother's entry into his house, but she becomes indomitable: she will even, in some versions, cut off her feet to gain admittance, and eventually succeeds, only to find that her children are now disguised, or renamed, or turned into pies or other delicacies, and about to be eaten. Nevertheless, by skill or luck, she identifies and releases them. Usually the game ends in a chase.

When little girls, dressed in T-shirts and jeans, are seen playing this acting-game at the end of the street, it is difficult to believe that they are not making up the incidents as they go along. Only when comparison is made, scene by scene, with nineteenth-century recordings, does it become apparent that they are following an old and international script dictated by

[1] A further connecting link is apparent in the game 'Cripple Chirsty' played in the Lorne district of Argyllshire (*Folk-Lore*, vol. xvi, 1905, p. 220). Cripple Chirsty came limping along leaning on a stick, and the protecting player, a hen, addressed her: 'Hey, Cripple Chirsty, what do you want with me today?'

'A beck and a bow and I would thank you for your eldest daughter.'

The hen gave the witch a curtsy and a bow, but refused to give up her eldest daughter, or any other of the children who were hanging on in a line at her back. The witch then attacked the hen, trying to reach the furthermost player.

folk memory. It seems best, therefore, with this game, to bring past and present descriptions together; and the following report is based on fifty-nine English-language accounts: twenty-seven contemporary, and thirty-two old, mostly nineteenth century, including eight from the United States (Newell, 1883, 1890, and 1903), and three from Australia (Howard MSS.).[1]

In most accounts the principal character is a mother who is going to market or 'going washing' or, in the present day, 'going shopping'. Occasionally she asks the children what they would like her to bring home, for instance in Cornwall, 1887: 'The mother says she is going to market and will bring home for each the thing that she most wishes for. Upon this they all name something.' These presents can have significance in the story, for when the mother is handing them out on her return she finds that she has one more present than she has children, and it is then that she realises one of her children is missing. Ordinarily she knows how many children she has by naming them after the days of the week, 'Monday, Tuesday, Wednesday . . .' (e.g. Northumberland, 1846; Liss, 1964). Before she goes out she counts them by reciting the days of the week, or gets the maid or eldest daughter to do so. In the nineteenth century it was usually the 'eldest daughter', sometimes named 'Sue', who was put in charge; today, although this may seem unrealistic, the person who looks after them for her is 'the maid', but the whole play is unrealistic. Thus it was in verse that the mother used to caution her children before she went out:

I charge my daughters every one
To keep good house while I am
    gone.
You and you, but specially Sue,
Or else I'll beat you black and blue.
          1849. Similar Sussex, 1882

I am going into the garden to gather
    some rue,
And mind old Jack-daw don't get you.
Especially you, my daughter Sue,
I'll beat you till you're black and
    blue.   Ipswich, 1893

Now all you children stay at home,
And be good girls while I am gone . . .
Especially you, my daughter Sue,
Or else I'll beat you black and blue.
        Hertford, Connecticut, 1883

Go down the garden and get a bit of
    rue,
Mind old Jack-a-Bed don't take you:
Especially you, my daughter Sue,
Or I'll beat you black and blue.
      Wrecclesham, Surrey, c. 1897

[1] Printed sources: Brockett, North Country Words, 1846, under Keeling; Halliwell, Popular Rhymes, 1849, p. 131; Folk-Lore Record, vol. v, 1882, p. 88; Shropshire Folk-Lore, 1883, p. 520; Newell, Games of American Children, 1883, pp. 215–21, and 1903, pp. 258–63; Folk-Lore Journal, vol. v, 1887, pp. 53–4; Journal of American Folk-Lore, vol. iii, 1890, pp. 139–48 (Newell); Northall, Folk Rhymes, 1892, p. 391; County Folk-Lore: Suffolk, 1893, p. 62; Gomme, Traditional Games, vol. i, 1894, pp. 396–401 and vol. ii, 1898, pp. 187–9, 215, 391–6, 449; Rodger, Lang Strang, 1948, pp. 43–4 (Forfar, 1910).

Manuscript versions in our files, other than present-day accounts, are quoted with the date when the game was played.

Today, after counting them, she merely warns the maid not to lose them, or, pointing to 'Sunday' (who is usually the special child), says, 'That's my very best chicken, and don't you let her go' (Stromness), 'Don't you let my Sunday go' (Alton), or, pointing to a child who is kneeling to represent the youngest one, says, 'Don't you dare lose my best Blacking Topper' (West Ham).

The evil visitor who comes to the door is often a witch. In the United States in 1883 she was sometimes called 'Old Mother Cripsy-crops' or 'Old Mother Hippletyhop'. At Crickhowell in Breconshire, in the present day, she is named 'Heckedy Peg', a lame old woman who announces herself 'I'm Heckedy Peg, I've lost my leg'. In Langholm, today, she is 'Jenny from o'er the hill' a wicked old woman who hobbles along supported by a stick. In 1849 she was a gipsy, as also in east Devon in 1922. In Derbyshire, 1910, she was a 'Peggy Woman' (i.e. a seller of pegs, traditionally a gipsy), and in Somerset, 1922, a 'Pins and Needle-Woman'. The child-snatcher may also be a 'Bogie-man' (Taunton, 1922), 'Black Man' (Hanley, 1890), 'Old Man' (South Elmsall, 1935); and in the present day a 'Tramp' (Westerkirk, Dumfriesshire), a 'Beggar' (Swansea), a 'Baker' (Dublin), a 'Funnyman' (Burslem), 'Black Jack' (Whalsay), 'Jack in the Chimney' (Roe). In the south and west the thief may be a fox or wolf (Cornwall, 1898; Chard, 1922; Abergavenny, c. 1930; Alton and St. Peter Port in the present day). In the west country the 'children' were often 'sheep'. They may also, even today, be 'chickens', 'scones', or 'pots of jam' (cf. the game 'Jams', pp. 286–8).

It is noticeable that in four out of the ten contemporary accounts in which the visitor begs at the door, he or she asks for something connected with fire. For instance in Oxford the witch asks for a match; in Alton the fox begs a box of matches, in Radnorshire the children are called 'Matchsticks'. This accords with the nineteenth century. At Deptford, in 1898, the witch begged politely, 'Please you, give me a match'; in London, in the same period, the Old Woman promised the eldest daughter a gay ribbon if she would give her a light; in Boston, Massachusetts, 1890, the witch begged 'Give me fire; I'm cold', and when refused, tempted the eldest daughter with a necklace, offering it 'All for one lighted sod and one fat child'.[1]

In the St. Thomas district of Swansea, today, the 'Man' asks, 'Can I light my pipe?' The maid at first says 'No'. The man spits on the

[1] In The Journal of American Folk-Lore, vol. iii, 1890, pp. 142–4, Newell deduces from parallel Swedish, Italian, and Catalan games, and by references to superstition in Ireland, that 'a demand for fire, or for a light, on the part of a stranger, constitutes ground for suspicion of witchcraft, and that such a request must not be complied with'.

'carpet'. 'How dare you dirty my carpet!' exclaims the maid, and rubs the spit in the ground. 'Look the cake's burning', says the man. The maid goes to look at the cake, and while she is away the man steals a child. This incident, too, accords with nineteenth-century practice. In Dronfield, Derbyshire, 1894, the Old Witch begs, 'Please, can I light my pipe?' Children: 'Yes, if you won't spit on t'hearth.' The witch pretends to light her pipe, but maliciously spits on the hearth, and this seems to give her the necessary power, for only then does she run off with the girl called Sunday. Likewise at Sneyd Green, Stoke-on-Trent, in the present day, the 'funnyman' who has begged a drink of water, does not attempt to enter the house and seize a child until he has smashed the cup which holds the water. Similarly at Whalsay in Shetland, in the present day, Black Jack is apparently able to seize the 'youngest child' without hindrance after he has spilt water on the hearth and ruined the bland.[1] There may also be significance and a connection between the ruined bland, the burning cake, and the boiling kettle. At Knighton the witch comes in and says to the maid, 'Quick, the kettle's boiling', and when she goes to see runs off with a 'matchstick'. Formerly the cry was that the pot was boiling over. Among the *dramatis personae* in London in 1894 was a child who played the part of the pot, and at the appropriate time made a 'hissing and fizzing' noise.[2] Sometimes, when the mother was supposed to be a washer-woman, and worked outside the house but within hailing distance, the daughter called out: 'Mother, Mother, the pot boils over' (London, 1894).

The mother replied: 'Take the spoon and skim it.'

Daughter: 'Can't find it.'

Mother: 'Look on the shelf.'

Daughter: 'Can't reach it.'

Mother: 'Take the stool.'

Daughter: 'The leg's broke.'

Mother: 'Take the chair.'

Daughter: 'Chair's gone to be mended.'

Mother: 'I suppose I must come myself.'

This dialogue was also reported in 1846 as known in Northumberland, one girl saying, 'Mother, Mother, the pot's boiling over', the mother answering, 'Then get the ladle and keel it'; and the girl objecting succes-

[1] Bland, made with equal parts of hot water and whey, remains a popular drink in Shetland, although not drunk perhaps to the extent it was in 1701, when it was reported to be so ordinary there were 'many people in the countrey who never saw ale or beer all their lifetime'.

[2] Just as the boys did who represented burning biscuits in the game called 'Jack, Jack, the Biscuit Burns'—*School Boys' Diversions*, 1820, pp. 10–12.

sively that the ladle was 'up a height', that the 'steul' wanted a leg, and that the joiner was either sick or dead.

In some versions the maid appears to deserve the chastisement that is coming to her. In Alton, when the fox knocks at the door and says 'I want a box of matches, please', the maid says 'Take one', and the fox takes Sunday. This may be a contraction of the story, but at Westerkirk in Dumfriesshire, where the children are said to be scones, the maid actually summons the tramp and gives him the best ones. This is precisely as the game was played in Forfar about 1910. The servant had been told to watch all the bannocks and 'pey affa gude attention tae this ane. O' a' my bannockies, this is my favourite'. No sooner is the mistress away than the servant goes to the 'Deil's' hiding-place and cries:

> 'Deil, Deil, come doon the lum,
> And wash yer face afore ye come.'

The maid is also clearly an accomplice in the game played today in Stromness. When the mistress goes out for a short walk the maid pushes 'the best chicken' to the Old Woman, who hides it behind her back. When the mistress returns, and demands in a rage, 'Where is my best chicken?' the maid replies she does not know, and when the maid is told she must go 'up the lane and down the lane' until she finds it, she only pretends to look for it.[1] In Dublin, too, in the present day, the children themselves seem to entice if not summon the evil visitor. As soon as the mother is absent they give notice of it by singing out all together, 'I am stirring my chicken, my chicken', and the villain comes running to take one of them.

In many versions the mother's return to her family provides moments of comic suspense. Everyone knows that a child is missing, and that the maid is about to be smacked for her negligence. But the mother acts as if nothing was wrong until the children have been counted, or their names recited, 'Monday, Tuesday, Wednesday . . .', or until she has handed the children their presents and there is one present too many. It almost seems that this repeated counting before and after the loss of each child is to ensure that a primitive audience shall understand what has happened. In

---

[1] This appears to be in accord with the game as played in Sweden. There, the old woman, who also hobbles on a stick, points at one of the children and asks 'May I have a chicken?' Although at first refused she is eventually given each of the 'chickens' and does not have to steal them—A. I. Arwidsson, *Svenska Fornsånger*, vol. iii, 1842, p. 437. In Spain the Mother leaves Marquilla in charge of the brood, with directions, if the wolf comes, to fling him the smallest chicken—K. L. Bates, *Spanish Highways and Byways*, 1900, p. 317. In German versions of the game, under the name 'Frau Rose', the children are begged rather than stolen—Wilhelm Mannhardt, *Germanische Mythen*, 1858, pp. 273–83. The plot may here have been influenced by the otherwise unrelated European game of 'Rich and Poor' in which children are begged.

Scalloway, when the mother asks where the missing child has gone, the other children say 'She has gone to the well', 'She has gone to a dance', 'She has gone for a walk', and the mother at first believes them. At Evenjobb in Radnorshire, when the 'matches' have been counted, and one of them is found missing, the master cross-examines the maid, 'Where is the other one gone?' The maid replies, 'I don't know.'

'Where was you?'

'Down the cellar.'

'What doing?'

'Making leather.'

'What for?'

'To tie your nose and mine together', replies the maid.

Whereupon the master gives the maid 'a good whacking', although whether for her cheekiness or her carelessness in allowing a match to be stolen is not clear.[1] In London, in 1898, the daughter in charge of the children made them promise not to tell their mother. When the mother returns and asks 'Are all the children safe?' the eldest daughter says 'Yes'. 'Then let me count them.' The children stand in a row, and the mother, who is blind, counts them by placing her hands on their heads. When the eldest daughter has been counted, she runs to the other end of the line to be counted again, and the mother does not at first realize that a child is missing. A similar ruse is attempted today in West Ham. When the maid finds that the favourite child, 'Blacking Topper', is missing, she puts herself in the line with the children to make up the right number. However the mother discovers her trick. 'Where's my Blacking Topper?' she cries. 'I don't know', says the maid. Then the mother spanks the maid for losing her 'Blacking Topper'. This spanking is one of the highlights of the game, and takes place six or seven times. Yet in some places when the mother asks about the missing child there is no prevarication. The children shout 'The witch has took her' (or, in Hanley, 1890, 'The Black Man took her for a basin of broth'), and the maid is promptly spanked. But whatever the form of the kidnapping and its discovery, it is ritualistically repeated for each child. In West Ham, for example, a new 'Blacking Topper' (or 'Mackintosh') is chosen each time, and kneels down to be the littlest one, so that each time when the mother returns she can make the same complaint, 'Where's my Blacking Topper?', and each time, even on the seventh occasion, the maid tries to hide the fact that another 'Blacking Topper' is missing.[2]

---

[1] 'I call this a very interesting game', stated our informant, a girl aged 12.

[2] It was no surprise when one small girl remarked, 'Sometimes we never get to the end of this game.'

When the last child has been stolen, and sometimes the maid too, or she has run away, or has accepted the fox's invitation to a party (Alton), the mother seeks out the evil visitor's house. At Alton the mother says to the fox, 'Where's my children gone?' The fox says 'Up the road and round the corner.' The mother goes and looks and, of course, she does not find them. In Whalsay, Shetland, she asks Black Jack if he has seen any children. Black Jack says he saw one go to the bridge, and carefully describes one of the children, as he is well able to do. The mother goes there and cannot find the child, so she comes back and Black Jack tells her another place where he says he saw another child. In Cornwall in 1887, when the Old Witch was asked if she had seen the children, she said 'Yes, I think by Eastgate.' The mother went and looked by Eastgate, and not finding them, the witch suggested 'I think by Westgate', and sent the mother on another futile search. In London in 1894 the mother met the witch and asked, 'Is this the way to the witch's house?' The witch replied 'There's a red bull that way'. The mother said 'I think I'll go this way', and the witch warned, 'There's a mad cow that way'. This is echoed today in Dublin. When the mother asks the Baker (who has run off with the children) where they are, the Baker gives her false directions. Then the children, who are hidden in the Baker's 'den', set up a cry 'There are snakes over there', and the mother comes running back. At Taunton, in 1922, the 'Bogie' declared that the children were gone to London. 'How many miles?' asked the mother. 'Forty-two' (or some such number), replied the Bogie, and the mother took forty-two steps for miles and then came back: 'They're not there.' So the Bogie named further towns, stating their distance. In Swansea today, when the mother demands of the kidnapper 'Where are the children?' the 'Man' replies: 'Gone to school.'

Mother: 'How many miles?'
Man: 'Eight' (or any number).
The mother, and the maid who is with her, take eight steps backwards.
Mother: 'They are not here.'
Man: 'They've gone to bed.'
Mother: 'I want to see them.'
Man: 'Your shoes will dirty the carpet.'
Mother: 'I'll take them off.'
Man: 'The wool will come off your stockings.'
Mother: 'I'll take off my stockings.'
Man: 'Your feet are dirty.'
Mother: 'I'll cut them off.'

Man: 'The blood will go on the carpet.'

Mother: 'I'll fly up.'

And in one Swansea version, while man and mother have their strange altercation, the stolen children stand in a row, swaying from side to side as if trying to work a magic spell, chanting 'Mother, the cake is burning! Mother, the cake is burning!'

Similar dialogues were recorded in the south and west of England in the nineteenth century, and also in the United States. For instance in 1883, when the mother asked to be let into the witch's house, the witch refused in exactly the same style, saying, 'No, your shoes are too dirty.'

Mother: 'But I will take off my shoes.'

Witch: 'Your stockings are too dirty.'

Mother: 'Then I will take off my stockings.'

Witch: 'Your feet are too dirty.'

Mother: 'I will cut off my feet.'

Witch: 'That would make the carpet all bloody.'[1]

Dialogues such as this, in which the formalities of polite speech are maintained despite the desperateness of the situation, are not uncommon in folk humour. In the Cornish version of 1898, social decorum is observed even when the shepherd has arrived at the wolf's house and smelt meat being cooked. He says, 'May I go up and taste your soup?' and the wolf blandly replies, 'You can't go upstairs, your shoes are too dirty.' Likewise at Chaffcombe, Somerset, in 1922, after a horrifying argument between the Old Man and the mother, in which it is finally agreed that a million blankets will staunch the blood from the mother's bleeding stumps (she has cut off her feet), the Old Man says 'I have some lovely jam for sale', and the mother inquires, as any housekeeper might be expected to do in other circumstances, 'What kind have you?' 'Some plum and apple', he replies. 'Can I have 4 lb. please?' she says. However, the mother may here be showing presence of mind. In several accounts it is apparent that the kidnapper is an ogre, and has cooked or intends to cook the children, and turn them into pies, jams, or other confections in which the ingredients are transformed. In New York, in 1883, the children were made into pies which were actually served up to the mother to eat. She tries one and

[1] Little girls, today, in Krombach, Westphalia, skip to a similar fancy:

'Aprikose Liese, geh' mal auf die Wiese.'	*'Apricot Liz, go to the meadow.'*
'Mutter, ich hab' kein Schuh.'	*'Mother, I have no shoes.'*
'Zieh des Vaters Pantoffl'n an.'	*'Put your father's slippers on.'*
'Mutter, die sind zu klein.'	*'Mother, they are too small.'*
'Schneid' ein Stück von der Ferse ab.'	*'Then cut a bit off your heel.'*
'Mutter, dann gibt es Blut.'	*'Mother, then it will bleed.'*

exclaims: 'This tastes like my Monday.' This reanimates Monday, who is then spanked and sent home. In London in 1898 the children had their faces covered when their mother arrived (little girls' faces could easily be covered at this time by having their pinafores pulled over them), and they were served to the mother as various kinds of meat: beef, mutton, or lamb. In Langholm, today, the mother or shopkeeper arrives in the nick of time to save the children from being put into Jenny's pot. In South Elmsall, *c.* 1935, where the Old Man who stole the children was a butcher, the mother went to the butcher's shop and asked for a leg of mutton. One of the children stuck her leg through the Old Man's legs, and the mother exclaimed, 'That's not mutton, that's one of my children.' In Stoke-on-Trent, today, after the mother has been sent a 'hundred miles' in one direction, and 'fifty miles' in another, and found her children are not there, she says 'What's that between your legs? Why, it's one of my children's legs!' Thereupon the kidnapper cuts off the child's leg, and cuts off the legs of the other children too, and they all cry out 'Oh mother, I've lost my leg!' In Guernsey, too, where juvenile lore is largely derived from the north country (owing to evacuation during the war), the children line up behind their captor, the wolf, and the first child puts a leg forward between the wolf's legs. The mother has to recognize whose leg it is, and say correctly 'That's my Wednesday', or whatever is the name of the child. 'Of course she can see who they are', remarked an informant, aged 9. 'She can see them sticking up behind the wolf.' Again at Westerkirk in Dumfriesshire, the children seem to be renamed (as they were in the United States in 1883), for the mistress does not recognize her scones, and calls them by the names of animals, until the quisling maid runs out of the witch's house, whereon 'the mistress calls her something nasty and she hits her'.

Indeed the game can end in a number of ways, most of them incongruous. In some places the mother merely asks for the children, and they are handed over. At Waunerllwyd, near Swansea, the mother has to buy her children from the Beggar with kisses, one kiss for each child and two for the maid. ('It is best if the beggar can be a boy.') In several places the mother chases the children. In others, mother and children chase the thief, and when he is caught, put him in jail. This was the mildest of the fates awaiting the malefactor in former times, when wolves had their heads chopped off, old men were hung, and witches burned. It is recorded for instance in Cornwall, in 1887, that when the Old Witch had been caught the children pretended to bind her hand and foot, and put her on the pile, and then they burnt her, 'fanning the imaginary flames with their pinafores'.

Names in present day: 'Mother, the Cake is Burning' (Swansea and district); 'Amy and Witchie' (Scalloway); 'Fairies and Witches' (Liss); 'Jenny Came O'er the Hill' (Langholm); 'Heckedy Peg' (Crickhowell); 'Jack in the Chimney' (Roe, Shetland); 'Black Jack' (Whalsay, Shetland); 'Blacking Topper' (West Ham); 'Mackintopper' (Isle of Wight); 'Mackintosh' (West Ham);[1] 'Got a Match, Jack?' (Oxford); 'Boxes of Ma ches' (Alton and Herriard); 'Matchsticks' (Knighton and Evenjobb); 'Black Ram' (Chudleigh, south Devon); 'Scones' (Westerkirk); 'Chickens' (Stromness, Orkney); 'I am Stirring my Chicken' (Dublin); 'Run Away Children' (Berry Hill, Forest of Dean); 'Days of the Week' (Swansea).

Names previously current: 'Limpie, limpie, the pot's boiling owre' (listed by Robert Chambers, *Popular Rhymes*, 1826, p. 299); 'Keeling the Pot' (Northumberland, 1846); 'Mother, Mother, the Pot Boils Over' (London and Dronfield, 1894); 'Mother, the Kettle's Boiling' (Abergavenny, c. 1920); 'Witch' or 'Old Witch' (Cornwall, 1887; Dartmouth, 1894); 'Pins and Needle Woman' (Somerset, 1922); 'Shepherd and Sheep' (Oswestry, 1883); 'Count my Sheep, Jack' (Luppitt, near Honiton, 1922); 'Black Ram' (Dalwood and Alfington, east Devon, 1922); 'Steal the Pigs' (Fraserburgh, 1898); 'The Children' (Chaffcombe, 1922); 'Box of Matches' (Taunton, 1922); 'Bannockies' (Forfar, 1910).

In Australia: 'Mother, Mother, the Kettle's Boiling' (Powelltown, Victoria, c. 1935); 'Blackie, Blackie, Come Here' (Footscray, Victoria, c. 1915); 'Diggley Bones' (Toowoomba, Queensland, c. 1905). In New Zealand: 'Old Digley Bones (Bay of Plenty, 1885); 'Blacksmith' (Kaitaia, Westland, 1949)—Sutton-Smith, *Games of New Zealand Children*, 1959, pp. 33–5.

** The absence of any recordings of this farce before the nineteenth century is disappointing, but hardly lessens the likelihood of its being old. Games are played throughout Europe in which children are stolen, begged, or bought one by one; in which passages of comic dialogue occur closely matching the British texts; and in which the child-taker limps with a defective foot, as did the man-eating hag Empusa, whose name scared little children in classical times. In the German acting game 'Lange Elen', described in 1836, before any account in English of 'Mother, the Cake is Burning', a mother goes off leaving her children (or lengths of cloth) in

---

1 A 10-year-old girl told us confidently 'Blacking Topper or Mackintosh is a foreign name for a favourite child, and this is how the game gets its name'. The name 'Blacking Toppers' (also spelt 'Topas' and 'Topa') appeared in 1910 as the name of a game played by children in London elementary schools (*Notes and Queries*, 11th ser., vol. i, p. 483). Norman Douglas, *London Street Games*, 1916, p. 48, listed 'Black In Topper'. Our own records show the game being called 'Blacking Topper' in London c. 1930, c. 1947, and 1960.

one person's care, and a thief comes to the house and snatches one of them. Immediately the guardian's cry is not that there has been a thief but that the broth boils over: 'Moder, Moder, de Brei kaakt aver!' The mother calls back, 'Throw a bit of salt in it', and, as in Britain, the guardian protests her inability to carry out a simple action, so that the mother is induced to return and learn what has really happened (H. Smidt, *Kinder- und Ammenreime in plattdeutscher Mundart*, 1836, cited Böhme, 1897, pp. 583–4). In the game 'Stoff Verkaufen', played today in Austria, that part of the English text is enacted where the evil visitor gains each child by a trick. The witch's response each time her plea for a length of cloth is refused, is to say to the mother: 'Run quickly to your neighbour, her cow has calved and she needs help' (Kampmüller, *Kinderspiele*, 1965, pp. 164–5). Likewise in the cloth-stealing game played in post-war Berlin, the thief obtains each length of cloth by saying, for instance, 'Your milk is boiling over'; and the human personality of the cloth becomes apparent as the game proceeds, for when the thief makes a suit out of the cloth, the suit scolds the thief saying 'You are bad, you are bad', or 'Wicked Emma, Wicked Emma' (Peesch, *Berliner Kinderspiel*, 1957, pp. 47–8). In the game 'Moder o Moder wo ist Kindken bliewen?' described about 1820 in a letter to the Grimm brothers, two buyers come to the mother again and again to buy a lamb, and it is significant that the child they ask for each time, and which the mother most strenuously refuses them, is the youngest child. In this game when the last and youngest child has eventually been obtained, the buyers hide him. Then the other children cry out 'O Mother, O Mother, where is the little child that has eaten the snakes and toads?' and the mother goes off to search for him (*Zeitschrift für Volkskunde*, N.F., vol. ii, 1931, p. 147). In the once widely popular game 'Frau Rose' or 'Mutter Rose' (a source of recreation, also, to German mythologists), it is noticeable that the evil character who approaches with the intention of obtaining the children (whether in the guise of chickens, flowers, or plasters for a sore foot) walks with a limp, and that although the mother gives her children away, the evil character has to wheedle them from her, one by one, with stories that are palpably false. In this game, however, the fun lies in the struggle which follows when the mother has to wrest the children from the witch's grasp by sheer strength (Böhme, *Deutsches Kinderspiel*, 1897, pp. 539–41). Likewise, in the game 'Brot backen' or 'Backofenkraucher', only that part of the play is enacted where the children, in the guise of loaves or buns, are cooked or overcooked in an oven, and the point of the game is the strenuous action which ensues when the loaves are pulled from their place, as in the old game in England of

'Drawing the Oven' or 'Jack, Jack, the Biscuit Burns', much described in the nineteenth century.[1]

Thus, despite the fact that many individual points of similarity occur between 'Mother, the Cake is Burning' and the games played on the Continent, it appears that nowhere is the full drama preserved so effectively as in Britain. On the Continent the acting is often of comparatively short duration, it merely sets the scene for a physical contest; and the games are the equivalents not so much of 'Mother, the Cake is Burning' as of 'Honey Pots', 'Jams', 'Limpety Lil', and 'Cripple Chirsty' (qq.v.). Yet there is one exception. In Sicily, and in the foot of Italy, playlets are enacted whose similarity to the British witch-play are such that they would scarcely seem out of place if found being performed in the Scilly Islands or in Cornwall.

In the game 'Le galline', played today at Forenza, near Potenza, in Basilicata, one child is a fox, another an old dame, a third a cock, and the rest are hens. The old dame tells the cock to look after the hens while she goes to church, and warns him about the fox. As soon as she has gone the cock calls the fox and gives him a hen. The other hens cry out 'Coccodè, coccodè', and the old dame comes running back. 'Is the fox here?' 'No', replies the cock. 'Beh', says the old dame, 'pruvamm a cuntà'. The old dame counts the hens, and cries 'There's one missing!' 'How, one,' replies the cock, 'let me count them.' The cock counts them and cheats as he counts, so that the number comes right. The old dame is pacified, and sets off again to church, repeating her previous injunction about the fox. This goes on until there are no hens left and the cock can no longer trick the dame into thinking that she still has the right number. Yet the dame goes again to church, telling the cock to be careful or the fox will get him too; and this, of course, is what happens. She returns from church to find the cock has gone. Only now does she set out in search of her hens. On the road she meets some neighbours who tell her that various things have happened to her hens: some have been lamed, some have gone blind, some lost a wing. The children who are playing the parts of the hens then show

---

[1] In this game Idle Jack and his master, after some playful preliminaries, had to go up to the oven and draw out each of the loaves, who were boys seated on the ground in single file, each boy holding on to the one in front. Descriptions appear in *School Boys' Diversions*, 1820, pp. 10–12; *Boy's Own Book*, 4th ed., 1829, p. 37; *Every Boy's Book*, 1841, p. 30; *Alphabet of Sports*, 1866 ('Draw the Batch'); and Seán O Súilleabháin's *Irish Wake Amusements*, 1967, p. 82 ('Drawing the Bonnavs'). Northall, *English Folk-Rhymes*, 1892, p. 390, states that in Warwickshire the baked bread used to attract attention by chanting:

> Jack, Jack, the bread's a-burning,
>   All to a cinder,
> If you don't come and fetch it out,
>   We'll throw it through the winder.

themselves in the distance: some limp, some shut an eye, some bend an arm. The old dame thinks she recognizes them and scatters chicken food, calling 'Cosc, cosc, cosc mei'. But when the chickens follow her, and she sees the pitiful state they are in, she no longer thinks they are hers, and shouts 'Sciò, sciò, sciò, sciò, ca nun sit l mei'. However, the hens recover, and the old dame takes them back, calling 'Cosc, cosc, cosc, sit le mei' (you are mine).

Likewise, in the game 'Mamma caduta dal monte' played at Paceco in Sicily, a mother puts her children in the charge of her eldest daughter while she goes to buy vegetables. While the mother is away an old woman comes and tells the eldest daughter that her mother has fallen from a balcony. When the eldest daughter goes to see the old woman takes a child; and she repeats this trick for each child, telling the eldest daughter that her mother has fallen from one high place and then another. When the mother returns home from shopping she asks the eldest daughter where the children are, and the eldest daughter confesses that the old woman has taken them. The Sicilian mother, however, is apparently as simple-minded as the British mother; she leaves the eldest daughter and goes to buy some *pasta*. This time the old woman snatches the eldest daughter. The theft of the last child, as in Britain, seems to bring the mother to her senses. When she returns from her shopping, she realizes *subito* that the old woman is to blame. She goes to the old woman's house and asks her not whether she has her children, but (as in Britain, apparently realizing that they are to be eaten) whether she has a hen. The old woman replies that she has a hen, but only one, which she is going to eat herself. The mother implores her to give it up, and makes her way into the hen-house. The children who are in the hen-house cry out, and, in the words of the child who gave this description to Signora Maroni Lumbroso, 'the old woman is going to finish badly because mother and daughter give her so many blows that she will remember them all her life' (*Giochi*, 1967, pp. 143–4 and 188–9).

Thus do children in Sicily, and the Shetland Isles, although separated by race, language, and history, not to mention about two thousand miles, apparently share a common heritage.

12

Pretending Games

'Some times I kill Some One and Some One kills me but my men release me and I release them Back.'

Boy, c. 6, Peckham Rye

CHILDREN, as Stevenson observed, 'take their enjoyment in a world of moonshine'. Each day, when the spell is upon them, they skip through a labyrinth of pretendings, and we suppose them to be imaginative, ignoring the evidence that the young do not commonly invent, merely imitate. We overlook, perhaps, when they amuse us with their oddities, that what passes as original is due not to art but to artlessness, to mishearings, to imperfect understanding, to the three-foot-high viewpoint which sees a palace roof in a table top, and fears a thunderstorm when the dog snores. Thus their pretending games turn out to be little more than reflections (often distorted reflections) of how they themselves live, and of how their mothers and fathers live, and of the books they read, and the TV programmes they watch. Whatever has latest caught their fancy is tested on their perpetual stage. A class of 10-year-olds are taken round a post office, and the craze for the next week is playing 'Telephone Exchange'. A teacher uses the John o' Groats to Land's End walking race in a geography lesson, and the children arrive home from school breathless and bedraggled because they, too, have been walking from John o' Groats. They re-present, as if they were newsreels, the more spectacular national events: Coronations and rescues at sea; the climbing of Mount Everest, Princess Alexandra's wedding, and the Great Train Robbery. In Berlin, when the wall was built (1961), West German children began shooting at each other across miniature walls. In the United States, after the death of President Kennedy (1963), children were found playing 'Assassination'. Throughout time, it seems, juvenile performances have varied only as their surroundings have varied. In classical Rome, where the law was administered in public, Seneca observed Roman boys playing at judges and magistrates, 'the magistrates being accompanied by little lictors with fasces and axes' (*De constantia sapientis*, xii). In the sixteenth century Bruegel, in his painting of children amusing themselves in the street, shows one group playing at weddings, another group at christenings, and a third group taking part in a pretence religious procession. In eighteenth-century France Noël-Antoine Pluche noted:

'Children mimick by Turns the Processions of the Church, the March of Soldiers, the Attack of Places, the driving of a Coach, the several Attitudes of Trades-People, in short, whatever they see.'

(*Spectacle de la Nature*, vol. vi, 1748, p. 97). At the beginning of the twentieth century Norman Douglas came upon the smaller boys and girls in the streets of London playing:

'Mothers and Fathers, for instance, and Teachers, and Schools, and Soldiers, and Nurses, and Hospitals, and Carts and Horses, and Shops, and Convicts and Warders, and Railway Stations, and games of that kind.'

And during the Second World War children in Auschwitz concentration camp, well aware of the reality, were seen playing a game that proved the most terrible indictment ever made against man, a game called 'Going to the Gas Chamber'.

MOTHERS AND FATHERS

The 6-year-old child who plays 'Mothers and Fathers' re-enacts the common incidents of his everyday life with what seems tedious exactness, until one realizes that there is a thrilling difference: he has promoted himself, he is no longer the protesting offspring being scrubbed to bed, but is the father or mother; and the 6-month-old doll is the one being scolded for not getting into the bath.

> The little actor cons another part;
> Filling from time to time his 'humorous stage'
> With all the persons, down to palsied age,
> That life brings with her in her equipage;
>   As if his whole vocation
>   Were endless imitation.

Even beyond Infant School the girls sometimes play 'Mothers and Fathers', and 'Houses' and 'Tea Parties' and 'Shopping' and 'Supermarkets' and 'Brides'. ('When I play "Brides",' says a 7-year-old, 'and pretend to be married I have to make my husband's tea and find his slippers'); and sometimes the boys cannot resist the fascination of the game. They ask teacher 'Can I go over and play with the girls?' and are usually welcome. In domestic dramas the male role is not a popular one;

in some young eyes (an East Dulwich 10-year-old's, for example) the father is little more than a figure of fun:

'My friend and I play "Husband and Wife". I am the wife, I have to wash-up, and wash the floor or clean the windows, while my friend sleeps.'

This revealing drama is apparently also stock repertory in the United States where, according to Brian Sutton-Smith, the domestic hero enters the home, clasps the 'mother' in his arms, gives her a loud kiss, stretches, and says 'Well, I guess I'll take a nap'. Indeed it seems to be traditional for the breadwinner to be portrayed as self-indulgent. In 1903 Edwin Pugh reported in *Living London* that in the game 'Mothers and Fathers' it was not uncommon for the boys to 'reel about the pavement in a dreadful pantomime as "father"'.

The marvellous pleasure of playing 'Houses' lies first in making the house:

'We make little squares with dead grass,' writes a Scots girl in the Border Country, 'and leave a little doorway in one side. We put dead grass in a circle against one wall as a fire and put rowans in it for burning coal. We hunt stones to serve as sideboards and dressers. A piece of heather and brown bracken is put in a corner as an ornament. We take up our old dolls as babies. Then we pretend a baker arrives in a van. We go out and buy a cake, bread, and biscuits. Sometimes we go and visit one of our neighbours, make up a row, and not speak for a few days.'

Idyllic surroundings are not, however, essential to their make-believe. City children are fully competent, when the desire is upon them, to lay their hands on boards, wire netting, old curtains, and packing-cases, for building material. In 1962 amidst the dust of demolition in the University area of Manchester little girls were seen being very busy round a 'house' made of a large plush Victorian armchair. We asked them (7–9 years old) what they were playing, and their proud reply was 'Bombed Houses'.

When children speak of dressing-up their accounts become long and detailed. Ankle-length dresses, large hats, veils, high heels (one youngster thought they were 'high hills'), cigarettes, powder and make-up, represent the joys of adulthood. 'One of my favourite tricks to get my mother's make-up out of the house,' recalls a 14-year-old, 'was to shout and tell my mother that my little sister was crying. When she went to see, I would sneak out of the back door. My friends thought that I was the best rouger in our street. They all wanted me to put on their rouge, which was a very great honour. The powder they liked to put on themselves.' However, when rouge and mascara are not to be had, coal dust, black-board chalk, and tomato sauce seem to do as well. A group of girls running a 'Beauty Salon' were seen to be quite happy (in January) dipping their fingers in puddles for lotions and putting real mud-packs on their foreheads. To be like teenagers they loosen their hair, adopt affected voices, and push their fists inside their jerseys to make bosoms, for the emphasis is on realism as they see it. Northerner II in *The Yorkshire Post* (November 1965) reported a kindly old gentleman watching some children in the street playing 'Mums and Dads'. Across the road, with his feet in the gutter, sat a lonesome 2-year-old. 'Why isn't he playing with you?' asked the old gentleman. 'Oh he is', said a 10-year-old Mum. 'He's a baby waiting to be born.'

PLAYING SCHOOLS

Perhaps the most popular of the mundane pretences is playing 'Schools'. Boys as well as girls play 'Schools' when they get home from school, and on Saturdays when they do not go to school, and even in playtime at school.

'The most favourite game played in school is "Schools",' says an Edinburgh 9-year-old. 'Tommy is the headmaster, Robin is the school-teacher, and I am the naughty boy. Robin asks us what are two and two. We say they are six. He gives us the belt. Sometimes we run away from school and what a commotion! Tommy and Robin run after us. When we are caught we are taken back and everybody is sorry.'

Clearly playing 'Schools' is a way to turn the tables on real school: a child can become a teacher, pupils can be naughty, and fun can be made of punishments. It is noticeable, too, that the most demure child in the real classroom is liable to become the most talkative when the canes are make-believe.

Playing schools has, in fact, long been ritualized. In the following game called 'Johnny Green', described by a Langholm 12-year-old, the cheeky replies are traditional, and this game is quite as much an 'acting game' as mimicry:

'You are all supposed to be in school when the teacher comes in. She says to Johnny Green, who was absent yesterday, "What was wrong with you yesterday, Johnny Green?"

He says, "I killed a fly and had to go to its funeral."

Johnny Green gets a thrashing and while he is still out on the floor after his thrashing, she says, "What was wrong with you the day before yesterday?"

Johnny Green replies, "My mother made an apple tart and I had to stay at home and get a piece."

He gets another thrashing for that. After giving him the thrashing the teacher sends him to his seat. "Who will go for some caramels for me?" she asks.

Johnny Green says he will. He comes back and says "The camels' humps won't come in the door."

There are roars of laughter. He gets another thrashing and gets the rest of the children into trouble for laughing.'

We have also been audience to this act in Glasgow; and in J. T. R. Ritchie's *The Golden City*, 1965, pp. 15–16, there is an account of the game as known in Edinburgh. A similar mad playlet, which seems to be a perennial in Aberdeen, concerns a child named 'Silly Sally', who is continuously unhelpful when her mother receives guests, answering the telephone instead of the front door, putting washing powder into the biscuit mixture, tripping over the visitor's legs and pouring tea on their heads.

ROAD ACCIDENTS

Some pretending games portray events which might happen, but have not happened yet. As a child grows older he becomes increasingly aware of the world around him, and needs to explore its possibilities. 'He is not sure how one would feel in certain circumstances; to make sure, he must come as near trying it as his means permit' (Stevenson again). Thus all over Britain children play 'Road Accidents':

'When we run about the playground pretending that we are driving cars', writes a 10-year-old boy, 'we pretend that we are drunk, and go wobbling about the roads we make out of shingle. When someone crashes we sent out a breakdown lorry. Sometimes the girls make a hospital in the G.P. room, and we get an ambulance to come out to fetch the injured drivers.'

Similarly when girls play 'Hospitals', as all little girls seem to do at one time or another, they are not being morbid (as an adult would be, acting such a fantasy), they are expressing healthy and commendable emotions: the feminine urge to soothe, to put right, to have everybody happy; and they are looking for reassurance that should they themselves happen to be stricken they, too, will be looked after and made well. Over and over again it is evident that 'Hospitals' is a comforting and even joyful game.

'I play hospitals when I have some time to spare. I pretend I am a very strict Matron, and make the nurses work hard. When my friends come to play with me, we become surgeons and do operations and make blood out of red paint and water. We use imaginary instruments and use small sweets for pills. Sometimes mummy lets us take some orange juice to the hospital, which we have outside. Mummy is making me a nurse's uniform, and when it is finished I shall play hospitals much more often.'   *Girl, 10, Whetstone*

Not infrequently their patients are their dolls:

'I sometimes pretend in my tent with my dolls that I am a nurse in a far away country. I had a hospital set for Christmas once, so I give my dolls operations. My two favourite dolls are Josephine and Peggy-Sue. Josephine has blonde coloured hair and Peggy-Sue has a blouse and skirt because she is a "teenage" doll. She has high heeled shoes and stockings. When my dolls get better, I give them a tea-party, so on a sunny day I think I am one of the happiest girls in the world.'   *Girl, 10, Finchley*

The boys seem to prefer causing injuries, or being injured, to being those who heal. In the playground they pretend to be jeeps, trains, buses, fire engines, motorbikes, jets, or ships.

'Me and David play "Jeeps" in the playground every day. We bash through people knocking them over and tripping them over. We have lots of fun together.'   *Boy, 9, Ipswich*

They play 'Trains', forming up in lines and holding on to each other's coat-tails:

'Last Friday i and my frend played trains and we keep braking them other trains to peses and wen we have dun one train we go and do a nuther.'
                                                        Boy, 8, Ipswich

Except in the game of 'Cops and Robbers' (see hereafter), only girls seem to think it fun to be upholders of the law:

'My Friend and I play "Police Women". We go around the streets petending to tell poeple off. We say for exampiall stop brarking bottels or get that child off your cross-bar.'   *Girl, 10, Fulham*

PLAYING HORSES

Amongst little girls 'Playing Horses' is almost as popular as playing 'Hospitals', and popular with some boys too. Whether or not the game can be associated with the sense of animism which W. H. Hudson experienced when about 8 years old is for the analysts to say; and depends on whether a child is pretending to possess a horse (the common dream-wish), or to be a horse himself. When a child pretends to be a horse or other animal he often becomes, as in no other pretending game, almost unconscious that he is pretending. 'There is something mystical about the game', a professor observed, recalling his youth. 'I remember once when I was playing horses with the other boys, I was serving a mare, and a grown-up came over and stopped me, he said I was being indecent.'

And the professor, fifty years later, still clearly resented the charge, adding—as if the plea was necessary—'I was completely innocent.'

This confusion between horse and self is often evident in the children's accounts:

> 'I play horses with two other girls, I am the horse and the other two are the masters. First we find a hook in the wall. Then one of the girls takes off her belt and fixes one end to me and the other to the hook. When the horse wants her master she must neigh. When the master comes she puts the belt round my waist and takes me round the playground and brings me back to the stable and ties me up again. Then I am ready for the next trip.'   *Girl, 7, Wilmslow*

If children's pretending is extrovert, they speak of it readily; if the pretending is introvert they will not admit to it under any circumstances. One of us remembers when she was a rabbit, being astonished at her nurse's stupidity when the nurse chided her for lagging behind on a walk. How could a little rabbit be expected to keep up? And even less, how could a rabbit explain that rabbits did not keep up?

Yet in most make-believe games, for instance when children are playing with toys, it is apparent that they are fully aware of the difference between play and reality. Korney Chukovsky, in his book От двух до пяти (part-translated by Miriam Morton, 1963, p. 26), tells of a 4-year-old playing with her wooden horse who whispered: 'The horsie put on a tail and went for a walk.' Her mother overheard and interrupted, saying: 'Horses' tails are not tied to them; they cannot be put on and taken off.'

'How silly you are Mommie!' replied the child, 'I am just playing!'

A child is likewise 'just playing' when he mounts a hobby-horse. He is not deceived by the stick between his legs into thinking that he is actually riding. He is content to be pretending. He is engaged upon one of the oldest of juvenile pastimes, and one of the most frequently depicted in previous centuries; and it is interesting to find that despite the superior mobility of bicycles, pedal cars, and roller-skates, the appeal remains today.

> 'My way for passing the time is playing with my home-made hobby-horses. I have three horses, two grey ones and one white horse. The grey ones are called Pixie and Frosty, the white one is Snowball. My friend and I made our hobby-horses by collecting all the old, worn-out socks, and stuffing them with rags. After that, we sew buttons on for eyes and then we make ears with white or grey cloth. We then find quite a short stick and put it in the sock, then we tie string or wool around it to make it secure. Then our horses are nearly finished but not quite, we have still the bridles to make. This is done by getting wool (and sometimes plaiting it to make it strong). Then we tie the wool to form a bridle and put it on the horse.'   *Girl, 10, Weybridge*

STORYBOOK WORLD

Some of the children's attempts to project themselves into special situations are so prosaic that only natural affection restrains us from terming them ludicrous.

'We play "Hotels". When we play "Hotels" you have a room, and one person goes round and collects the money for staying in the room.'
Girl, 9, Camberwell

Often, however, the story enacted is beyond ordinary experience; it is a vision of improbable possibility, a tale from contemporary mythology. They pretend, for example, to be giving a concert or a fashion show, and 'stars' are discovered during the performance; or it is a beauty contest, and there are judges who 'pick the first, second, and third, and tell them what their prizes are, like a weekend in Paris'. There are also the games in which, while still being themselves, they allow their circumstances to alter—usually for the worse:

'We play "Runaways". We pretended our father was horrible and made us work like slaves. One night we ran away with only a few shillings and a sackful of food taken from the pantry, and took a ride in the back of a hay lorry.'
Girl, 12, Oxford

They seem to have a particular predilection for being orphans ('We play sisters who are very poor, and have no mother and father and nobody to love them'), although it is noticeable that they are not unwilling to triumph over their misfortunes, either by their own exertions or Cinderella-wise, or by a judicious compound of both.

'My favourite game is to play at circuses. I play at starting with being poor. We are orphons. Then we find wild animals and train them. Then we get a present from our uncle which is a lot of money.'
Girl, 9, Wilmslow[1]

Girls also pretend to be princesses, or Red Riding-Hood, or a film star, or a fairy queen ('The fairies have wings and flutter about. I like this game because it does not end up in a fight or a battle'). The boys have

---

[1] Psychologists doubtless equate such make-believe with playing 'Hospitals'. The child wishes to assure herself that even if something disastrous occurred in her home-life she would still be able to manage. The gifted Ann and Jane Taylor, towards the end of the eighteenth century, were other little girls who indulged in these fancies. 'We most frequently personated two poor women making a hard shift to live; or we were "aunt and niece", Jane the latter and I the former; or we acted a fiction entitled "the twin sisters", or another, the "two Miss Parks". And we had, too, a great taste for royalty, and were not a little intimate with various members of the royal family. Even the two poor women, "Moll and Bet", were so exemplary in their management and industry as to attract the notice of their Royal Highnesses the Princesses ("when George the Third was King").'—*Autobiography of Mrs Gilbert, formerly Ann Taylor*, vol. i, 1874, p. 29.

mock adventures, rather formless games of 'Pirates and Castaways', 'Deep Sea Diving', or 'Tarzan in the Trees'. Groups of two or three may be seen crossing deserts, searching for hidden treasure (girls as well as boys), and exploring unknown jungles: a game that adults, less fancifully minded, usually know as 'going for a walk'.

'My best friend and I often play at explorers in the Hermitage of Braid. We pretend to hack our way through the undergrowth with broad knives. After that we pretend to be attacked by savage natives and when one of us is wounded we often start an argument. Then we make our way down to the stream and try to jump across.'

Boy, 9, Edinburgh

'If we were always to judge from reality,' Tolstoy commented, 'games would be nonsense. But if games were nonsense what else would there be left to do?'

WAR GAMES

There is a noteworthy difference between playing at 'Soldiers', and playing at 'War' with two opposing sides. Many old prints show children playing at soldiers, dressing up in uniforms and drilling with pretence muskets, or marching in procession with a little officer or sergeant strutting in front; but they do not, as far as we know, show children having mock battles. Pugh, in *Living London*, 1903, tells how during the Boer War wonderfully drilled regiments of juvenile soldiers were to be seen parading the streets: 'It was a memorable spectacle to see these bands of little ones, to whom some tiny *vivandières* were usually attached, marching along in perfect step through the mire and dust of the road, wearing their helmets and tunics, carrying their weapons, also an "ambulance", beating their drums and blowing their toy trumpets, with that dignified gravity of which only children know the secret.' Such a parade (a photograph of a back street shows a troop with flags flying and rifles shouldered) would be improbable today, and not only because soldiers are less often seen on parade. Children today are nearer to the realities of war: there is no parading and much killing.[1]

---

[1] Boys are shown parading as soldiers in Jacob Cats's *Emblemata*, 1622; in Stella's *Jeux de l'enfance*, 1657; and frequently in juvenile literature, for instance in *A Little Pretty Pocket-Book*, 1744. Husenbeth recalled that when he was at Sedgly Park School during the Napoleonic Wars, the boys formed a troop of soldiers, each paying nine pence for a wooden gun, and three pence for his pasteboard cap, so that they made a fine show on parade (*History of Sedgley Park*, 1856, pp. 109–10). Thomas Miller, born 1807, played at 'French and English', enlisting his recruits with broken bits of white pot to represent the King's Shilling (*Country Year Book*, 1847, p. 18). There is a tale that when Elizabeth I was concerned about the effects of her policies on the Scottish nation, she would inquire how the boys were amusing themselves. If they were playing soldiers, she took it as a warning that it was time for her to arm.

Occasionally when they play a war game there is only one side, the enemy is non-existent, the battle an illusion:

'Some nights in the summer the boys in our avenue go in and fetch their guns and war helmets, and go to the dark field and make a camp. When we are ready two boys go off, and when they are coming back they start running, and when they get to us they say there is an enemy convoy going over the warren and we get our guns, fill the pistols with caps, and put on our helmets. Then we go to the warren and put old hairnets over our helmets and stick ferns in them. When we have done that we start firing our guns, and some of us reckon to be killed.'

Boy, 10, Annesley

Usually, however, the fighting is only partially make-believe.

'First you pick two sides, one English and one German. One side which is always the English creeps up to the camp of the Germans, and one person which is usually me goes behind a German guard and hits him on the head. Then we take the leader of the Germans so that they won't have a leader to show them what to do. I myself have been captured but have always got away. I have been wounded and been taken to hospital.'

Boy, 8, Wilmslow

And sometimes the game is more rules and planning than imagination. The players 'dip' to see who shall be the commanders of the opposing forces; the commanders pick their sides; an agreement is made on how many stones each soldier may carry for grenades; and a coin is tossed to decide which side shall go out first. 'There is much fighting in this game therefore girls cannot play, but if you have had enough you say "I'm quitting" and then you have to go back to the headquarters of your side' (Boy, 11). In other accounts, however, action and fantasy merge into Walter Mitty heroics, and an adult scarcely knows whether the young historian is more concerned with the facts of the game or with self-aggrandizement.

'My game is cowboys. Now I am Cheyenne, I play it in the school yard I am fighting the indians with my gun, I killed 20 of them and the chief, that's why all the others ran away, but if their chief was not dead they would keep charging till they kill me so that when I was just ready to get on my knees all the troops came then all the other indians came but not from the same camps they had funny faces they were like lions like in yellow stone kelly so we all had a fight they had lots of indians but we still killed them they was only about 40 of us left.'

Boy, 9, York

Some boys manage to be both romantic and down-to-earth at the same time. Outside the Students' Union in Bristol a lad aged about 10 was organizing his playfellows for a battle and trying to be fair to both sides. When one player demurred at not being on the English side, the leader was overheard saying emphatically: 'I tell you none of us is English, the English always wins, that's why!'

Names: The following are some of the battles that are regularly fought on British soil in the second half of the twentieth century, 'Germans and English', 'British and Reds' (Chinese), 'Commandoes and Japs', 'Nips and Yanks' ('Nips and Yanks' explained a 10-year-old 'is really another game of "Germans and British"'), 'Foreign Legion and Arabs', 'Redcoats and Scottish Highlanders', 'Greeks and Romans'.

'Cowboys and Indians' remains an outstanding favourite (as also on the Continent[1]), and according to the TV programme of the moment, may be named: 'Raw Hide', 'Wagon Train', 'Pony Express', 'Totem Pole', 'Cavalry and Indians', 'Gun Law', 'Apache Warpath' ('If you kill the cowboys the game is finished'), 'Cheyenne', 'Laramie', 'Wells Fargo', 'Lone Ranger', or 'Cisco Kid'. (' "Cowboys and Indians" does not obey any set of rules and as often as not ends up in a free fight.')

Surprisingly popular (TV influence again) are 'Knights in Armour', 'Knights of Old', 'King Arthur and His Knights', 'Sir Lancelot', 'Medieval Men', 'Robin Hood', 'Ivanhoe', and 'William Tell'.

Of a slightly more domestic character, but mostly TV inspired, are 'Cops and Robbers' (see hereafter), 'Biggles and Crooks', 'Z-Cars', 'Highway Patrol' (on bikes), 'Black Riders', 'Stage Coaches and Robbers' (at Hounslow, once a notorious district for highwaymen), 'Russian Spies', 'Dangerous Mission', 'Security Police', and 'Man from U.N.C.L.E.'

Amongst the fantastical: 'Super Car', 'Superman', 'Batman', 'Captain Marvel', 'Thunderbirds', 'Robots' ('They pretend to be robots gone mad', reports a headmaster), 'Daleks' and 'Spacemen'. But spacemen are not as popular as they were in the 1950s, when astronauts were still fabulous, and, led by Dan Dare of *Eagle*, they had to contend in outer space with green-hued foes of mighty intellect. By the mid-sixties spacemen had become commonplace, and the scope for the wonderful seems to have been correspondingly reduced.

COPS AND ROBBERS

Of all the perennial characters which 9-year-olds adopt to enliven a chase or give point to a brawl, the most popular are cops and robbers. Year in and year out boys enact the battle between crime and the law,

---

[1] In France children play 'Le Cowboy et l'Indien', in Germany 'Cowboy und Indianer', in Italy 'Caw-boys e Indiani'.

passionately engaging themselves to whichever side they find themselves on, without apparent preference for right or wrong. On occasion the game is an elaborate make-believe, with stones for gold, pieces of paper for bank notes, guns, handcuffs, stocking masks, and almost a story.

'You have a bank and a bank manager and the robbers rob the bank. Then the cops chase the robbers, the robbers go to their hide-out until the cops have gone. Then the robbers rob another bank and the cops chase the robbers and they go to their hide-out, and it goes on like that until the cops find the robbers hide-out and catch them.'

Sometimes the make-believe is no more than an occasion for a fight. 'The robbers knowk the cops out and the cops knok the robbers out and so on. If anybody falls out we call them big babys.' 'When a convict is caught he can struggle until he is put into the den. When he is in the den he is tied to a pole and is asked who the leader is. If he does not talk with this method he is tied to the merry-go-round.' (The merry-go-round is a type of swing placed in the recreation ground by the local council.) Quite often the game is played on bikes. 'You are on fast bicks. Every person must have a spud gun and a packet of caps. You must not hit the Robber with the spud but you must hit just in frunt of his bick. When he is caught you have a jail. You have to give him a chance to get away and if he does you start all over again.' And, popularly, the game is played at night, and the chasing side has torches. As a practical-minded 12-year-old remarked:

'You have to play in the dark because torches are no good in the daytime. It is more exciting also. Because if they chase you through the gardens they can see where you go but at night they need torches to find you. You can easily give them the slip (nip into the gardens) and lose them.'

Indeed when the game is played at night, in the unrestricting darkness, it often takes on the form of 'Release' or 'Relievo' (q.v.), the den being the jail, the seeking side the cops, and if 'the free robbers touch the caught ones without being caught themselves they can release them'.

Names: 'Bobbies and Thieves', 'Convicts', 'Cops and Burglars', 'Cops and Robbers' (much the most common name), 'Escaped Prisoners', 'Kidnappers', 'Policemen and Robbers', 'Prisoners and Warders'. In Wilmslow, 'River Police and Smugglers'. In the Outer Hebrides 'Smugglers and Customs'.

** In the past highwaymen and press-gangs were amongst the *dramatis personae*. Edward Moor, *Suffolk Words*, 1823, p. 238, gives simply 'Robbers'. *Notes and Queries*, 18 June 1910, p. 483, gives 'Robbers and Travellers' and 'Robbers and Policemen', as played by children at

London elementary schools. In Forfar, *c.* 1910, it was 'Snouts and Robbers' and 'Takkies and Thieves'. Norman Douglas, *London Street Games*, 1916, p. 36, gives 'Robbers and Coppers' and he, too, equates the game with 'Release'.

FAIRIES AND WITCHES

In the war games that the girls play the protagonists are almost invariably supernatural. They are fairies and witches, roles which give 5- to 8-year-olds infinite scope for pantomime: the fairies flapping their arms and quickly frightened, the witches who pursue them holding their fingers as if they were claws, bending their bodies as if they were aged, and having coats or pieces of cloth flowing from their shoulders for cloaks. The witches generally emerge from a den, and the fairies too sometimes have a palace or place of safety. Often 'there are not as many witches as there are fairies because we do not like being witches'. And they like to play in the dark. In Langholm the fairies dance round together in the darkness until a voice cries 'Witches!' and the fairies flutter off into the night. In Bristol the fairies sing in small shrill voices:

> Wicked old witch are you hungry today?
> If you are then we will all run away.

In Radnorshire the witch has a magic wand that 'turns us into a frog, or something like that'. In fact the enjoyable part of the game is what happens to the fairies after they have been caught. Generally, as in Accrington, 'they are put in a stewpot and eaten for the witches' tea'. The more grisly the details, the more the girls enjoy the game. On the outskirts of Gloucester we found the boys playing conkers, studiously ignoring a group of giggling screaming little girls playing, as they informed us, 'Fairies and Witches'.

'How do you play that?' we asked.

'There's a witch and she catches you', they said.

'And then?'

'Takes you to her corner.'

'And then?'

'*Stews* you.'

'And then?'

'*Eats* you.'

'And then?'

'*Throws the bones away*.'

Hence 'Fairies and Witches' is sometimes known as 'Stewpot' (Camber-well) and 'Witches' Cauldron' (Knottingley). In some places the evil one is a wizard, a wolf, or a ghost, but the fate of those caught is the same, to be boiled in a stewpot and eaten. Even when they play 'Invisible Man', as girls do in Wolstanton, those unlucky enough to be caught end up in the pot:

'The first person the Invisible Man touches becomes cook. The cook pretends to stir a cauldron of boiling water. In the meantime the Invisible Man is still catching other people. When all the players are caught the cook goes in the pot. Then, when the people in the pot are supposed to be cooking nicely the Invisible Man does all kinds of horrid things to you like hanging you or putting you in a fridge at freezing point. Thus the game goes on.'   *Girl, 9*

And in the fields around Liss, where they play 'Dragons', and 'you are not allowed to run out of the field when the dragon chases you', the fate of those caught is equally unpleasant and final:

'The dragon takes the person he has caught to the haunted tower where he chains them to the wall, and then he presses a button. Suddenly a whole lot of spiders drop down on the prisoner and make him scream. When the spiders have woven cobwebs all over the prisoner the dragon puts the prisoner in the fire and burns him. This game is nearly always played in Autumn and Winter.'   *Girl, 9*

At this point, it seems, we reach the boundary of juvenile invention.

Index of Games and Game-Rhymes